Malliavin Calculus in Finance

Chapman & Hall/CRC Financial Mathematics Series

Aims and scope:
The field of financial mathematics forms an ever-expanding slice of the financial sector. This series aims to capture new developments and summarize what is known over the whole spectrum of this field. It will include a broad range of textbooks, reference works and handbooks that are meant to appeal to both academics and practitioners. The inclusion of numerical code and concrete real-world examples is highly encouraged.

Handbook of Financial Risk Management
Thierry Roncalli

Optional Processes
Stochastic Calculus and Applications
Mohamed Abdelghani, Alexander Melnikov

Machine Learning for Factor Investing
Guillaume Coqueret and Tony Guida

Malliavin Calculus in Finance
Theory and Practice
Elisa Alos, David Garcia Lorite

Risk Measures and Insurance Solvency Benchmarks
Fixed-Probability Levels in Renewal Risk Models
Vsevolod K. Malinovskii

For more information about this series please visit: https://www.crcpress.com/Chapman-and-HallCRC-Financial-Mathematics-Series/book-series/CHFINANCMTH

Malliavin Calculus in Finance

Theory and Practice

Elisa Alòs

David García Lorite

Foreword by Dariusz Gatarek

CRC Press is an imprint of the
Taylor & Francis Group, an **informa** business
A CHAPMAN & HALL BOOK

First edition published 2021
by CRC Press
6000 Broken Sound Parkway NW, Suite 300, Boca Raton, FL 33487-2742

and by CRC Press
2 Park Square, Milton Park, Abingdon, Oxon, OX14 4RN

CRC Press is an imprint of Taylor & Francis Group, LLC

Library of Congress Cataloging-in-Publication Data

Names: Alòs, Elisa, author. | Lorite, David Garcia, author.
Title: Malliavin calculus in finance : theory and practice / Elisa Alòs, David Garcia Lorite.
Description: First edition. | Boca Raton : Chapman & Hall/CRC Press, 2021. | Series: Chapman and Hall/CRC financial mathematics series | Includes bibliographical references and index.
Identifiers: LCCN 2021000977 (print) | LCCN 2021000978 (ebook) | ISBN 9780367893446 (hardback) | ISBN 9781003018681 (ebook)
Subjects: LCSH: Finance--Mathematical models. | Malliavin calculus. | Stochastic analysis.
Classification: LCC HG106 .A54 2021 (print) | LCC HG106 (ebook) | DDC 332.01/51922--dc23
LC record available at https://lccn.loc.gov/2021000977
LC ebook record available at https://lccn.loc.gov/2021000978

ISBN: 978-0-367-89344-6 (hbk)
ISBN: 978-0-367-86325-8 (pbk)
ISBN: 978-1-003-01868-1 (ebk)

Typeset in Nimbus
by KnowledgeWorks Global Ltd.

To Xavi, Àlex and Dídac
To Nerea and Martín

Contents

Foreword xv

Preface xix

SECTION I A primer on option pricing and volatility modelling

CHAPTER 1 ▪ The option pricing problem 3

1.1 DERIVATIVES 4

1.1.1 Forwards and futures 4

1.1.2 Options 5

1.2 NON-ARBITRAGE PRICES AND THE BLACK-SCHOLES FORMULA 7

1.2.1 The forward contract 7

1.2.2 The price of a European option as a risk-neutral expectation 9

1.2.3 The price of a vanilla option and the Black-Scholes formula 11

1.3 THE BLACK-SCHOLES MODEL 14

1.3.1 From the Black-Scholes formula to the Black-Scholes model 14

1.3.2 Option replication and delta hedging in the Black-Scholes model 16

1.4 THE BLACK-SCHOLES IMPLIED VOLATILITY AND THE NON-CONSTANT VOLATILITY CASE 18

1.4.1 The implied volatility surface 18

1.4.2 The implied and spot volatilities 19

1.5 CHAPTER'S DIGEST 21

CHAPTER 2 ■ The volatility process 25

2.1 THE ESTIMATION OF THE INTEGRATED AND THE SPOT VOLATILITY 25
2.1.1 Methods based on the realised variance 25
2.1.2 Fourier estimation of volatility 26
2.1.3 Properties of the spot volatility 29
2.2 LOCAL VOLATILITIES 32
2.2.1 Mimicking processes 33
2.2.2 Forward equation and Dupire formula 34
2.3 STOCHASTIC VOLATILITIES 38
2.3.1 The Heston model 40
2.3.2 The SABR model 41
2.4 STOCHASTIC-LOCAL VOLATILITIES 46
2.5 MODELS BASED ON THE FRACTIONAL BROWNIAN MOTION AND ROUGH VOLATILITIES 47
2.6 VOLATILITY DERIVATIVES 49
2.6.1 Variance swaps and the VIX 50
2.6.2 Volatility swaps 51
2.6.3 Weighted variance swaps and gamma swaps 53
2.7 CHAPTER'S DIGEST 53

SECTION II Mathematical tools

CHAPTER 3 ■ A primer on Malliavin Calculus 57

3.1 DEFINITIONS AND BASIC PROPERTIES 57
3.1.1 The Malliavin derivative operator 58
3.1.1.1 Basic properties 59
3.1.2 The divergence operator 61
3.2 COMPUTATION OF MALLIAVIN DERIVATIVES 62
3.2.1 The Malliavin derivative of an Itô process 63
3.2.2 The Malliavin derivative of a diffusion process 63

3.2.2.1 The Malliavin derivative of a diffusion process as a solution of a linear SDE 64
3.2.2.2 Representation formulas for the Malliavin derivative of a diffusion process 65
3.3 MALLIAVIN DERIVATIVES FOR GENERAL SV MODELS 68
3.3.1 The SABR volatility 68
3.3.2 The Heston volatility 69
3.3.3 The 3/2 Heston volatility 71
3.4 CHAPTER'S DIGEST 72

CHAPTER 4 ▪ Key tools in Malliavin Calculus 73

4.1 THE CLARK-OCONE-HAUSSMAN FORMULA 73
4.1.1 The Clark-Ocone-Haussman formula and the martingale representation theorem 73
4.1.2 Hedging in the Black-Scholes model 76
4.1.3 A martingale representation for spot and integrated volatilities 79
4.1.3.1 The SABR volatility 79
4.1.3.2 The Heston volatility 80
4.1.4 A martingale representation for non-log-normal assets 82
4.2 THE INTEGRATION BY PARTS FORMULA 84
4.2.1 The integration-by-parts formula for the Malliavin derivative and the Skorohod integral operators 84
4.2.2 Delta, Vega, and Gamma in the Black-Scholes model 85
4.2.2.1 The delta 85
4.2.2.2 The vega 86
4.2.2.3 The gamma 87
4.2.3 The Delta of an Asian option in the Black-Scholes model 88
4.2.4 The Stochastic volatility case 89
4.2.4.1 The delta in stochastic volatility models 90

4.2.4.2 The gamma in stochastic volatility models 90
4.3 THE ANTICIPATING ITÔ'S FORMULA 94
4.3.1 The anticipating Itô's formula as an extension of Itô's formula 94
4.3.2 The law of an asset price as a perturbation of a mixed log-normal distribution 97
4.3.3 The moments of log-prices in stochastic volatility models 101
4.3.4 Some applications to volatility derivatives 104
4.3.4.1 Leverage swaps and gamma swaps 104
4.3.4.2 Arithmetic variance swaps 106
4.4 CHAPTER'S DIGEST 108

CHAPTER 5 ■ Fractional Brownian motion and rough volatilities 111

5.1 THE FRACTIONAL BROWNIAN MOTION 111
5.1.1 Correlated increments 113
5.1.2 Long and short memory 113
5.1.3 Stationary increments and self-similarity 114
5.1.4 Hölder continuity 115
5.1.5 The p-variation and the semimartingale property 115
5.1.6 Representations of the fBm 116
5.2 THE RIEMANN-LIOUVILLE FRACTIONAL BROWNIAN MOTION 117
5.3 STOCHASTIC INTEGRATION WITH RESPECT TO THE FBM 118
5.4 SIMULATION METHODS FOR THE FBM AND THE RLFBM 122
5.5 THE FRACTIONAL BROWNIAN MOTION IN FINANCE 124
5.6 THE MALLIAVIN DERIVATIVE OF FRACTIONAL VOLATILITIES 126
5.6.1 Fractional Ornstein-Uhlenbeck volatilities 127
5.6.2 The rough Bergomi model 128
5.6.3 A fractional Heston model 129

5.7 CHAPTER'S DIGEST 129

SECTION III Applications of Malliavin Calculus to the study of the implied volatility surface

CHAPTER 6 ▪ The ATM short-time level of the implied volatility 133

6.1 BASIC DEFINITIONS AND NOTATION 134
6.2 THE CLASSICAL HULL AND WHITE FORMULA 135
6.2.1 Two proofs of the Hull and White formula 135
6.2.1.1 Conditional expectations 135
6.2.1.2 The Hull and White formula from classical Itô's formula 136
6.3 AN EXTENSION OF THE HULL AND WHITE FORMULA FROM THE ANTICIPATING ITÔ'S FORMULA 138
6.4 DECOMPOSITION FORMULAS FOR IMPLIED VOLATILITIES 145
6.5 THE ATM SHORT-TIME LEVEL OF THE IMPLIED VOLATILITY 146
6.5.1 The uncorrelated case 147
6.5.2 The correlated case 152
6.5.3 Approximation formulas for the ATMI 165
6.5.4 Examples 168
6.5.4.1 Diffusion models 168
6.5.4.2 Local volatility models 172
6.5.4.3 Fractional volatilities 176
6.5.5 Numerical experiments 178
6.6 CHAPTER'S DIGEST 182

CHAPTER 7 ▪ The ATM short-time skew 185

7.1 THE TERM STRUCTURE OF THE EMPIRICAL IMPLIED VOLATILITY SURFACE 185
7.2 THE MAIN PROBLEM AND NOTATIONS 187
7.3 THE UNCORRELATED CASE 189
7.4 THE CORRELATED CASE 190

7.5 THE SHORT-TIME LIMIT OF IMPLIED VOLATILITY SKEW 192
7.6 APPLICATIONS 198
7.6.1 Diffusion stochastic volatilities: finite limit of the ATM skew slope 198
7.6.1.1 Models based on the Ornstein-Uhlenbeck process 199
7.6.1.2 The SABR model 199
7.6.1.3 The Heston model 200
7.6.1.4 The two-factor Bergomi model 200
7.6.2 Local volatility models: the one-half rule and dynamic inconsistency 201
7.6.3 Stochastic-local volatility models 203
7.6.4 Fractional stochastic volatility models 203
7.6.4.1 Fractional Ornstein-Uhlenbeck volatilities 203
7.6.4.2 The rough Bergomi model 204
7.6.4.3 The approximation of fractional volatilities by Markov processes 205
7.6.5 Time-varying coefficients 206
7.7 IS THE VOLATILITY LONG-MEMORY, SHORT-MEMORY, OR BOTH? 207
7.8 A COMPARISON WITH JUMP-DIFFUSION MODELS: THE BATES MODEL 208
7.9 CHAPTER'S DIGEST 212
CHAPTER 8 ■ The ATM short-time curvature 215
8.1 SOME EMPIRICAL FACTS 215
8.2 THE UNCORRELATED CASE 216
8.2.1 A representation for the ATM curvature 216
8.2.2 Limit results 220
8.2.3 Examples 223
8.2.3.1 Diffusion stochastic volatilities 223
8.2.3.2 Fractional volatility models 225

8.3 THE CORRELATED CASE 227
8.3.1 A representation for the ATM curvature 227
8.3.2 Limit results 230
8.3.3 The convexity of the short-time implied volatility 238
8.4 EXAMPLES 239
8.4.1 Local volatility models 239
8.4.2 Diffusion volatility models 241
8.4.3 Fractional volatilities 243
8.4.3.1 Models based on fractional Ornstein-Uhlenbeck processes 243
8.5 CHAPTER'S DIGEST 246

SECTION IV The implied volatility of non-vanilla options

CHAPTER 9 ▪ Options with random strikes and the forward smile 249

9.1 A DECOMPOSITION FORMULA FOR RANDOM STRIKE OPTIONS 250
9.2 FORWARD-START OPTIONS AS RANDOM STRIKE OPTIONS 253
9.3 FORWARD-START OPTIONS AND THE DECOMPOSITION FORMULA 256
9.4 THE ATM SHORT-TIME LIMIT OF THE IMPLIED VOLATILITY 258
9.5 AT-THE-MONEY SKEW 264
9.5.1 Local volatility models 269
9.5.2 Stochastic volatility models 269
9.5.3 Fractional volatility models 271
9.5.4 Time-depending coefficients 271
9.6 AT-THE-MONEY CURVATURE 272
9.6.1 The uncorrelated case 272
9.6.2 The correlated case 275
9.7 CHAPTER'S DIGEST 279

CHAPTER 10 ▪ Options on the VIX 281

10.1 THE ATM SHORT-TIME LEVEL AND SKEW OF THE IMPLIED VOLATILITY 283
10.1.1 The ATMI short-time limit 284
10.1.2 The short-time skew of the ATMI volatility 287
10.2 VIX OPTIONS 290
10.2.1 The short-end level of the ATMI of VIX options 290
10.2.2 The ATM skew of VIX options 293
10.3 CHAPTER'S DIGEST 301
Bibliography 303
Index 323

Foreword

Louis Bachelier's 1900 Ph.D. thesis *Theory of Speculation* was the first publication on both theory of stochastic processes and quantitative theory of finance, so it could be a revolutionary result and start of two important branches of mathematics. It did not, the thesis was neglected and rediscovered several decades later. The theory of stochastic processes was rather inspired by physics, namely two seminal papers by Albert Einstein and Marian Smoluchowski in 1905 and 1906 respectively.

It was Norbert Wiener, in 1923, who first produced a mathematically rigorous construction of Brownian motion. First paper on stochastic integration was published by Kiyoshi Itô in 1944, followed then by a series of papers on stochastic differential equations and what is now known as Itô formula creating the branch of stochastic analysis.

It took several decades for the stochastic analysis to grow and become fully operational and produce tools such as Feynman-Kac formula, Kolmogorov equations, and Girsanov transformation finding various applications in engineering such as Kalman filter and Bellman control principle.

The calculus of variations of stochastic processes was introduced in 1976 on a symposium in Kyoto by Paul Malliavin and is also called Malliavin calculus after him. The initial aim of the research was to provide a probabilistic proof of the Hörmander criterion of smoothness of the solutions of a second-order partial differential. It quickly exceeded its initial functions and became a big and sophisticated theory.

During half a century the theory of stochastic process grew from an obscure semi-mathematical field to a rich and mature research area. Call it a revolution, if you want. I do.

Another revolution in parallel happened in the area of finance. The option is a possibility but not an obligation and there are plenty of options in all areas of life in general (what is probably related to the existence of the free will) and in finance in particular. Financial specialists were always aware that options have some value and wanted to have an objective pricing tool. That need became sudden after the introduction of option exchanges. As Cox, Ross, and Rubinstein state it in their 1979 paper: *'Options have been traded for centuries, but remained relatively obscure financial instruments until the introduction of a listed options exchange in 1973.'*

It was Paul Samuelson who rediscovered the Wiener process as a good model of stock prices in 1965 with one slight modification – in order to avoid negative stock prices it was not just Brownian motion but its exponent called the geometric Brownian motion. This framework was accepted by three authors of the Black-Scholes-Merton formula. Their celebrated results were published the same year options started to be publically listed – in 1973. An accident or fate? Decide yourself.

The approach of Fischer Black and Myron Scholes was via the notion of equilibrium while Robert Merton introduced self-financing strategies and hedging. The second approach requires correct calculations of sensitivities of the underlying stock price with respect to various parameters (in finance called Greeks) in order to have proper hedges. To this end, the stochastic calculus of variations (Malliavin calculus) may be very useful, in particular the integration by parts.

Beyond this one there are two other books on Malliavin calculus in finance: *Stochastic Calculus of Variations in Mathematical Finance* by Paul Malliavin himself and Anton Thalmaier and *Malliavin Calculus for Lévy Processes with Applications to Finance* by Giulia Di Nunno, Bernt Øksendal, and Frank Proske. Both are of very high quality rigorously written by eminent specialists. Despite that they do not exhaust the topic of Malliavin calculus in finance.

I first met Elisa Alos Alcalde 20 years ago when she was a young and enthusiastic researcher and she still is young and enthusiastic – although already an eminent expert in both finance and stochastic analysis. David García Lorite is a mathematically educated senior quant within the banking industry with more than 10 years of experience and who is familiar with many practical issues.

That is the reason to read this book – a clever synthesis of academic knowledge with practical experience. You get a stereoscopic view of the subject – with academic rigour but dealing mostly with practically important issues, a real bridge from theory to practice. The authors explain why some models can reproduce (or not) some properties observed in real market data, collecting some previous results in a simple and general way. Malliavin calculus is a practical tool.

There are also several issues not covered by other books like rough volatility i.e. driven by the fractional Brownian motion. The book includes a Python code repository to facilitate the reader to understand the different concepts in the book with numerical examples. Enjoy and don't panic.

Dariusz Gatarek
Warsaw, November 2020

Preface

Malliavin calculus, introduced in the 70s by P. Malliavin, has had a profound impact on stochastic analysis. Originally motivated by the study of the existence of smooth densities of certain random variables (such as solutions of stochastic differential equations), it has proved to be a useful tool in several other probability problems. In particular, it has found applications in quantitative finance, such as in the computation of hedging strategies or the efficient estimation of the Greeks.

Despite its potential applicability, Malliavin calculus is often seen by practitioners as a sophisticated theory, far away from daily market problems. Our objective in this book is to fill this gap by offering a bridge between theory and practice. We see how Malliavin calculus is an easy-to-apply tool that allows us to recover, unify, and generalise several previous results in the literature on stochastic volatility modelling related to the vanilla, the forward, and the VIX implied volatility surfaces. It can be applied to local, stochastic, and also to rough volatilities (driven by a fractional Brownian motion) leading to simple and explicit results.

This book includes several examples and numerical experiments. It is oriented to practitioners, as well as to Ph.D. students with a background in stochastic analysis that are interested in quantitative finance problems. We remark that this book is not a textbook on Malliavin Calculus. Several excellent references (see Di Nunno, Øksendal and Proske (2009), Malliavin and Thalmaier (2006), Nualart (2005), or Sanz-Solé (2005)) offer a more complete and rigorous approach to this theory. Here, our focus is on practical issues, and we aim to show how these issues can be addressed simply through Malliavin calculus.

The text is organised as follows. The first part (Chapters 1 and 2) is oriented to Ph.D. students and researchers that need an introduction to the main concepts in option pricing and volatility modelling. In Chapter 1 we discuss some general concepts such as risk-neutral prices, the Black-Scholes model, and implied volatilities. Chapter 2 is devoted to the estimation and modelling of volatility. In particular, we present some classical models such as local, stochastic, and stochastic-local volatilities, as well as rough volatilities.

The second part (Chapters 3, 4, and 5) focuses on the mathematical tools we need in our analysis. Chapter 3 is a primer on the main concepts and tools of Malliavin calculus, like the derivative and the divergence operators and their basic properties. In Chapter 4 we deep on the main tools of Malliavin calculus that we use through the book, like the Clark-Ocone-Haussmann, the integration-by-parts, or the anticipating Itô's formulas. We also study some applications of these tools. More precisely, we see how to use the Clark-Ocone-Haussmann formula in hedging, the integration by parts formula in the computation of the Greeks, or the anticipating Itô's formula in the analysis of the law of asset prices. In Chapter 5 we introduce the fractional Brownian motion, a process that is the basis of rough volatility models.

In the third part (Chapters 6, 7, and 8) we make use of Malliavin calculus techniques to analyse the at-the-money short-end of the implied volatility. In Chapter 6 we study the short-term implied volatility level and its relationship with the volatility swap and the variance swap. In Chapter 7 we focus on the at-the-money skew and we compare its short-term behaviour for local, stochastic, and rough volatilities, as well as for jump-diffusion models. A similar analysis for the at-the-money curvature is carried out in Chapter 8.

The last part (Chapters 9 and 10) is devoted to non-vanilla smiles. In chapter 9 we study the short-term level, skew, and curvature of the forward smile, while in Chapter 10 we analyse the VIX skew and we study under which conditions this skew is positive, as observed in real market data.

Finally, the book contains a Github repository (that can be accessed here: https://bit.ly/2KNex2Y) with the Python library corresponding to the numerical examples in the text. The library has been implemented so that the users can re-use the numerical code for building their examples. We have split the next way:

- **Examples:** Here the reader can find the numerical experiments performed for each chapter of the book.
- **Monte Carlo Engines:** Where the reader can find the different

models used in the examples (SABR, rBergomi, Heston,..). We have used reduction variance techniques like control variates or antithetic variables.

- **PDE solver engine:** We have included a one-dimensional PDE solver. Under the local volatility model, this numerical is so stable and fast. The user can use generic boundary conditions and solve the PDE using explicit, implicit, or the theta method.

- **Analytic tools:** In this module, the reader can find two numerical methods, Fourier inversion and COS method, to price options under dynamics with known characteristic function. Currently, we have implemented the characteristic function for the Heston and the Merton models. Also, we have included the different approximations for implied volatility, vol swap, variance swap,... that we have obtained using Malliavin tools.

- **Instruments:** In this module, the reader can find the different instruments that we have used in the book. These instruments are independent of the model. For this reason, we cause the same instrument for different MC engines.

- **Tools:** This module contains generic functionalities (random number generator, numba functions, ...) that are used in several parts of the library.

All details of the code can be found in PyStochasticVolatility. To end, we remark that this library does not pretend to be industrial but a tool from where the reader can build examples using the last modules that we have described and a included brief introduction to Python code without scripts.

ACKNOWLEDGEMENTS

We would like to express our gratitude to the people who have helped us in this project. In particular, we would like to thank Dariusz Gatarek for his constant feedback and key suggestions that allowed us to substantially improve our original project. To Jorge A. León, Matt Lorig, Maria Elvira Mancino, Raúl Merino Fernández, Frido Rolloos, and Kenichiro Shiraya for their encouragement, careful reading of previous versions of the manuscript, and fruitful discussions. We are grateful to CaixaBank and especially to Miquel Campaner Perelló for his help with the market data and deep explanations about the equity market. As well, we are indebted to Miguel Canteli Fermández for his smart appraisals about numerical tricks and for sharing his extensive knowledge of exotic derivatives. We would also like to thank our editors Callum Fraser and Mansi Kabra for their help and support. All remaining errors are ours.

Elisa Alòs and David García-Lorite
Barcelona and Madrid, December 2020

I

A primer on option pricing and volatility modelling

CHAPTER 1

The option pricing problem

Financial markets are complex. Not only assets, but also some other instruments such as futures, options, or swaps are traded. Most of these instruments are called **derivatives**, because their value depends on the value of other underlying entities. In general, they are agreements between two parties that promise to sell or to buy something, under some conditions, at (or before) some future moment T that we call the **maturity time**.

From the historical point of view, derivative products appear in the context of harvests. In particular, the first organised market for derivatives was the Dojima Rice Exchange in Osaka (17th century), while the oldest operating derivatives market, the Chicago Board of Trade (CBOT, currently part of the CME group), started as an association of grain merchants (see Algieri (2018)). Nowadays, the underlying asset of a derivative can be a commodity, but also a stock, or the value of an index, or even something more complex, for example the realised variance. These derivatives can be traded on an organised market (that we call **exchange**) or out of these markets (then we say that the derivative is traded **over the counter (OTC)**).

Pricing derivatives is not a trivial problem. One could think that the value of a financial instrument is the expectation of its future value, but this is not true because of several reasons. The first one is the *time value* of money. 10 euros today is not the same as 10 euros in one year, and this means that we have to consider the effect of interest rates. The second one is more relevant: the value of a financial instrument does not only rely on its expected value, but investors take into account other factors as the risk. In general, pricing a derivative from the statistical analysis of the underlying asset is not straightforward. Also, practitioners do not focus on the construction of a perfect model for assets, but their main objective is to *consistently* price derivative instruments.

What does 'consistently' mean? That means, in such a way that we do not allow for **arbitrage** opportunities. We do arbitrage when we construct a zero-risk strategy with a positive benefit. For example, if a portfolio ϕ composed of three financial instruments A, B, C has the same final value (even random) as another portfolio composed of just one derivative D (that is, $\phi_T = D_T$), the price of both portfolios has to be the same. If not, we can construct an arbitrage by simply buying the cheapest and selling the expensive.

The portfolio ϕ is called a **replicating portfolio** for the derivative D. Usually, it is composed of bonds, assets, and some other (simpler) derivatives. Derivatives pricing theory is based on the construction of this replicating portfolio ϕ, and the value of D is defined as the value of ϕ.

Now, we introduce the basic types of derivatives, and we present the mathematical formulation of the option pricing problem.

1.1 DERIVATIVES

The basic types of derivatives are forwards and options. In this section, we define them and introduce the basic concepts related to these instruments.

1.1.1 Forwards and futures

A **forward contract** is an agreement where one party promises to buy an asset from another party at some specified time in the future T and some specified price F. We say that the party that agrees to buy takes the **long position**, while the party that agrees to sell takes the **short** position. We refer to T as the **maturity time** of the contract.

The value at maturity of the contract is called the **payoff**. If we denote by S_t the asset price at time t, this payoff, for the long contract, is equal to

$$S_T - F.$$

That is,

- If $S_T > F$, there is a benefit of $S_T - F$, while
- If $S_T < F$, there is a loss equal to $F - S_T$.

We can write this payoff as $h(S_T)$, where $h(x) := (x - F)$. Then we say that h is the **payoff function** of the long forward contract. Similar reasoning gives us that the payoff of a short forward contract is equal to $F - S_T$, and the payoff function is equal to $h(x) = F - x$.

When entering into a contract, the delivery price F is computed such that the value of the contract is zero for both the long and the short positions (that is, such a that we consider the contract 'fair'). This fair delivery price is called the **forward price**. The value of the contract changes from day to day, being equal to the corresponding payoff at maturity. We will see in Section 1.2 how to compute the forward price F, as well as the price of the forward contract at any time before maturity.

A **futures contract** has the same payoff as a forward contract. The only difference is that forwards are traded OTC, while we call them *futures* when these derivatives are traded on an exchange. The delivery price of a futures contract is called the **future price**.

A particular example of a forward contract is the **variance swap**, where the underlying value is the realised variance of an asset. Similarly, a **volatility swap** is defined as a forward contract on the realised volatility of an asset.

1.1.2 Options

An **option** is a contract that gives its owner the right (but it takes out the obligation) of buying or selling the underlying asset at a prefixed delivery price K that we call the **strike price**. If it gives the holder the right to buy, we say it is a **call** option, and if it gives the right to sell, we say it is a **put** option. The holder of the option is said to take the **long** position, while the one who sells the option is said to take the **short** position.

Options that can be exercised at a fixed maturity T are called **European**. If they can be exercised at any time before maturity, they are called **American**. European and American calls and puts are the simplest options, and we refer to them as **vanilla** options. Options that are not vanillas are called **exotics**. Some common examples include

- **Forward-start options**, where the strike is fixed at a future moment before maturity,
- **Path-dependent options**, where the final payoff depends not only on the final value of the asset but on the whole path of the asset price before maturity. This class includes *barrier* options (the option can be exercised if and only if the price reaches (or not) a specific level), *lookback* options (where the payoff depends on the maximum or minimum asset price during a period), and *Asian* options (the payoff depends on the mean of the asset price in a fixed time interval).
- **Spread options**, whose payoff depends on the difference between two asset prices.
- **Basket options**, whose payoff depends on the value of a set of assets.

European options are defined by their payoff. As a European call option is exercised only if the final value of the asset S_T is greater than K, its payoff is calculated as follows:

$$(S_T - K)_+.$$

That is, the payoff function of a European call is given by $h(x) = (x - K)_+$. A European put option is exercised only when the final value of the asset S_T is less than K and its payoff can be expressed as

$$(K - S_T)_+.$$

Then, the payoff function of a European put is $h(x) = (K - x)_+$.

The **intrinsic value** of a European option at any given time $t < T$ is expressed as $h(S_t)$, where S_t is the price of the asset at time t and h is the corresponding payoff function. If this intrinsic value is positive, we say that the option is **in-the-money** (ITM). If an option has no intrinsic value, we say that the option is **out-of-the-money** (OTM). We define **at-the-money spot** (ATMS) options as having a strike equal to the spot price of the underlying asset, while **at-the-money forward**

(ATMF) options as having a strike equal to the forward price of the underlying. In the sequel, **at-the-money** (ATM) will denote ATMF.

1.2 NON-ARBITRAGE PRICES AND THE BLACK-SCHOLES FORMULA

Let us see how to price a derivative instrument. For the sake of simplicity, let us assume first that the market is composed of a non-risky asset (a bond) β and a non-paying dividend stock S. The price of the bond satisfies that

$$\begin{cases} d\beta_t &= r\beta_t dt, \\ \beta_0 &= 1, \end{cases}$$

for some interest rate r that we assume to be constant. That is, $\beta_t = e^{rt}$. The price of the stock S is a random process in a probability space $(\Omega, \mathcal{F}, P)$. For the moment, we impose no hypotheses on S. We only assume that the market is **viable**. That is, there is no possibility of constructing an arbitrage strategy with bonds and stocks.

In this context, let us first consider the case of a forward contract.

1.2.1 The forward contract

In a long forward contract, the payoff is given by

$$S_T - F.$$

What is the 'fair' value of F? Let us see how to answer this question using non-arbitrage arguments.

We said that, if there is a replicating portfolio ϕ of bonds and stocks with the same payoff as the forward contract, the price of the forward contract is defined as the value of the portfolio ϕ. Observe that this portfolio ϕ can be constructed in a very simple way by taking:

- A long position in the asset (that is, we buy an asset), and
- A short position in $e^{-rT}F$ bonds (that is, we borrow bonds)

As the final value of the asset is S_T and the final value of the bond is e^{rT}, the final payoff of this portfolio is equal to $S_T - F$. That is, ϕ is a portfolio composed of bonds and stocks that replicates the forward contract. The value of this portfolio at any time $t < T$ is given by

$$S_t - e^{-rT} F e^{rt} = S_t - e^{-r(T-t)} F.$$

In particular, when the contract is entered into ($t = 0$), the price is equal to

$$S_0 - e^{-rT} F. \tag{1.2.1}$$

If we want this price to be zero at $t = 0$, we simply need to take

$$F = S_0 e^{-rT}, \tag{1.2.2}$$

which gives us the **non-arbitrage forward price** when the underlying asset is a non-paying dividend stock. We remark that the forward price (1.2.2) is model free, in the sense that we need not assume a specific model for S (but only very simple non-arbitrage arguments) to compute it.

Remark 1.2.1 *The above formula is not valid when the underlying asset provides an income, which can be positive (as in dividend-paying stocks) or negative (as the storage costs of a commodity). In these cases, the computation of the forward price has to be modified taking into account this income (see for example Hull (2016)).*

Remark 1.2.2 (Static and dynamic replication) *The above construction of ϕ is an example of* **static** *replication, where the portfolio is simply constructed at $t = 0$, with no need of rebalancing it before maturity. In the replication of most derivatives, the portfolio has to be reorganised, changing the weight of bonds and assets at different moments before maturity. Then, we say the replication is* **dynamic**.

Remark 1.2.3 (The call-put parity) *Consider a portfolio consisting of a long European call and a short European put on a non-paying dividend stock. The value of this portfolio is equal to $c - p$, where c and p denote the price of the call and the put option, respectively. On the other hand, the payoff of this portfolio is given by*

$$(S_T - K)_+ - (K - S_T)_+ = S_T - K.$$

The same arguments as before allow us to replicate this payoff with a portfolio composed of a long asset and $e^{-rT}K$ *bonds. As the price of this replicating portfolio is equal to* $S_0 - e^{-rT}K$*, we get*

$$c - p = S_0 - e^{-rT}K. \tag{1.2.3}$$

Equation (1.2.3) is a model-free relationship called the call-put parity (see for example Bronzin (1908) and Stoll (1969)).

1.2.2 The price of a European option as a risk-neutral expectation

Consider now the case of a European option with a payoff given by $h(S_T)$, for some real function h.

Let us see first that a static replication is not possible in this context. In fact, in every static replication, the value of the portfolio at every moment $t \in [0, T]$ is given by

$$\alpha\beta_t + \gamma S_t,$$

where α denotes the number of bonds and γ denotes the number of assets. Then, at $t = T$, the value of the portfolio is given by $\alpha\beta_T + \gamma S_T$. If this portfolio replicates the European option, one should have

$$\alpha\beta_T + \gamma S_T = h(S_T),$$

for every possible value of S_T. As the left-hand side is a linear function of S_T while h is non-linear, it is not possible to construct a static replication for a vanilla option.

Let us try to construct a dynamic replication. This means that, fixed $n > 1$, we readjust the portfolio at some moments $\{0 = t_0 < t \cdots t_n = T\}$ of the time interval $[0, T]$, and then α and β change with time. That is, the portfolio ϕ is composed, at every time $t \in [0, T]$, of α_t bonds and γ_t assets, and the price $V_t(\phi)$ of the portfolio at time t is given by

$$V_t(\phi) = \alpha_t\beta_t + \gamma_t S_t.$$

Notice that by changing α_t and β_t at every moment $t_i, i = 1, .., n$, we are just rebalancing the portfolio, without bringing or consuming any wealth.That is,

$$\alpha_{t_i}\beta_{t_i} + \gamma_{t_i}S_{t_i} = \alpha_{t_{i+1}}\beta_{t_i} + \gamma_{t_{i+1}}S_{t_i}.$$

This expression is equivalent to the next one

$$\alpha_{t_{i+1}}\beta_{t_{i+1}} + \gamma_{t_{i+1}}S_{t_{i+1}} - \alpha_{t_i}\beta_{t_i} - \gamma_{t_i}S_{t_i} = \alpha_{t_{i+1}}(\beta_{t_{i+1}} - \beta_{t_i}) + \gamma_{t_{i+1}}(S_{t_{i+}} - S_{t_i}),$$

that can also be written as

$$V_{t_{i+1}}(\phi) - V_{t_i}(\phi) = \alpha_{t_{i+1}}\Delta\beta_{t_i} + \gamma_{t_{i+1}}\Delta S_{t_i},$$

where $\Delta\beta_{t_i} := \beta_{t_{i+1}} - \beta_{t_i}$ and $\Delta S_{t_i} := S_{t_{i+}} - S_{t_i}$. Now, letting $n \to \infty$ the above expression reads

$$dV_t(\phi) = \alpha_t d\beta_t + \gamma_t dS_t = r\alpha_t\beta_t dt + \gamma_t dS_t. \tag{1.2.4}$$

A portfolio ϕ satisfying (1.2.4) is called a **self-financing portfolio.** Notice that the *discounted* value of this portfolio, $\tilde{V}_t(\phi) := \beta_t^{-1}V_t(\phi)$, satisfies that

$$d\tilde{V}_t = \gamma_t d\tilde{S}_t,$$

where $\tilde{S}_t = \beta_t^{-1}S_t$ is the discounted price of the asset. This means, in particular, that

$$\tilde{V}_t(\phi) = \tilde{V}_0(\phi) + \int_0^t \gamma_u d\tilde{S}_u. \tag{1.2.5}$$

Now our aim is to construct a self-financing portfolio with the same final value as the option (that is, $V_T(\phi) = h(S_T)$), and to compute its value at time $t = 0$. It is not obvious that such a portfolio exists. Markets where this is possible for every payoff function h are called **complete** markets. In general, as we will see latter, markets are not complete, but let us assume for the moment that our market is complete.

If the market is complete, there exists a self-financing portfolio ϕ such that $V_T(\phi) = h(S_T)$ and then $\tilde{V}_T(\phi) = \beta_T^{-1}h(S_T)$. This, jointly with (1.2.5) implies that, for every $t \in [0, T]$

$$\beta_T^{-1}h(S_T) = \tilde{V}_t(\phi) + \int_t^T \gamma_u d\tilde{S}_u. \tag{1.2.6}$$

If $\tilde{S}$ is a martingale, $\tilde{V}(\phi)$ will also be a martingale and $\tilde{V}_t(\phi)$ will be simply the conditional expectation (up to time t) of $\beta_T^{-1}h(S_T)$. Obviously we do not know if $\tilde{S}$ is a martingale, but imagine there exists an equivalent probability P^* under which $\tilde{S}$ is a martingale. Then, $\tilde{V}_t(\phi) = \beta_T^{-1}E_t^*(h(S_T))$, where E_t^* denotes the conditional expectation up to time t with respect to the probability P^* and, consequently

$$V_t(\phi) = \beta_T^{-1}\beta_t E_t^*(h(S_T)) = e^{-r(T-t)}E_t^*(h(S_T)). \tag{1.2.7}$$

Now, the key question is: does this probability exist? The answer is given by the First and Second Fundamental Theorems of Asset Pricing (see Harrison and Pliska (1981), Kreps (1981), or Delbaen and Schachermayer(1994)). These results prove the existence and uniqueness of an equivalent probability measure in viable and complete markets. We state here these two theorems as in Lamberton and Lapeyere (2008).

Theorem 1.2.4 (First Fundamental Theorem of Asset Pricing) *The market is viable if and only if there exists a probability measure P^* equivalent to P such that the discounted prices of assets are P^*-martingales.*

Theorem 1.2.5 (Second Fundamental Theorem of Asset Pricing) *A viable market is complete if and only if there exists a unique probability measure P^* equivalent to P such that the discounted prices of assets are P^*-martingales.*

The probability P^* is commonly called the **risk-neutral probability**. This is because, under it, the discounted prices of assets are martingales, and then asset prices are just the expectation (discounted) of their value at maturity. That is, the risk is not considered in the computation of prices, and this price is the one a risk-neutral investor would compute.

Remark 1.2.6 (Incomplete markets) *If the market is viable but not complete, the probability measure P^* is not unique. This is the case when the dynamics of asset prices S depend on another process Y that has a random component on its own. Then, the option cannot be replicated by just bonds and stocks. Following Fouque, Papanicolau, and Sircar (2000), we assume through this book that the market selects a unique probability P^* under which derivative contracts are priced. Then, option prices are still given by*

$$V_t = e^{-r(T-t)} E_t^*(h(S_T)).$$

Under this approach, the common practice is not to estimate the parameters of the model by econometric methods, but to estimate them from derivative data (usually vanilla option prices). Then, vanilla prices are not computed but observed in the market, and they are the inputs in the calibration of our models. Then, these calibrated models are used to price exotics.

1.2.3 The price of a vanilla option and the Black-Scholes formula

We have identified the price of a vanilla option as a risk-neutral expectation. But we do not know about the distribution of S under this risk-neutral measure. To look for an adequate model for S, let us see the problem from another point of view. Assume again a viable and complete market. The price of the replicating portfolio (and then the price of our vanilla option) at any time moment t is given by

$$V_t(\phi) = \alpha_t \beta_t + \gamma_t S_t,$$

where α_t and γ_t denote the number of bonds and stocks, respectively. Notice that then, the price of the option is given by $f(t, S_t)$, for some real function $f : \mathbb{R}^2 \to \mathbb{R}$. That is,

$$f(t, S_t) = V_t(\phi)$$

which implies that

$$d(f(t, S_t) - V_t(\phi)) = 0. \tag{1.2.8}$$

Now, we use this relationship to get information about the pricing formula f and the composition of the replicating portfolio. As $\tilde{S}$ is a martingale under P^*, S is a semimartingale. Assume that S is continuous[1]. Then, a direct application of Itô's formula to $f(t, S_t)$, jointly with (1.2.4), allows us to write

$$\begin{aligned}
&d(f(t, S_t) - V_t(\phi)) \\
&= d(f(t, S_t) - (\alpha_t \beta_t + \gamma_t S_t)) \\
&= \frac{\partial f}{\partial t}(t, S_t)dt + \frac{\partial f}{\partial x}(t, S_t)dS_t + \frac{\partial f}{2\partial x^2}(t, S_t)d\langle S, S\rangle_s \\
&-\alpha_t d\beta_t + \gamma_t dS_t.
\end{aligned} \tag{1.2.9}$$

In order to cancel the terms of first order in dS_t, we can choose $\gamma_t = \frac{\partial f}{\partial x}(t, S_t)$. This derivative is the sensitivity of the price with respect to the asset price S_t, and it is commonly called the **delta** of the option. Then, taking into account that

$$\begin{aligned}
\alpha_t &= \beta_t^{-1}(V_t(\phi) - \gamma_t S_t) = \beta_t^{-1}(f(t, S_t) - \gamma_t S_t)) \\
&= \beta_t^{-1}\left(f(t, S_t) - \frac{\partial f}{\partial x}(t, S_t)S_t\right),
\end{aligned}$$

[1] The case of non-continuous models for asset prices is briefly discussed in Chapter 7. We refer to Cont and Tankov (2009) for a complete introduction to this field.

(1.2.9) reads

$$
\begin{aligned}
&d(f(t,S_t)-V_t(\phi))\\
&=\frac{\partial f}{\partial t}(t,S_t)dt+\frac{\partial f}{2\partial x^2}(t,S_t)d\langle S,S\rangle_s\\
&-\beta_t^{-1}\left(f(t,S_t)-\frac{\partial f}{\partial x}(t,S_t)S_t\right)d\beta_t\\
&=\frac{\partial f}{\partial t}(t,S_t)dt+\frac{\partial f}{2\partial x^2}(t,S_t)d\langle S,S\rangle_s\\
&-r\left(f(t,S_t)-\frac{\partial f}{\partial x}(t,S_t)S_t\right)dt.
\end{aligned}
\tag{1.2.10}
$$

As $d(f(t,S_t)-V_t(\phi))=0$, our pricing formula $f(t,S_t)$ should satisfy the following partial differential equation (PDE)

$$
0=\left(\frac{\partial f}{\partial t}(t,S_t)-r\left(f(t,S_t)-\frac{\partial f}{\partial x}(t,S_t)S_t\right)\right)dt+\frac{\partial f}{2\partial x^2}(t,S_t)d\langle S,S\rangle_t.
$$

The properties of the quadratic variation allow us to write

$$
d\langle S,S\rangle_t=S_t^2 d\langle X,X\rangle_t,
$$

where $X:=\ln S$ denotes the log-price. Empirical observations give us that $\langle X,X\rangle_t=O(t)$. If we assume that $d\langle X,X\rangle_t=\sigma^2 dt$ for some constant σ that we call the **volatility**, the above PDE can be written as

$$
0=\frac{\partial f}{\partial t}(t,S_t)-r\left(f(t,S_t)-\frac{\partial f}{\partial x}(t,S_t)S_t\right)+\frac{\sigma^2}{2}\frac{\partial^2 f}{\partial x^2}(t,S_t)S_t^2.
\tag{1.2.11}
$$

Then, $f(t,S_t)$ is the solution of a PDE, with the final condition $f(T,S_T)=h(S_T)$. Equation (1.2.11) is the well-known **Black-Scholes equation** (see Black and Scholes (1973) and Merton (1973)). When $h(x)=(x-K)_+$, the solution is the **Black-Scholes price** for a call

$$
c(t,S_t)=S_tN(d_1)-Ke^{-r(T-t)}N(d_2),
$$

where N denotes the normal distribution function and

$$
\begin{aligned}
d_1&=\frac{\ln S_t-\ln K+r(T-t)}{\sigma\sqrt{T-t}}+\frac{\sigma\sqrt{T-t}}{2},\\
d_2&=\frac{\ln S_t-\ln K+r(T-t)}{\sigma\sqrt{T-t}}-\frac{\sigma\sqrt{T-t}}{2}.
\end{aligned}
$$

The corresponding put price is given by

$$
p(t,S_t)=Ke^{-r(T-t)}N(-d_2)-S_tN(-d_1).
$$

1.3 THE BLACK-SCHOLES MODEL

After deriving the Black-Scholes option pricing formula, our objective is to find a model such that the discounted risk-neutral expectation $\beta_T^{-1} E_t^*(h(S_T))$ coincides with the corresponding Black-Scholes price. Towards this end, we use the relationship between PDEs and probability expectations.

1.3.1 From the Black-Scholes formula to the Black-Scholes model

The Feynman-Kack formula (see for example Borodin (2000)) links the solution of a PDE like (1.2.11) with expectations. In particular, one can see that $f(t, S_t) = \beta_T^{-1} E_t^*(h(S_T))$ where, under our risk-neutral probability P^*,

$$dS_t = rS_t dt + \sigma S_t dW_s, \tag{1.3.1}$$

for some P^*-Brownian motion W. That is, the Black-Scholes price $f(t, S_t)$ matches the discounted expectation $\beta_T^{-1} E_t^*(h(S_T))$ under the risk-neutral model (1.3.1). We remark that we have arrived at this model by simply applying non-arbitrage arguments and assuming that the volatility is constant. Notice also that this is not a model for stock prices in the real world, but simply a risk-neutral model that matches these Black-Scholes prices.

If we assume the stock prices to follow (under the real probability) a geometric Brownian motion of the type

$$dS_t = \mu S_t dt + \sigma S_t dB_s, \tag{1.3.2}$$

for some constant μ and some P-Brownian motion B, Girsanov theorem (see for example Musiela and Rutkowski (2005)) gives us that the stock price follows, under a risk-neutral probability P^* and for some P^*-Brownian motion W, the model (1.3.1). When asset prices follow a geometric Brownian motion, we say they satisfy the **Black-Scholes model.** Notice that, even the drift coefficients μ and r are different, the volatility σ is the same under both probabilities. Moreover, as option prices are computed under the risk-neutral probability, the drift coefficient μ does not affect them.

The solution of Equation (1.3.2) can be expressed explicitly. A direct application of Itô's formula gives us that

$$d \ln S_t = \left(\mu - \frac{1}{2}\sigma^2\right) dt + \sigma dB_t. \tag{1.3.3}$$

Then,

$$\ln S_t = \ln S_0 + \left(\mu - \frac{1}{2}\sigma^2\right) t + \sigma B_t,$$

which implies

$$S_t = S_0 \exp\left(\left(\mu - \frac{1}{2}\sigma^2\right) t + \sigma B_t\right). \qquad (1.3.4)$$

That is, S_t is the exponential of a Brownian motion with drift. As S_t is the exponential of a normal random variable, it is log-normally distributed. Moreover, notice that the **returns** $\frac{dS_t}{S_t}$ satisfy

$$\frac{dS_t}{S_t} = \mu dt + \sigma dB_s$$

and then they are Gaussian. Equation (1.3.3) gives us that the *log-returns* $d \ln S_t$ are also Gaussian, but notice that the mean of the log-returns $\left(\left(\mu - \frac{1}{2}\sigma^2\right) dt\right)$ does not coincide with the mean of the returns (μdt).

Obviously, under the risk-neutral probability, the above computations apply, taking $\mu = r$. Then the expected returns are equal to rdt, while the expected log-returns are given by $\left(r - \frac{1}{2}\sigma^2\right) dt$.

The Black-Scholes model is not able to reproduce the complexity of real market data. In particular, we can see that empirical returns are not independent Gaussian variables (see Example 1.3.1). Several extensions of this model have been proposed in the literature. The most common alternative in practice (and the one that we will study in this book) is to allow the volatility to be a stochastic process. Some other interesting approaches are based on the inclusion of jumps in asset prices (see for example Cont and Tankov (2004)) or in the replacement of the Brownian motion by another adequate process (see Mandelbrot (2001)).

Example 1.3.1 (Fat tails) *Figure 1.1 corresponds to the Q-Q plot of the daily log-returns of EURO STOXX 50 from August 10, 2010 to August 7, 2020 (data courtesy of CaixaBank). This analysis indicates the presence of fat tails, suggesting that the probability of extreme events is higher than for a Gaussian distribution. The existence of fat tails is one of the main departures from normality of asset returns, and it has been largely observed in the literature (see for example Aldrich, Heckenbach, and Laughlin (2015)).*

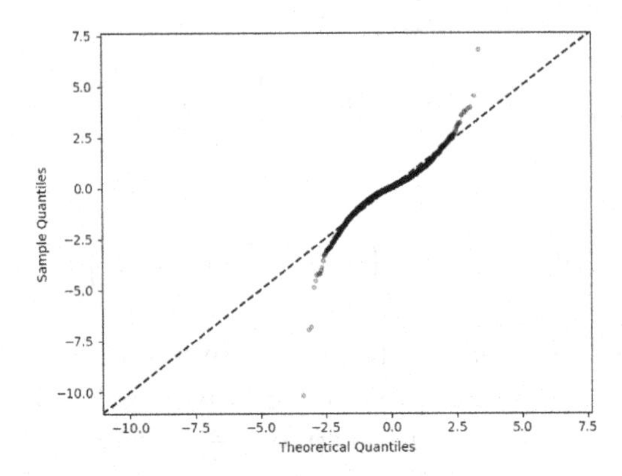

Figure 1.1 Q-Q plot of the daily log-returns of EURO STOXX 50 from August 10, 2010 to August 7, 2020. Data courtesy of CaixaBank.

Remark 1.3.2 *Brownian motion was first used in market modelling in Bachelier (1900), where the first theory of option pricing was presented. Bachelier model for asset prices assumes the returns to be normal (instead of log-normal), and it is used when the underlying asset can have negative prices, as in the case of spread options. Recently, during the COVID-19 crisis, the oil prices fell significantly, reaching for the first time in history negative values in April 2020. This motivated the change of oil options models from Black-Scholes to Bachelier by the CME group. A comparison between Bachelier and Black-Scholes prices can be found in Schachermayer and Teichmann (2007).*

1.3.2 Option replication and delta hedging in the Black-Scholes model

The results in Section 1.2.3 imply that, under the Black-Scholes model, a vanilla option can be replicated by a portfolio composed of bonds and stocks, where the number of assets at every moment t is given by the corresponding delta ($\frac{\partial c}{\partial S_t}$ in the case of a call and $\frac{\partial p}{\partial S_t}$ in the case of a put). Notice that, as delta changes at every time t, the composition of the portfolio has to be adjusted continuously. In real market practice, this adjustment cannot be done continuously but only at some discrete moments. This leads to a hedging error due to the discretisation, depending on the rebalancing frequency. The impact of this discrete-time hedging has been studied by several authors (see for example Hayashi and Mykland (2005), Cai, Fukasawa, Rosenbaum, and Tankov (2016),

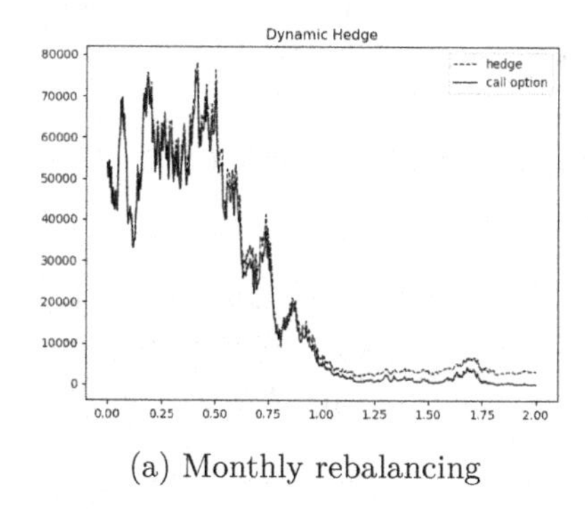

(a) Monthly rebalancing

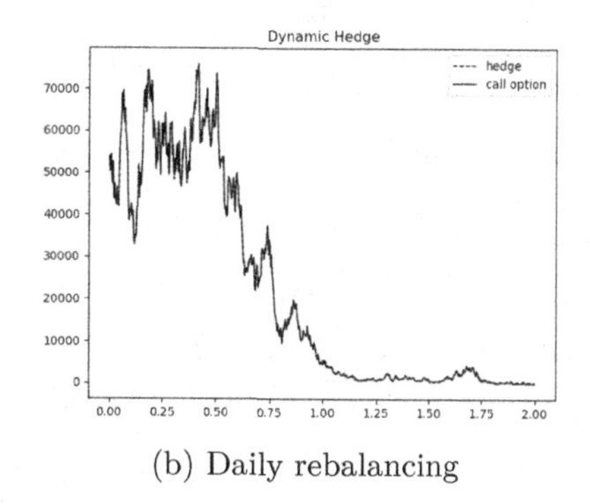

(b) Daily rebalancing

Figure 1.2 Monthly rebalanced vs daily rebalanced replicating portfolios for a call option with $K = 100$ and $T = 2Y$ under a Black-Scholes model with $r = \mu = 0.02$, and $\sigma = 0.4$.

Alòs and Fukasawa (2020), and the references therein). We can observe this phenomenon in the following example.

Example 1.3.3 *Consider a call option with $K = 100$ and $T = 2Y$ under a Black-Scholes model with $r = \mu = 0.02$, and $\sigma = 0.4$. In Figure 1.2, we have shown a simulation of the evolution of the call option price and we compare it with a) the corresponding monthly rebalanced replicating portfolio and b) the daily adjusted replicating portfolio. We can see clearly that the hedging error decreases as the rebalancing frequency increases.*

In the construction of the replicating portfolio, we have hedged the risk associated with changes in the asset price by buying delta units of the underlying asset. This technique is called **delta hedging**. Notice that then, the delta (that is, the derivative with respect to the asset price) of the hedged portfolio composed of

- the option and
- delta units of the asset

is equal to zero. This means that it is insensitive to small movements in the underlying asset price. We remark that in incomplete markets options cannot be replicated by just bonds and stocks, and then delta hedging is not enough to hedge the option. In this case, some common alternatives associated consist of hedging with the underlying and a vanilla option, in such a way we make zero not only the delta of the hedged portfolio but also its **gamma** (the second derivative with respect to the asset price) or its **vega** (the derivative with respect to the volatility). These techniques are called **gamma** and **vega** hedging, respectively.

1.4 THE BLACK-SCHOLES IMPLIED VOLATILITY AND THE NON-CONSTANT VOLATILITY CASE

The implied volatility of a traded vanilla option I_t is defined as the unique volatility parameter one should put in the Black-Scholes formula to get the market option price V_t. That is, the quantity I_t such that

$$BS(t, S_t, I_t) = V_t,$$

where $BS(t, S_t, \sigma)$ denotes the corresponding Black-Scholes price with volatility σ and for some payoff $h(S_T)$. As famously stated by Rebonato (1999), *the implied volatility is the wrong number to put in the wrong formula to get the right price.* This section is devoted to present the empirical properties of the implied volatility, as well as its connection with the *spot* volatility σ_t. Excellent introductions to this field can be found in Gatheral (2006) and Lee (2005).

1.4.1 The implied volatility surface

An empirical analysis shows that, at every time moment t, the computed implied volatility of a vanilla is not constant but is a function of the strike K (or in other words, a function of the **log-moneyness** $\ln\left(\frac{F}{K}\right)$[2], where F denotes the asset forward price) and time to maturity. The corresponding plot is called the **implied volatility surface**. Fixed a time to maturity T, the implied volatility tends to exhibit a U-shaped pattern across different strikes. If it is symmetric, we refer to this as the **smile effect**. If asymmetric, we talk about the **skew effect**. In equity markets, the skew is the more typical pattern after 1987.

Empirical smiles and skews change with time to maturity T, being very pronounced for short maturities and flattening as T increases. A popular rule-of-thumb (see Lee (2002)) states that skew slopes decay with T approximately as $\frac{1}{\sqrt{T}}$. Moreover, these skews become more symmetric as time to maturity decreases. We can observe these phenomena in the following example.

[2]The term *moneyness* is not completely standardised in the literature and it may refer to $\frac{F}{K}$ or $\frac{K}{F}$.

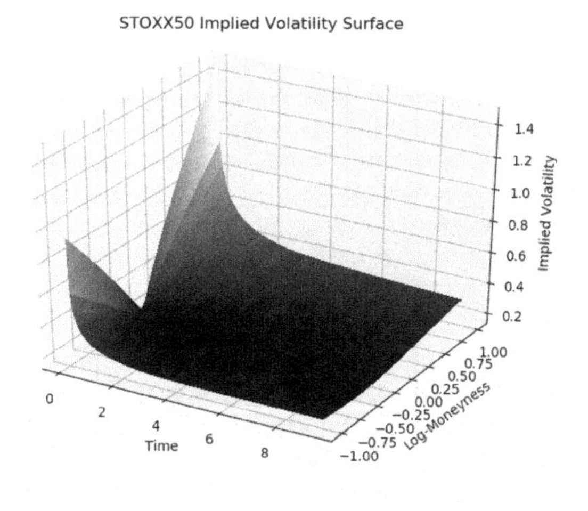

Figure 1.3 EURO STOXX 50 implied volatility surface as of August 17, 2020. Data courtesy of CaixaBank.

Example 1.4.1 *In Figure 1.3 we can see the implied volatility surface of EURO STOXX 50 as of August 17, 2020 (data courtesy of CaixaBank). We can see the typical behaviour of an implied volatility surface as a function of time to maturity and the log-moneyness* $z = \ln(\frac{F}{K})$. *We observe a strong skew/smile effect for short maturity options, followed by a flattening as time to maturity increases. In Figure 1.4 we plot the implied volatility as a function of the log-moneyness for some fixed maturities. Notice that the skew flattens as time to maturity increases. Moreover, the skews become more symmetric for short maturities.*

One of the main goals in option pricing is the development of a model for the spot volatility, in such a way it reproduces the main properties of the observed implied volatilities. This is a complex problem that we will address through this book. The next section, where we see that the implied volatility can be written in terms of weighted expectations of spot volatilities, is the first step in this direction.

1.4.2 The implied and spot volatilities

What is the relationship between the spot and implied volatilities? The following reasoning, gives us a representation of the implied volatility as a weighted mean of the future values of the volatility process. Let us denote by I_0 the implied volatility at $t = 0$. Obviously, $BS(t, S_t, I_0)$

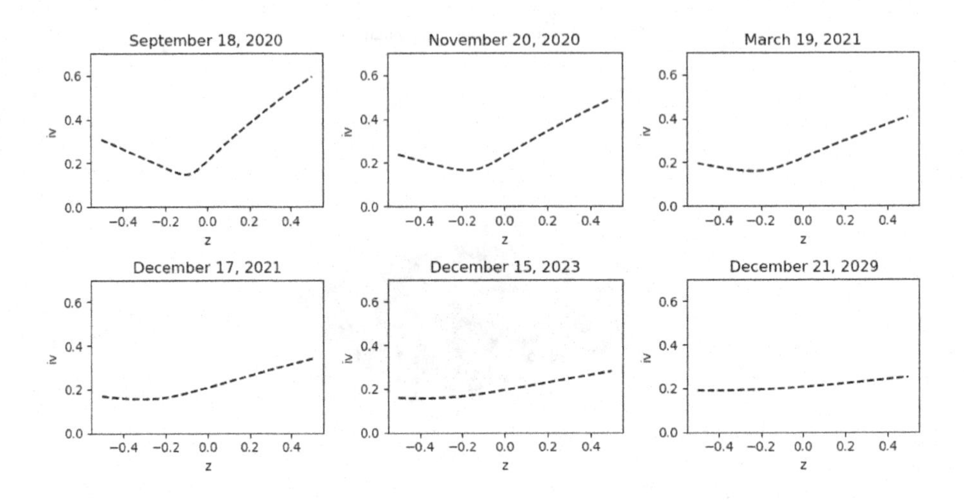

Figure 1.4 EURO STOXX 50 smiles for different maturities, as of August 17, 2020. Data courtesy of CaixaBank.

satisfies the Black-Scholes equation (1.2.11). Then, replacing σ by I_0

$$
\begin{aligned}
0 =& \frac{\partial BS}{\partial t}(t, S_t, I_0) - r\left(BS(t, S_t, I_0) - \frac{\partial BS}{\partial x}(t, S_t, I_0)S_t\right) \\
&+ \frac{I_0^2}{2}\frac{\partial^2 BS}{\partial x^2}(t, S_t, I_0)S_t^2. \qquad (1.4.1)
\end{aligned}
$$

On the other hand, Itô's formula applied to the process $e^{rt}BS(t, S_t, I_0)$ gives us that

$$
\begin{aligned}
&e^{-rT}BS(T, S_T, I_0) - BS(0, S_0, I_0) \\
&= \int_0^T e^{-rt}\left(\frac{\partial BS}{\partial t}(t, S_t, I_0) - rBS(t, S_t, I_0) + \frac{\sigma_t^2}{2}\frac{\partial^2 BS}{\partial x^2}(t, S_t, I_0)S_t^2\right)dt \\
&+ \int_0^T \frac{\partial BS}{\partial x}(t, S_t, I_0)dS_t. \qquad (1.4.2)
\end{aligned}
$$

Now, because of (1.2.7) and Remark 1.2.6, and taking into account the final condition for the Black-Scholes formula $BS(T, x, \sigma) = h(x)$, we have

$$
V_0 = e^{-rT}E^*(h(S_T)) = e^{-rT}E^*(BS(T, S_T, I_0)),
$$

for some risk-neutral probability P^*. Then, taking expectations

$$\begin{aligned}
&V_0 - BS(0, S_0, I_0) \\
&= E^* \int_0^T e^{-rt} \left(\frac{\partial BS}{\partial t}(t, S_t, I_0) - r(BS(t, S_t, I_0) + \frac{\partial BS}{\partial x}(t, S_t, I_0) r S_t \right. \\
&\left. + \frac{\sigma_t^2}{2} \frac{\partial^2 BS}{\partial x^2}(t, S_t, I_0) S_t^2 \right) dt. \qquad (1.4.3)
\end{aligned}$$

This, jointly with (1.4.1), leads to

$$\begin{aligned}
&V_0 - BS(0, S_0, I_0) \\
&= E^* \int_0^T e^{-rt} \left(\frac{\sigma_t^2 - I_0^2}{2} \frac{\partial^2 BS}{\partial x^2}(t, S_t, I_0) S_t^2 \right) dt. \qquad (1.4.4)
\end{aligned}$$

As from the definition of implied volatility, $V_0 - BS(0, S_0, I_0) = 0$, the term in the right-hand side of (1.4.4) is zero, and then

$$I_0^2 = \frac{E^* \int_0^T e^{-rt} \sigma_t^2 \frac{\partial^2 BS}{\partial x^2}(t, S_t, I_0) S_t^2 dt}{E^* \int_0^T e^{-rt} \frac{\partial^2 BS}{\partial x^2}(t, S_t, I_0) S_t^2 dt}. \qquad (1.4.5)$$

Under some general conditions, the above equality implies that I_0^2 tends to σ_0^2 as maturity T tends to zero. That is, the short-time limit of the ATMI is the spot volatility. Short-time limit results for the short-time behaviour of the skew $\frac{\partial I_0}{\partial K}$ and the curvature $\frac{\partial I_0^2}{\partial K^2}$ have been studied for several authors in the context of different models with random volatilities (see for example Alòs, León, and Vives (2007), Alòs, León, Pontier, and Vives (2008), Bergomi (2009b), Forde and Jacquier (2009, 2011), Forde, Jacquier, and Lee (2012), Durrleman (2010), Alòs and León (2015), and Fukasawa (2017), among others). Malliavin calculus techniques will allow us to find analytical expressions for the short-end behaviour of the ATM level, skew, and curvature, as we will see in Chapters 6, 7, and 8.

1.5 CHAPTER'S DIGEST

Forward contracts are agreements between two parties calling for the sale of an asset on a future date (the maturity time), at some specified price that we call the forward price. Futures are similar contracts as forwards, but they are traded on an exchange, while forwards are traded OTC. Particular cases of forward contracts are variance swaps and volatility swaps, where the asset is a measure of the variability of asset prices.

In an option contract, the owner has the right (but not the obligation) to buy or sell some asset on a future date at some specified price that we call the strike price. A call option gives the right to buy, while a put option gives the right to sell. European options can be exercised only at maturity, while American options can be exercised at any time before the expiration date. European and American calls and puts are the simplest examples of options, and we usually refer to them as vanilla options, while non-vanillas are called exotics. Some examples of exotics include forward-starting options, path-dependent options (lookback, barrier, or Asian, among others), spread options, and basket options.

European options are defined by their payoff function h. At any time t, the intrinsic value of an option is given by $h(S_t)$. If $h(S_t) > 0$, the option is in-the-money, while if there is no intrinsic value, the option is out-of-the-money. In ATM spot options, the strike is equal to the asset price, while in ATM forward (or simply, ATM) options the strike is equal to the forward price of the asset.

The forward price and the value of a forward contract can be obtained by simple non-arbitrage arguments, based on the concept of replicating portfolio. As a portfolio composed of a long call and a short put has a linear payoff, it is easy to see that call and put prices satisfy the so-called call-put parity.

Using again non-arbitrage arguments, we see that European option prices can be expressed as the discounted expectation of the payoff under a risk-neutral probability. The First Fundamental Theorem of Asset Pricing guarantees that this risk-neutral probability exists if the market is viable (that is, if there are no arbitrage opportunities), and the Second Fundamental Theorem of Asset Pricing gives us that this probability is unique if the market is complete (that is, if every option can be replicated by just bonds and stocks). For continuous asset prices and under the hypotheses of constant volatility, Itô's formula allows us to see that these prices have to satisfy the so-called Black-Scholes partial differential equation, whose solution is given by the Black-Scholes formula. Then, by Feynman-Kack lemma, we see that the model for asset prices that fits the Black-Scholes formula is a geometric Brownian motion. Asset prices that follow a geometric Brownian motion are said to satisfy the Black-Scholes model. Under this model, delta hedging allows us to replicate the payoff of European options.

The Black-Scholes formula is linked to the concept of implied volatility, defined as the volatility value one should put in the Black-Scholes formula to get the market option price. The plot of the implied volatility as a function of the strike and time to maturity is called the implied volatility surface. This surface tends to exhibit some common patterns, as the skew/smile effect, more pronounced and symmetric for short maturities. The ATM implied volatility can be written as the expectation of a weighted mean of future spot volatilities, and its short-time limit coincides with the current value of the spot volatility. Chapters 6, 7, and 8 will be devoted to study the short-end behaviour of the ATM implied volatility level, skew, and curvature.

CHAPTER 2

The volatility process

This chapter focuses on the study of the spot volatility defined as $\sigma_t^2 := \frac{d\langle X,X\rangle_t}{dt}$ (where $X := \ln S$). We see its main empirical properties, as well as the main volatility models proposed in the literature.

2.1 THE ESTIMATION OF THE INTEGRATED AND THE SPOT VOLATILITY

The volatility σ is a process. For a given $t > 0$, the random variable σ_t is called the **spot** or **instantaneous** volatility, while the integral $\int_0^T \sigma_s^2 ds$ of the squared spot volatility over a time interval $[0,T]$ is called the **integrated volatility** (or the **integrated variance**). Both the spot and the integrated volatility cannot be observed directly in the market, but they have to be estimated from asset prices. This chapter is devoted to present the main tools in this estimation.

2.1.1 Methods based on the realised variance

Consider a time interval $[s,t]$ and a partition $\{s = t_0, t_1, ..t_n = t\}$ of size n of this time interval. Then, the realised variance over this time interval $[s,t]$ is defined as

$$RV_n^{[s,t]} := \sum_{i=1}^{n} r_{s,t,i}^2, \tag{2.1.1}$$

where $r_{s,t,i} = X_{t_{i+1}} - X_{t_i} = \ln S_{t_{i+1}} - \ln S_{t_i}$ are the log-returns. Because of the definition of quadratic variation, we know that

$$\lim_{n\to\infty} RV_n^{[s,t]} = \langle X,X\rangle_t - \langle X,X\rangle_s.$$

Now, as $\sigma_t^2 dt = d\langle X, X\rangle_t$, it follows that

$$\lim_{n\to\infty} RV_n^{[s,t]} = \int_s^t \sigma_u^2 du.$$

Then, a natural estimator of the integrated volatility is given by the realised variance. Nevertheless, the goodness of the above approximation depends on the observability of the true price process. Observed asset prices are contaminated by microstructure noise (due to discretisation, asynchronicity, etc.), especially if intraday returns are sampled at high frequencies. Then, the application of techniques to mitigate the noise is interesting in practice. Some approaches introduced in the literature include preaveraging (Jacod, Li, Mykland, Podolskij, and Vetter (2007)), multiscale (Mykland and Ait-Sahalia (2005), and Zhang (2006)), and kernel-based estimators (Barndorff-Nielsen, Hansen, Lunde, and Shephard (2009)).

Suppose we want to estimate the daily spot volatility. Towards this end, we recall the definition

$$\sigma_t^2 := \frac{d\langle X, X\rangle_t}{dt}. \tag{2.1.2}$$

The same arguments as before give us that, for the t-th day, $d\langle X, X\rangle_t$ can be estimated as the realised variance RV_n^t defined as

$$RV_n^t = \sum_{i=1}^n r_{t,i}^2, \tag{2.1.3}$$

where $r_{t,i} = X_{t_{i+1}} - X_{t_i}$ are the intradaily log-returns on the t-th day. On the other hand, as the time is usually expressed in a year scale, we have $dt = 1/N$ years, where N denotes the number of trading days of the year (usually, N is around 253). Then, (2.1.2) leads to the following approximation for the spot volatility

$$\sigma_t^2 \approx N \times RV_n^t. \tag{2.1.4}$$

Again, we have to take into account that this estimation can be sensitive to noise. An alternative approach, based on Fourier analysis, is presented in the next section.

2.1.2 Fourier estimation of volatility

Here we present the Fourier estimation of volatility introduced in Malliavin and Mancino (2002). The key point of this methodology is to use the connection between the Fourier transform of the log-returns and the Fourier transform of the integrated volatility $\int_0^{2\pi} \sigma_s^2 ds$ over the time interval $[0, 2\pi]$. More precisely, it can be proved that

$$\int_0^{2\pi} \sigma_s^2 ds = (2\pi)^2 (\mathcal{F}(dX) *_B \mathcal{F}(dX))(0), \tag{2.1.5}$$

where $\mathcal{F}(dX)$ denotes the Fourier transform of the log-returns dX and $*_B$ is the Bohr convolution

$$(f *_B g)(k) = \lim_{N\to\infty} \frac{1}{2N+1} \sum_{s=-N}^{s=N} f(s)g(k-s).$$

Now, consider a partition $\{0 = t_0, ..., t_n = 2\pi\}$ of $[0, 2\pi]$ and the corresponding sample of log-returns

$$\delta(X)_{i,n} = (X)_{t_i} - (X)_{t_{i-1}}.$$

Then, the finite sample Fourier estimator of the integrated volatility is defined as

$$\hat{\sigma}^2_{n,N} = \frac{(2\pi)^2}{2N+1} \sum_{s=-N}^{s=N} (\mathcal{F}(dX)_n(s)_B \mathcal{F}(dX)_n)(-s), \tag{2.1.6}$$

where

$$\mathcal{F}(dX)_n(s) = \frac{1}{2\pi} \sum_{j=1}^{n} \exp(-ist_j)\delta_j(X).$$

Straightforward computations allow us to write the estimator (2.1.6) as

$$\hat{\sigma}^2_{n,N} = \sum_{j,l} D_N(t_j - t_l)\delta_j(X)\delta_l(X). \tag{2.1.7}$$

where $D_N(t_j - t_l)$ denotes the Dirichlet kernel

$$D_N(t) = \sum_{s=-N}^{s=N} \exp(ist) = \frac{1}{2N+1} \frac{\sin[(N+\frac{1}{2})t]}{\sin\frac{t}{2}}.$$

Remark 2.1.1 *Equation (2.1.7) can be written as*

$$\hat{\sigma}^2_{n,N} = RV_n + \sum_{j\neq l} D_N(t_j - t_l)\delta_j(X)\delta_l(X), \tag{2.1.8}$$

where $RV_n = \sum_j \delta_j^2(X)$ denotes the realised variance. The cross-terms contribute to the robustness of the estimator to microstructure noise (see for example Zhou (1996) and Barndorff-Nielsen, Hansen, Lunde, and Shephard (2008)). Numerical analysis show, that the Fourier estimator $\hat{\sigma}^2_{n,N}$ is more robust than the realised variance RV_n estimation under the presence of noise and it is asymptotically unbiased if $\frac{N^2}{n}$ tends to zero (see for example Mancino and Sanfelici (2008)).

Because of the Fourier-Fejér inversion formula, we can define the Fourier estimator of the spot volatility as

$$\hat{\sigma}^2_{n,N,M}(t) = \sum_{j,l} F_M(t-t_l)D_N(t_j-t_l)\delta_j(X)\delta_l(X), \tag{2.1.9}$$

where F_M is the Fejér kernel defined as

$$F_M(x) = \sum_{s=-N}^{s=N}\left(1-\frac{|n|}{N}\right)\exp(isx) = \frac{1}{N+1}\frac{\sin^2[(N+1)x/2]}{\sin^2(\frac{x}{2})}.$$

$\hat{\sigma}^2_{n,N,M}(t)$ is a global estimator, in the sense that it tends, as $n, N, M \to \infty$, uniformly to the spot volatility σ_t. Moreover, it relies on the integration of time series (not on differentiation), which makes it computationally stable. Nevertheless, the rate of convergence of this estimator depends on the choice of the cutting frequencies N and M (see Mancino, Recchioni, and Sanfelici (2017)). The optimisation of Fourier estimators in different scenarios is the object of ongoing research (see for example Wang (2014), Chen (2019), or Chang (2020)). An extension of this methodology that is robust against the presence of jumps can be found in Cuchiero and Teichman (2015).

Remark 2.1.2 *Equation (2.1.9) can be written as*

$$\begin{aligned}\hat{\sigma}^2_{n,N,M}(t) &= \sum_j F_M(t-t_j)\delta_j^2(X) \\ &+ \sum_{j\neq l} F_M(t-t_l)D_N(t_j-t_l)\delta_j(X)\delta_l(X).\end{aligned} \tag{2.1.10}$$

The first term in the right-hand side of the above equality behaves like the kernel-based estimators in Fan and Wang (2008) and Kristensen (2010). The second term, which contains the cross-terms, contributes again to the robustness to microstructure noise.

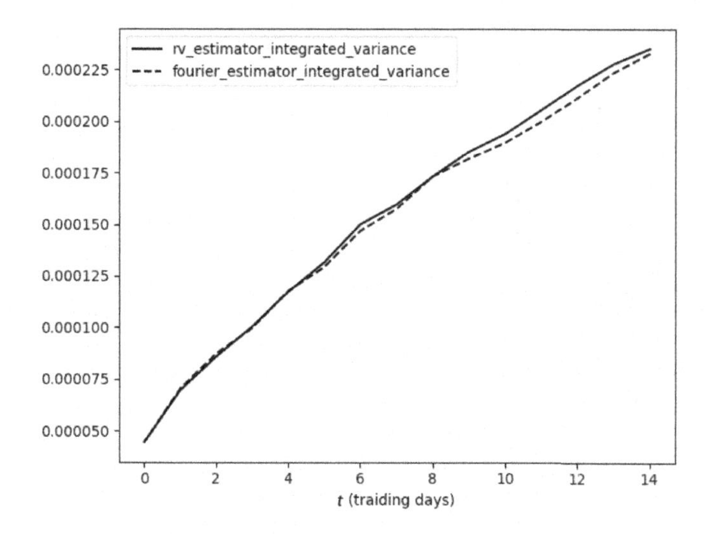

Figure 2.1 Integrated volatility estimators for the SPX index, from January 2, 2013 to January 23, 2013.

Example 2.1.3 *In Figures 2.1 and 2.2 we have considered a dataset containing the 1-minute prices of the SPX index, from January 2, 2013 to January 23, 2013 ($n = 6061$, source: FirstRate Data , https://firstratedata.com/). Then, we have applied the above-realised variance and Fourier estimators for the integrated and the spot volatilities. In the Fourier analysis, we have taken $N = n/2$ for the integrated variance and $N = 0.85n^{\frac{3}{4}}$ and $M = \frac{1}{16\pi}\sqrt{n^{\frac{3}{4}}}\ln(n^{\frac{3}{4}})$ for the spot volatility.*

2.1.3 Properties of the spot volatility

In Figure 2.3 we have shown the EURO STOXX 50 5-minutes subsampled realised variance, from January 3, 2000 to December 11, 2003, downloaded from Gerd, Lunde, Shephard, and Sheppard (2009). The statistical analysis of volatility paths has revealed some stylised facts:

- **The presence of jumps**, observed in empirical studies such as Eraker, Johannes, and Polson (2003) and Todorov and Tauchen (2011), among others.

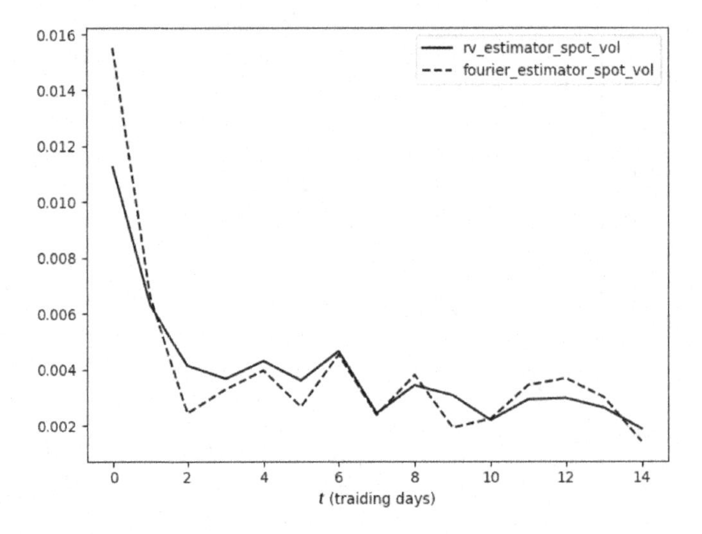

Figure 2.2 Spot volatility estimators for the SPX index, from January 2, 2013 to January 23, 2013.

- **The 'roughness' of volatility paths.** The volatility paths tend to be rougher than those we should expect for classical diffusion volatilities. In particular, they are more compatible with models based on the fractional Brownian motion (fBm)[1] with Hurst parameter $H < \frac{1}{2}$ (see Gatheral, Jaisson, and Rosenbaum (2018)). Girsanov theorem gives us that, under any equivalent martingale measure, the volatility is still rough (see Bayer, Friz, and Gatheral (2016)). These results fit the previous results in Alòs, León, and Vives (2007), where it was established that (risk-neutral) volatility models based on a fBm with $H < \frac{1}{2}$ fit the short-end blow-up of the implied volatility skew (see Chapter 7).

- **The volatility clustering**. As observed by Mandelbrot (1963): 'large changes tend to be followed by large changes, of either sign, and small changes tend to be followed by small changes.' This implies that the volatility tends to persist and that there are periods of low volatilities followed by periods of slow volatilities (see

[1]The fBm with Hurst parameter H is an extension of the Brownian motion whose paths are α-Hölder continuous for all $\alpha < H$. If $H = \frac{1}{2}$, it coincides with the classical Brownian motion. If $H < \frac{1}{2}$ its paths are rougher than for the Brownian motion and if $H > \frac{1}{2}$ we have smoother paths (see Chapter 5).

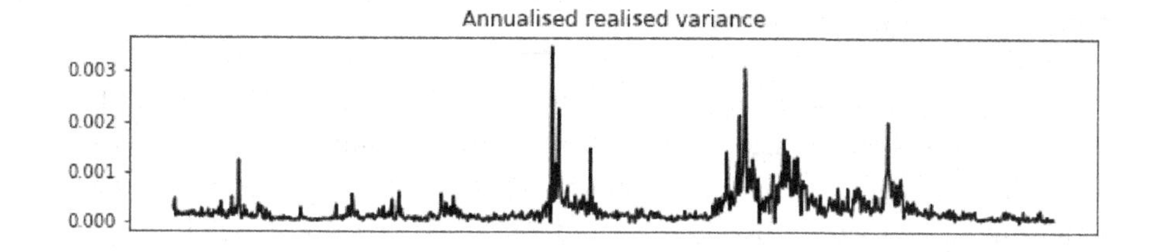

Figure 2.3 EURO STOXX 50 annualised realised variance (5-minutes subsampled), from January 3, 2000 to December 11, 2003 (data downloaded from Gerd, Lunde, Shephard, and Sheppard (2009)).

Baillie and Bollerslev (1989), Bollerslev (1987), Hsieh (1989), McCurdy and Morgan (1988), Ding and Granger (1996), Andersen and Bollerslev (1997), Andersen, Bollerslev, Diebold, and Labys (2003), and Cont (2007), among others).

- The volatility persistence is linked to **long-memory properties**, in the sense that the autocorrelation function of its increments decays slowly (see for example Harvey (1993) and Breidt, Crato, and De Lima (1993)). In Comte and Renault (1998) it was proved that long-memory volatility models are adequate to reproduce the slow flattening of the implied volatility skews and smiles observed in real market data.

The long-memory property of volatility has been a subject of controversy in recent years. From one side, some authors point out that the observed long memory in the volatility would be spurious (see Mikosch and C. Stărică (2000) and Gatheral, Jaisson, and Rosenbaum (2018)). On the other hand, the roughness of the volatility path is compatible with a fractional Brownian motion with a Hurst parameter $H < \frac{1}{2}$, a process that does not exhibit long-memory but short-memory properties, in the sense the autocorrelation function of its increments has a fast decay. We would like to notice that both properties (short and long memory) *are not incompatible.* A process can depend both on long-term and short-term components, and then each one of these components dominates at different scales and, in consequence, at different maturities in the implied volatility surface. We will discuss further this phenomenon in Chapter 5.

2.2 LOCAL VOLATILITIES

As we pointed out in Remark 1.2.6, we assume through this book that the market selects a risk-neutral probability measure P^* under which derivatives are priced. We do not know, a priori, which is this risk-neutral probability, and we do not know what is the corresponding risk-neutral model for asset prices. Our objective will not be to 'find' this model but to construct a (as simple as possible) model fitting option prices observed in the market.

We know from Chapter 1 that the Black-Scholes model is not able to replicate the vanilla option prices (or, equivalently, the implied volatility surface) observed in the market. If asset prices were according to the Black-Scholes model with volatility σ, the implied volatility would always be equal σ, independently on the strike and time to maturity, leading to a flat implied volatility surface.

The next simplest model we can consider is a diffusion of the form

$$dS_t = rS_t + \sigma(t, S_t)dW_t, \tag{2.2.1}$$

for some function σ. Such a model is called a **local volatility model**, and the function σ is the **local volatility function**. The question is: are these models able to reproduce the European vanilla options observed in the market?

The answer to this question depends on the (unknown) model that the market uses to price options. If this model is a diffusion satisfying some regularity conditions, Gyongy's lemma (see Theorem 2.2.1) gives us that the corresponding local volatility model reproducing vanilla option prices exists. Moreover, in this case, the local volatility function can be computed through the Dupire formula (2.2.12). Notice that, even when Dupire's formula fails in the non-diffusion framework, it is widely used by practitioners, with some ad-hoc adaptations (see for example De Marco, Friz, and Gerhold (2012), Friz, Gerhold, and Yor (2014), and Itkin (2020)).

In the next section, we study the existence of this local volatility model in the diffusion setting, where we assume a risk-neutral model of the form

$$dS_t = rS_t + \sigma_t dW_t, \tag{2.2.2}$$

for some random process σ.

2.2.1 Mimicking processes

Local volatilities are mimicking processes, in the sense that they are able to reproduce the marginal distributions of the process S (which is equivalent to reproducing vanilla option prices). A key result in the construction of mimicking processes is the following theorem.

Theorem 2.2.1 (Gyöngy's lemma) *Let $T > 0$. Consider a n-dimensional Itô process $\xi = \{\xi_t, t \in [0,T]\}$ of the form*

$$d\xi_t = \beta_t dt + \delta_t dW_t, \tag{2.2.3}$$

where $W = (W^1, ..., W^d)$ is a d-dimensional Brownian motion and β and $\delta = (\delta^1, ..., \delta^d)$ are $\mathbb{R}^n$-valued adapted and bounded processes such that $\delta_t \delta_t^T$ is uniformly positive definite. Then, there exists a stochastic differential equation

$$dx_t = b(t, x_t)dt + \sigma(t, x_t)dW_t, \tag{2.2.4}$$

where b is a $\mathbb{R}^n$-valued deterministic function and σ is a $n \times n$-valued deterministic matrix, which admits a weak solution x_t with the same one-dimensional probability distributions as ξ_t for every $t \in [0,T]$. Moreover, the coefficients b and σ satisfy that

$$\sigma(t,x)\,\sigma^T(t,x) = E\left(\delta_t \delta_t^T | \xi_t = x\right),$$

and

$$b(t,x) = E\left(\beta_t | \xi_t = x\right).$$

Proof. The idea of the proof can be summarised as follows. For the sake of simplicity, let us focus on the case $\beta = 0$ and take $n = d = 1$. Consider a function u in $C_0^\infty([0,\infty) \times \mathbb{R}$ (that is, u is zero for large $t + |x|$). Then, $d\xi_t = \delta_t dW_t$ and a direct application of Itò's formula to the process $u(t, \xi_t)$ gives us, after taking expectations

$$\begin{aligned} E[u(T,\xi_T)] - u(0,\xi_0) &= E\int_0^T \frac{\partial u}{\partial s}(s,\xi_s)ds + \frac{1}{2}E\int_0^T \frac{\partial^2 u}{\partial x^2}(s,\xi_s)\delta_s^2 ds \\ &= E\int_0^T \frac{\partial u}{\partial s}(s,\xi_s)ds + \frac{1}{2}E\int_0^T \frac{\partial^2 u}{\partial s^2}(s,\xi_s)E[\delta_s^2|\xi_s]ds, \end{aligned} \tag{2.2.5}$$

for all $T \geq 0$. This allows us to write

$$\int_0^\infty \int_{-\infty}^\infty u(t,x)\nu(dt,dx) = \int_0^\infty \int_{-\infty}^\infty \left[\frac{\partial}{\partial s} + \frac{1}{2}\sigma^2(t,x)\frac{\partial^2}{\partial x^2}\right] u(s,x)d\mu_s(x)dx, \tag{2.2.6}$$

where ν is the Dirac measure concentrated at $(0,0)$, $\sigma^2(s,x) := E[\delta_s^2|\xi_s = x]$, and $\mu_t(x) = P(\xi_t \in dx)$. Notice that (2.2.6) can be seen as an equation satisfied by the distributions $P(\xi_t \in dx)$. Now, consider a weak solution of the equation

$$x_t = \sigma(t, x_t)dW_t.$$

By Itô's formula, we can see that the distributions of the process x also satisfy (2.2.6), for every $u \in C_0^\infty([0,\infty) \times \mathbb{R})$. This suggests mimicking the process ξ by the process x. Nevertheless, we have to take into account that the hypotheses of the theorem do not imply uniqueness. To overcome this difficulty, we can make use of an approximation argument to prove the equality of both Green measures, which implies the equality of the one-dimensional distributions (see Gyöngy (1986)). Then, the proof is complete. ■

Remark 2.2.2 *The hypotheses of this result have been relaxed in Kurtz and Stockbridge (1998) and Brunick and Shreve (2013). The extension to the case of non-continuous semimartingales has been studied in Bentata and Cont (2009).*

This theorem is an example of *mimicking* result. Given a complex model for asset prices, we can construct another 'simpler' stochastic process x such that x_t and ξ_t have the same one-dimensional probability distribution for every t. As pointed out before, European option prices depend only on these marginal distributions, and both models 2.2.3 and 2.2.4 lead to the same vanilla option prices.

2.2.2 Forward equation and Dupire formula

Assume again the model (2.2.2). By Göngy's lemma, we know there exists a process $\hat{S}$ defined in some probability space $(\hat{\Omega}, \mathcal{F}, \hat{P})$, such that

$$d\hat{S}_t = r\hat{S}_t + \sigma(t, \hat{S}_t)\hat{S}_t d\hat{W}_t, \tag{2.2.7}$$

for some deterministic function σ and for some $\hat{P}$-Brownian motion $\hat{W}$, and such that S_t and $\hat{S}_t$ have the same one-dimensional distributions. Now our objective is to compute explicitly the values of the function σ from observed vanilla option prices. Towards this end, let us assume that S_t has a density $\phi(t,\cdot)$ (then $\hat{S}_t$ has the same density). Then, the price C of a call with strike K and time to maturity T is given by

$$C = e^{-rT}\int_{\mathbb{R}}(x-K)_+\phi(T,x)dx = e^{-rT}\int_K^\infty (x-K)\phi(T,x)dx,$$

which implies that

$$\begin{aligned}\frac{\partial C}{\partial T} &= -re^{-rT}\int_K^{\infty}(x-K)\phi(T,x)dx \\ &\quad +e^{-rT}\int_K^{\infty}(x-K)\frac{\partial \phi}{\partial T}(T,x)dx. \end{aligned} \tag{2.2.8}$$

The Fokker-Plank equation corresponding to (2.2.7) reads as

$$\frac{\partial \phi}{\partial t} = -r\frac{\partial}{\partial t}[x\phi(x,t)] + \frac{1}{2}\frac{\partial^2}{\partial x^2}[\sigma^2(t,x)x^2\phi(t,x)], \tag{2.2.9}$$

which, jointly with (2.2.8), implies that

$$\begin{aligned}\frac{\partial C}{\partial T} &= -re^{-rT}\int_K^{\infty}(x-K)\phi(T,x)dx \\ &+e^{-rT}\int_K^{\infty}(x-K)\left(-r\frac{\partial}{\partial t}[x\phi(T,x)] + \frac{1}{2}\frac{\partial^2}{\partial x^2}[\sigma^2(t,x)x^2\phi(T,x)]\right)dx. \end{aligned} \tag{2.2.10}$$

Then, integrating by parts (see for example Rouah (2008)) we obtain that

$$\begin{aligned}\frac{\partial C}{\partial T} &= rKe^{-rT}\int_K^{\infty}\phi(T,x)dx + \frac{1}{2}\sigma^2(T,K)K^2\phi(T,x) \\ &= -rK\frac{\partial C}{\partial K} + \frac{1}{2}\sigma^2(T,K)K^2\frac{\partial^2 C}{\partial K^2}. \end{aligned} \tag{2.2.11}$$

This implies that

$$\sigma^2(T,K) = \frac{\frac{\partial C}{\partial T} + rK\frac{\partial C}{\partial K}}{\frac{1}{2}K^2\frac{\partial^2 C}{\partial K^2}}. \tag{2.2.12}$$

Equation (2.2.12) is known as the **Dupire formula** (see Dupire (1994)). Notice that, as $C(0,K) = (S_0 - K)_+$, Equation (2.2.11) gives a forward equation for call option prices given a local volatility $\sigma(T,K)$.

Remark 2.2.3 *The computation of local volatilities from (2.2.12) is not numerically stable due to the fact that, for out of the money options, the term* $\frac{\partial^2 C}{\partial K^2}$ *can be very small. One way to overcome this difficulty is to write local volatilities not in terms of call prices but in terms of implied volatilities. More precisely, let us denote by* $I(T,K)$ *the implied volatility of a call option with time to maturity* T *and strike price* K*, and by* $w(T,K) = I^2(T,K)T$ *the total variance in the time interval*

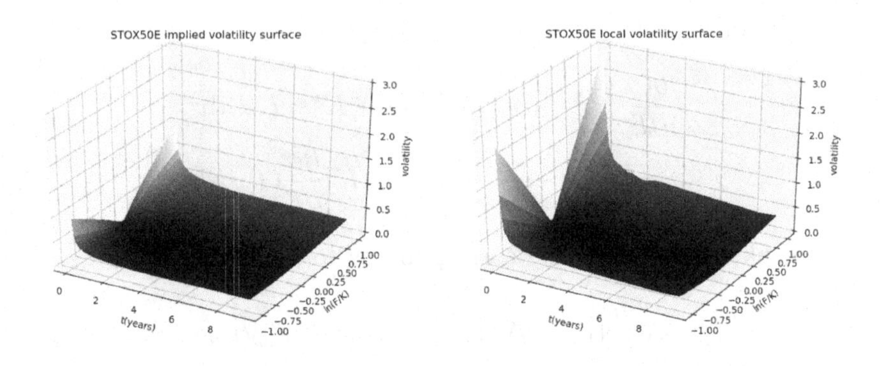

Figure 2.4 EURO STOXX 50 implied volatility vs local volatility surfaces, as of September 8, 2020 (data courtesy of CaixaBank).

$[0, T]$. *Consider also the variable* $y = \ln(K/F_T)$, *where* F_T *denotes the forward price* $F_T = S_0 e^{rT}$. *Then, taking into account the definition of the implied volatility, a direct application of the chain rule allows us to write Equation (2.2.12) as*

$$\sigma^2(T,K) = \left.\frac{\frac{\partial \omega}{\partial T}}{\left(1 - \frac{\partial \omega}{\partial y}\frac{y}{\omega} + \frac{1}{4}\left(\frac{\partial \omega}{\partial y}\right)^2\left(-\frac{1}{4} - \frac{1}{\omega} + \frac{y^2}{\omega^2}\right) + \frac{1}{2}\frac{\partial^2 \omega}{\partial y^2}\right)}\right|_{y=\ln(K/F_T)} \tag{2.2.13}$$

(see for example Gatheral (2006)).

Remark 2.2.4 *Given a local volatility model 2.2.7, there is only one risk-neutral probability under which discounted asset prices are martingales. Then, local volatility models are complete.*

Example 2.2.5 *In this example, we compute the implied volatility surface and the corresponding local volatility surface for EURO STOXX 50 index as of September 8, 2020, computed from 2.2.13 (data courtesy of CaixaBank). The results are shown in Figure 2.4. We can observe that the local volatility surface is similar to the implied volatility surface, but the short-end local skew and smile are stronger, as we study analytically in Chapters 7 and 8.*

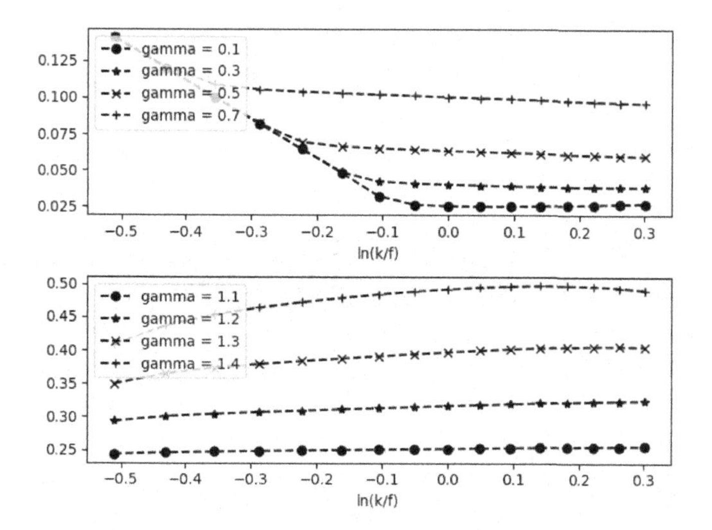

Figure 2.5 Implied volatility skews for the CEV model with $T = 1Y$, $\mu = 0.0$, $\sigma = 0.4$, $S_0 = 10$, and varying values of γ.

Example 2.2.6 (The CEV model) *The constant elasticity of variance model (CEV) for asset prices is an extension of the Black-Scholes model given by*

$$dS_t = \mu S_t dt + \sigma S_t^{\gamma} dW_t, \tag{2.2.14}$$

for some positive μ, σ, and γ. Notice that this is a local volatility model, where $\sigma(t, S_t) = \sigma S_t^{\gamma-1}$. When $\gamma < 1$, the volatility is negatively correlated with asset prices, as observed in real market data (see Cox (1996)). For this model the price of a call or put is usually computed by numerical methods based on PDEs or Monte Carlo simulations. It is widely known that for the CEV model we have downward-sloping and upward-sloping volatility skews for $\gamma < 1$ and $\gamma > 1$, respectively. That is, the implied volatility is a decreasing function of the log-moneyness $y = \ln(\frac{K}{T})$ if $\gamma < 1$ and it is increasing function if $\gamma > 1$. Moreover, the skew effect is stronger than the smile effect. We can observe this behaviour in Figure 2.5, corresponding to the simulations of the implied volatility for $T = 1Y$, $\mu = 0.0$, $\sigma = 0.4$, $S_0 = 10$, and varying values of γ.

Remark 2.2.7 *Local volatilities are one of the most popular tools in option pricing. Nevertheless, the dynamic behaviour of local volatility smiles and skews is not consistent with real market data. In the market-*

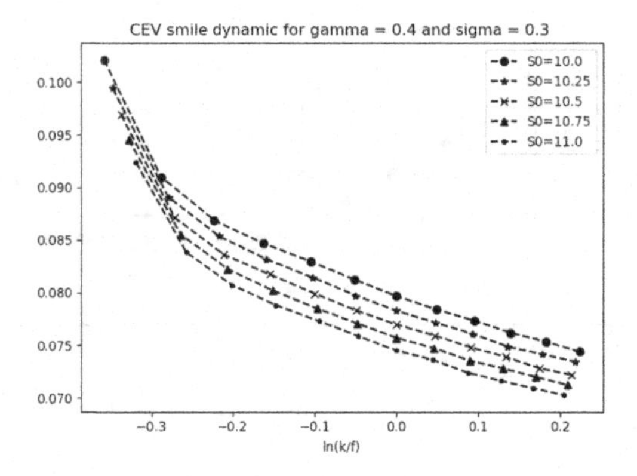

Figure 2.6 Smile dynamic for CEV model with $\sigma = 0.3, \gamma = 0.4$, and varying values of S_0.

place, asset prices and implied volatilities move in the same direction. When the asset price increases, the implied volatility increases, while when the asset price decreases, the implied volatility decreases. In local volatility models, we see the opposite behaviour, and asset prices and smiles move in the contrary direction, which may imply a poor performance of delta and vega hedging strategies (see for example Hagan, Kumar, Lesniewski, and Woodward (2002)). We can see this phenomenon in Figure 2.6, corresponding to the implied volatility curves (as a function of $y = \ln(\frac{K}{T})$*) of the CEV model with parameters* $\sigma = 0.3, \gamma = 0.4$*, and* $S_0 = 10.0, 10.25, 10.5, 10.75, 11$*. Notice that the implied volatility curves move in the opposite direction of the spot price.*

2.3 STOCHASTIC VOLATILITIES

Even when local volatility models can mimic the marginal distributions of asset prices S_t, they are not able to reproduce the joint distribution of the whole process S. In practice, this means that they may not be suitable for non-European options (see for example Chapter 3 in Bergomi (2016)). Moreover, as we pointed out in Remark 2.2.7, local volatilities present some dynamical inconsistencies with market data.

This leads us to consider general **stochastic volatility models** where the asset price S satisfies, under a risk-neutral probability, an equation of the form

$$dS_t = rS_t + \sigma_t S_t dW_t, \tag{2.3.1}$$

where r is the interest rate, W is a Brownian motion, and σ is a positive process that can be correlated with W. Even when local volatilities are a particular case of (2.3.1), the term *stochastic volatility* commonly refers to the case where σ is assumed to follow a diffusion process, driven by another Brownian motion B that can be correlated with W. In general, these models consider volatilities of the form $\sigma = f(Y)$, where f is a positive function and Y is a driving diffusion process satisfying a stochastic differential equation

$$dY_t = a(t, Y_t)dt + b(t, Y_t)dW_t \tag{2.3.2}$$

for some deterministic functions a and b. These functions are usually chosen for their analytical tractability rather than for economical reasons. Common choices for f include the square root, the exponential, or the absolute value, while some driving processes Y are: log-normal, mean-reverting Ornstein-Uhlenbeck (OU), or Cox-Ingersoll-Ross (CIR). Different combinations of the previous choices for f and Y lead to the most popular stochastic volatility models in the literature, such as Ball and Roma, the 3/2 model, Heston, Hull and White, SABR, Stein and Stein, Scott, and Schöbel and Zhu (see for example Ahn and Gao (1999), Fouque, Papanicolau, and Sircar (2000), and Hagan, Kumar, Lesniewsk, and Woodward (2002)).

When the Brownian motions W and B are uncorrelated, these models can explain the smile effect (see Renault and Touzi (1996)), while the skew effect is known to appear when W and B are correlated (see for example Lee (2002) and the references therein). Moreover, as observed in the marketplace, asset prices and smiles move in the same direction, which makes these models more consistent from the dynamical point of view.

Nevertheless, these models cannot reproduce the dependence of the implied volatility with respect to time to maturity discussed in Section 1.4.1 (see Lee (2002), Lewis (2000), and Medvedev and Scaillet (2007)). Then, stochastic volatility models cannot fit the whole implied volatility surface (in particular, and unlike local volatilities, they cannot reproduce all vanilla prices in the market), and their calibration is usually performed for each fixed maturity. We study in more detail two of the most popular stochastic volatility models (the Heston and the SABR) in the following subsections.

v_0	k	θ	ν	ρ	T
0.05	0.1	0.1	0.1	−0.75	0.1, 2

2.3.1 The Heston model

In the Heston model (Heston (1993)), the volatility σ is given by the square root of a CIR process. More precisely, $\sigma_t = \sqrt{v_t}$, where

$$dv_t = k(\theta - v_t)dt + \nu\sqrt{v_t}dB_t, \tag{2.3.3}$$

for some positive constants k, θ, μ, and where $d\langle W, B\rangle_t = \rho dt$, for $\rho \in (-1, 1)$. Notice that this is a mean-reverting process. The interpretation of the parameters is as follows: k is the rate of mean reversion, θ denotes the long-term variance, and ν is the volatility of the volatility (usually called the *vol-of-vol*). A common assumption is that $2k\theta \geq \nu^2$. This is often called the Feller condition. Given that $v_0 > 0$, this condition guarantees that the volatility process is always strictly positive. The Heston model has been very popular due to the existence of a quasi-closed form solution for European options (see Heston (1993)). In the next example, we analyse its corresponding implied volatilities.

Example 2.3.1 *Our goal in this example is to analyse the different smiles and skews that can be generated by the Heston model, as well as to see the effect of the parameters ρ, k, θ, and ν on these smiles/skews. Towards this end, we compute Heston option prices*[2] *and the corresponding implied volatilities for the following base parameters: In Figures 2.7 and 2.8 we can see the simulated implied volatility curves for $T = 0.1$ and $T = 2$, respectively, and for varying values for ρ, k, θ, and ν. In Figure 2.9 we plot the ATM implied volatility as a function of time to maturity. We can observe that:*

- *The correlation parameter ρ has a direct impact on the skew.*
- *The curvature is controlled by ν.*
- *The rate of mean reversion k controls the speed of the ATM implied volatility convergence to the reversion level θ.*
- *θ has a little effect on the smiles and skews. It is not relevant for short maturities.*

[2] Heston prices have been computed via the Fourier inversion of the corresponding characteristic function (see Rouah (2013)).

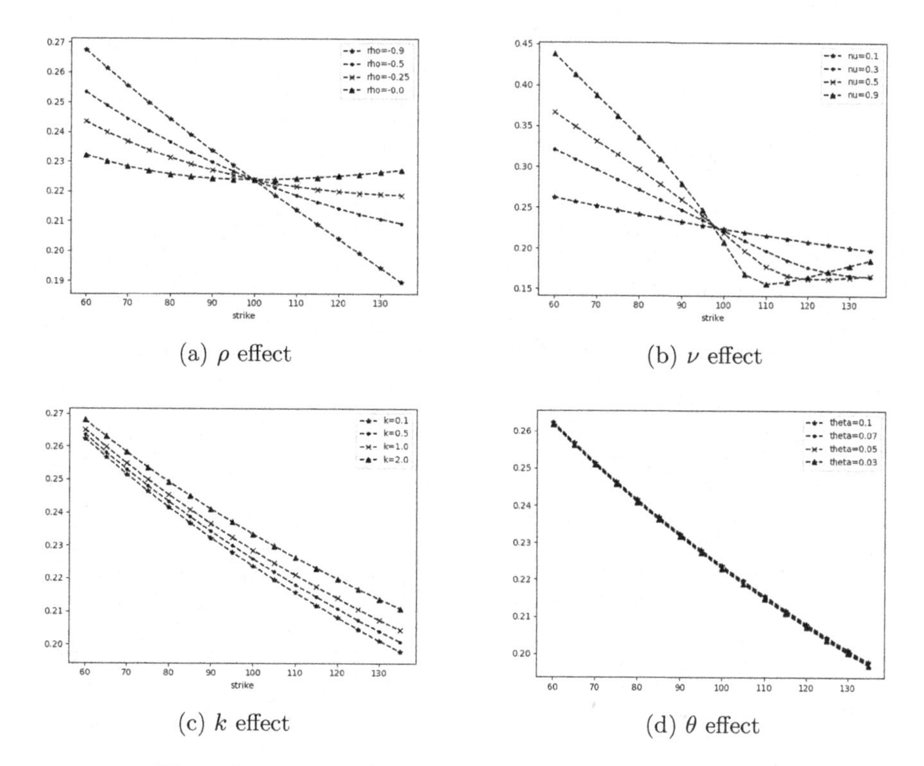

(a) ρ effect

(b) ν effect

(c) k effect

(d) θ effect

Figure 2.7 The effect of Heston parameters on smiles and skews for $T = 0.1$ and base parameters $v_0 = 0.05, k = 0.1, \theta = 0.1, \nu = 0.1$, and $\rho = -0.75$.

- *The ATM implied volatility is close to $\sqrt{v_0}$ in the short-end limit while it is near $\sqrt{\theta}$ for long maturities. Moreover, it depends strongly on all the parameters of the model.*

2.3.2 The SABR model

The Stochastic-Alpha-Beta-Rho (SABR) model for a forward price F is given by

$$\begin{cases} dF_t = \sigma_t F_t^{\beta} dW_t \\ d\sigma_t = \alpha \sigma_t dB_t, \quad t \in [0, T], \end{cases} \tag{2.3.4}$$

where we take $\beta \in [0, 1]$ and $\alpha > 0$. As the effect of correlation $\rho = \frac{d\langle W,B\rangle_t}{dt}$ and the effect of β are difficult to distinguish, it is usual to take $\beta = 1$. The popularity of the SABR model relies on the existence of a

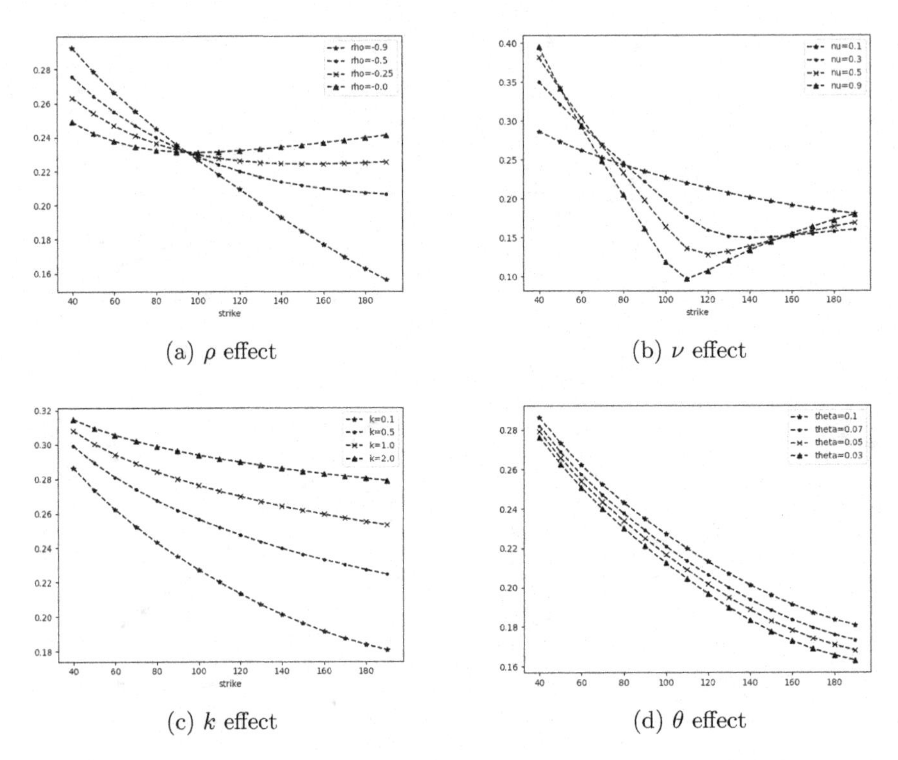

(a) ρ effect (b) ν effect

(c) k effect (d) θ effect

Figure 2.8 The effect of Heston parameters on smiles and skews for $T = 2$ and base parameters $v_0 = 0.05, k = 0.1, \theta = 0.1, \nu = 0.1$, and $\rho = -0.75$.

simple approximation formula for the corresponding implied volatility (see Hagan, Kumar, Lesniewski, and Woodward (2002)). For $\beta = 1$, this approximation reads (at $t = 0$) as

$$I(T, K) \approx \sigma_0 \left(1 + \left(\frac{\alpha\rho\sigma_0}{4} + \frac{2 - 3\rho^2}{24}\alpha^2\right) T\right) \frac{z}{x(z)}, \tag{2.3.5}$$

where $z = \frac{\alpha}{\sigma_0} \ln\left(\frac{F_0}{K}\right)$ and

$$x(z) = \ln\left(\frac{\sqrt{1 - 2\rho z + z^2} + z - \rho}{1 - \rho}\right).$$

In the particular case of at-the-money options ($F_0 = K$) we get

$$I(T, K) \approx \sigma_0 \left(1 + \left(\frac{\alpha\rho\sigma_0}{4} + \frac{2 - 3\rho^2}{24}\alpha^2\right) T\right). \tag{2.3.6}$$

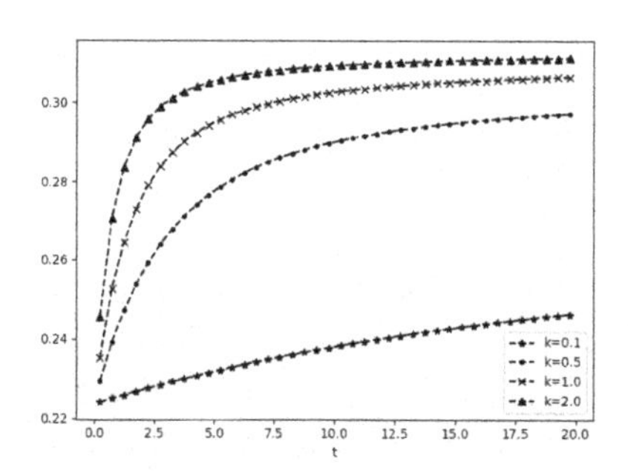

(a) k effect to ATM implied volatility in time

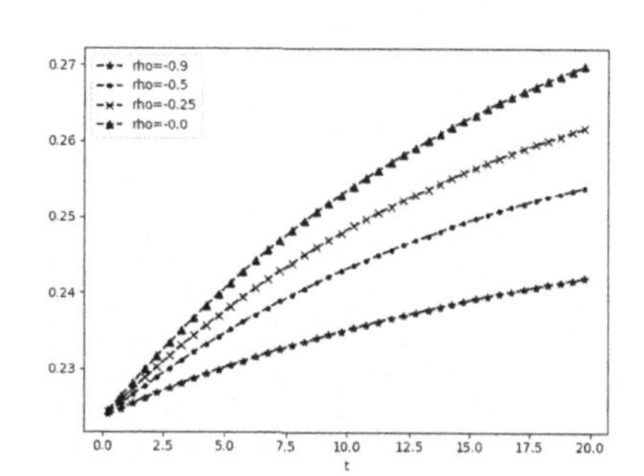

(b) ρ effect to ATM implied volatility in time

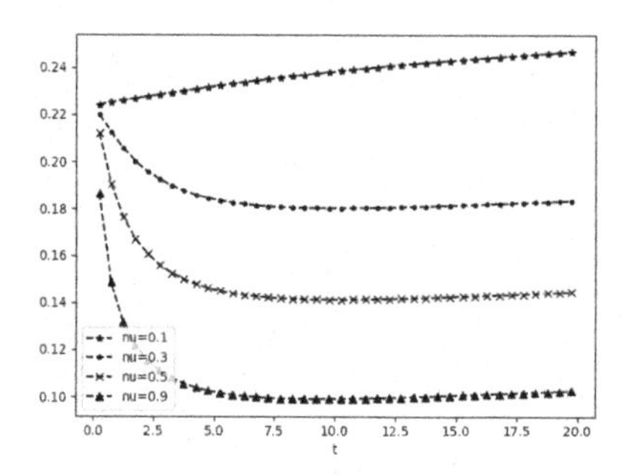

(c) ν effect to ATM implied volatility in time

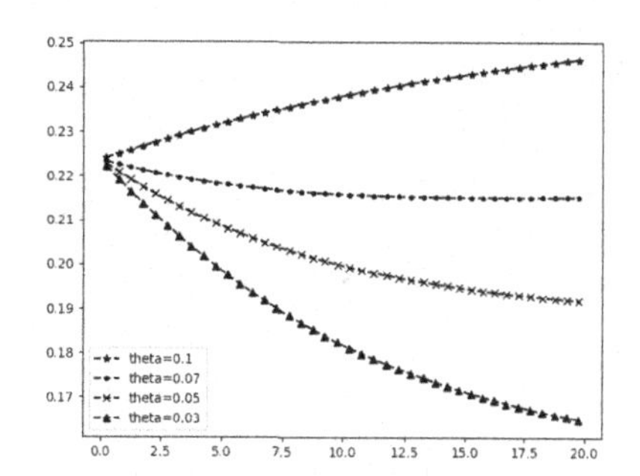

(d) θ effect to ATM implied volatility in time

Figure 2.9 The effect of Heston parameters on the ATM implied volatility with base parameters $v_0 = 0.05, k = 0.1, \theta = 0.1, \nu = 0.1$, and $\rho = -0.75$.

We notice that when the strike K is close to the forward price F_0, the expression (2.3.5) can be approximated as

$$I(T, K) \approx \sigma_0 + \frac{1}{2}\alpha\rho \ln\left(\frac{K}{F_0}\right) + \frac{(2 - 3\rho^2)\alpha^2}{12\sigma_0} \ln^2\left(\frac{K}{F_0}\right) \tag{2.3.7}$$

(see again Hagan, Kumar, Lesniewski, and Woodward (2002)). According to this expression, the skew is controlled by the product $\alpha\rho$, while the curvature depends on α^2 and ρ^2, being proportional to α^2. In the following example, we observe the effect of each parameter on the SABR implied volatility.

Example 2.3.2 *In this example, we analyse the behaviour of the approximated volatility curves defined by 2.3.5. The base parameters are as follows.*

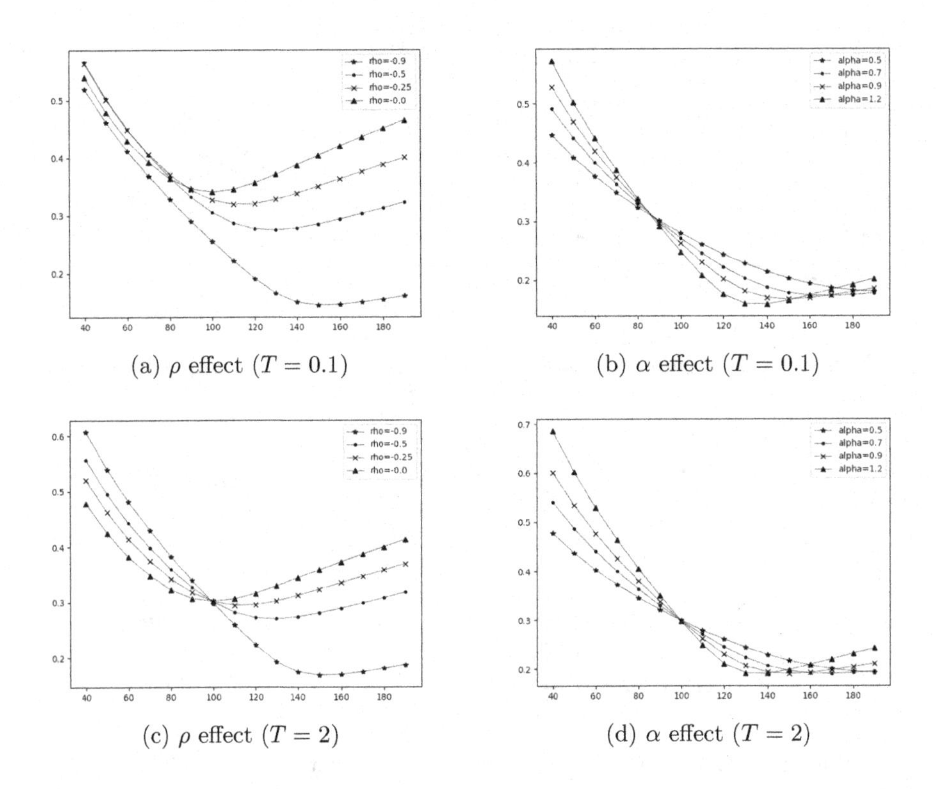

(a) ρ effect ($T = 0.1$)

(b) α effect ($T = 0.1$)

(c) ρ effect ($T = 2$)

(d) α effect ($T = 2$)

Figure 2.10 The effect of SABR parameters on skews and smiles with base parameters $F_0 = 100$, $\alpha = 1.5$, $\rho = -0.6$, $\sigma_0 = 0.4$, and $T = 2$.

F_0	α	ρ	σ_0	T
100.0	1.5	−0.6	0.4	2.0

In Figure 2.10, we can see the impact of ρ and α in the skew and curvature of the implied volatility curve, for both short and long maturities. We see again that the ATM skew is more pronounced for big values of $|\rho|$, and that the curvature increases with α, as predicted by 2.3.7.

Example 2.3.3 *Formula 2.3.5 is widely used by practitioners. There are two main reasons for this success. The first one is that one can explain the level, the skew, and the curvature of the implied volatility with only three parameters. The second one is that it is a simple and accurate approximation of the implied volatility and therefore is relatively easy to calibrate (for every fixed maturity) this model to market data. But the approximation 2.3.5 is not arbitrage-free (see Hagan, Kumar,*

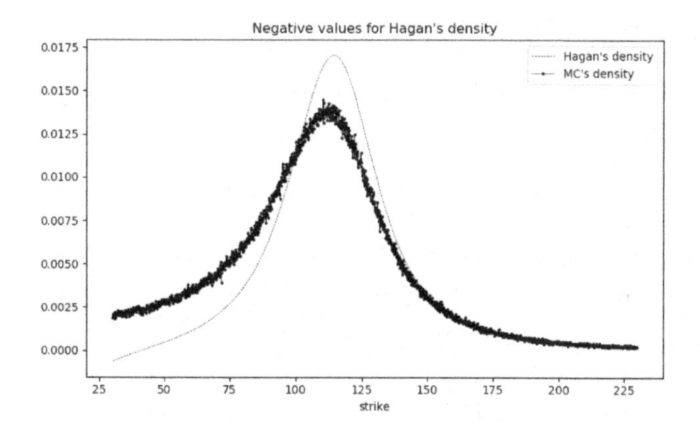

Figure 2.11 Negative density for low strikes in the formula of Hagan, Kumar, Lesniewski, and Woodward with $F_0 = 100$, $\alpha = 1.5$, $\rho = -0.6$, $\sigma_0 = 0.4$, and $T = 2$.

Lesniewski, and Woodward (2014)). In fact, it leads to an approximate density that takes negative values for some sets of parameters. To see this phenomenon, we first compute the price of a call option for different strikes both by Monte Carlo simulations and by 2.3.5. As

$$\partial^2_{K,K} C(T, K; F_0) = p_T(K), \tag{2.3.8}$$

where p_T denotes the probability density of F_T, we can estimate the density function corresponding to the computed Monte-Carlo call prices. On the other hand, we compute the density function corresponding to the prices computed from the approximation formula 2.3.5. In Figure 2.11 we compare these two densities. We can see that for low strikes the density of Hagan, Kumar, Lesniewski, and Woodward can get negative values.

We have seen that volatilities driven by a diffusion process can replicate skews and smiles. But as pointed out earlier, these models can not fit the whole implied volatility surface. One way to overcome this problem is to consider the so-called *stochastic-local* volatility models (LSV), where local volatilities and stochastic volatilities are combined.

2.4 STOCHASTIC-LOCAL VOLATILITIES

In a stochastic-local volatility model, asset prices are assumed to follow a model of the form

$$dS_t = rS_t + \sigma_t \lambda(t, S_t) S_t dW_t, \tag{2.4.1}$$

where r is the interest rate, W is a Brownian motion, σ is a positive diffusion driven by another Brownian motion B (that can be correlated with W), and λ is a deterministic function (see Alexander and Nogueira (2004)).

Notice that the model (2.4.1) has a corresponding local volatility function $\sigma_{loc}(t, x)$. According to Theorem 2.2.1,

$$\sigma_{loc}^2(t, x) = E[\sigma_t^2 \lambda^2(t, S_t)|S_t = x] = \lambda^2(t, x) E[\sigma_t^2 | S_t = x]. \tag{2.4.2}$$

Then, it follows that

$$\lambda(t, x) = \frac{\sigma_{LV}(t, x)}{E[\sigma_t^2 | S_t = x]}. \tag{2.4.3}$$

The numerator is the classic local volatility of Dupire, and the main difficulties rely on the denominator. The computation of this conditional expectation can be done solving the Kolmogorov forward PDE (see Engelmann, Koster, and Oeltz (2020)) or by non-parametric estimation (see Guyon and Henry-Labordere (2012, 2013) or Stoep, Grzelak, and Oosterlee (2013)), as we see in the following example.

Example 2.4.1 *In this example, we will use a non-parametric estimator kernel type for the conditional expectation (2.4.3). Assume for the sake of simplicity that $r = 0$, then we can approximate $E[\sigma_t^2 | S_t = x]$ from a simulation of size M in the following way*

$$E[\sigma_t^2 | S_t = x] \approx \frac{\sum_{i=1}^M \sigma_{t,i}^2 K_h(S_{t,i} - x)}{\sum_{i=1}^M K_h(S_{t,i} - x)}, \tag{2.4.4}$$

(see Guyon and Henry-Labordère (2013)), where $S_{t,i}$ and $\sigma_{t,i}$, $i = 1, ..., M$ are the simulations of the asset price and the spot volatility, respectively, and $K_h(\cdot)$ is a kernel function with bandwidth parameter $h > 0$. For example, we can take

$$K_h(x) = \frac{1}{\sqrt{2\pi h}} \exp(\frac{-x^2}{2h}). \tag{2.4.5}$$

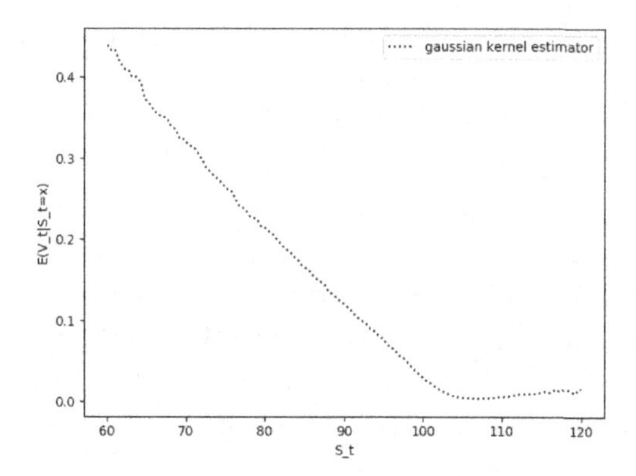

Figure 2.12 Gaussian kernel estimator for $E[\sigma_t^2|S_t = x]$ for a Heston model with $v_0 = 0.1$, $k = 0.1$, $\theta = 0.1$, $\nu = 0.9$, $\rho = -0.75$, and $t = 0.25$.

Now, consider the Heston model with parameters

S_0	v_0	k	θ	ν	ρ	t
100.0	0.05	0.1	0.1	0.9	-0.75	0.25

In Figure 2.12, we can see the results of the estimator 2.4.4, which we have obtained with bandwidth parameter $h = 1.5\sqrt{T}M^{-\frac{1}{5}}$ *(see again Guyon and Henry-Labordère (2013)).*

2.5 MODELS BASED ON THE FRACTIONAL BROWNIAN MOTION AND ROUGH VOLATILITIES

Rough volatilities are stochastic volatility models driven by the fractional Brownian motion (fBm). The fBm (Chapter 5) is an extension of the Brownian motion, where the increments are correlated. When this correlation is positive, the fBm is a long-memory process and this long-memory feature allows one to capture the observed persistence of implied volatilities when the time to maturity increases. When the increments are negatively correlated, the variance of the volatility in short-time intervals is greater than for the classical Brownian motion, and this leads to a better fit of the short-time behaviour of the implied volatility surface (see Alòs, León, and Vives (2007)). In particular, the corresponding skew slope blows up in the short-end, as observed in real market data (see Section 1.4.1). In this negatively correlated case, volatility paths are rougher than classical diffusion volatilities, and this is why these models are usually referred to as *rough volatilities*.

Fractional volatilities are not Markovian, which implies several technical difficulties in their implementation. Nevertheless, they succeed in explaining several properties of empirical implied volatilities, as we will see through this book. Due to their ability to describe the implied volatility surface, volatility models based on the fBm and other similar Gaussian processes have been deeply studied in the last years. In the next example, we see the short-end behaviour of the skews corresponding to rough volatilities.

Example 2.5.1 (The one-factor Bergomi and the rough Bergomi models) In the one-factor Bergomi model (see Bergomi (2005)) the volatility process satisfies that

$$\sigma_t^2 = \sigma_0^2 \exp\left(\nu Y_t - \frac{1}{2}\nu^2 < Y, Y >_t\right), t \in [0, T], \tag{2.5.1}$$

for some Ornstein-Uhlenbeck process and some positive real values σ_0^2 and ν. Let us extend this model, substituting the Ornstein-Uhlenbeck process by a Gaussian process defined by

$$Z_r := \int_0^r (r-s)^{H-\frac{1}{2}} dB_s,$$

for some $H < \frac{1}{2}$ and some Brownian motion B. Notice that the variance of Z is $O(t^{2H})$. In small time intervals, as $2H < 1$, this allows us to reproduce higher variances than in the case of the Brownian motion, whose variance is $O(t)$. Now, the rough Bergomi (rBergomi) model (see Bayer, Friz, and Gatheral (2016)) is defined as (under a risk-neutral probability)

$$dS_t = rS_t + \sigma_t S_t dW_t,$$

where W_t is a Brownian motion (that can be correlated with B) and the volatility process is given by

$$\sigma_t^2 = \sigma_0^2 \exp\left(\nu\sqrt{2H} Z_t - \frac{1}{2}\nu^2 t^{2H}\right), t \in [0, T], \tag{2.5.2}$$

for some positive real values σ_0^2 and ν. A direct computation gives us (see Bayer, Friz, and Gatheral (2016)) that, for all $s < t$

$$E(Z_t Z_s) = s^{H+\frac{1}{2}} t^{H-\frac{1}{2}} \frac{1}{H+\frac{1}{2}} {}_2F_1\left(\frac{1}{2} - H, 1, H + \frac{3}{2}, \frac{s}{t}\right) \tag{2.5.3}$$

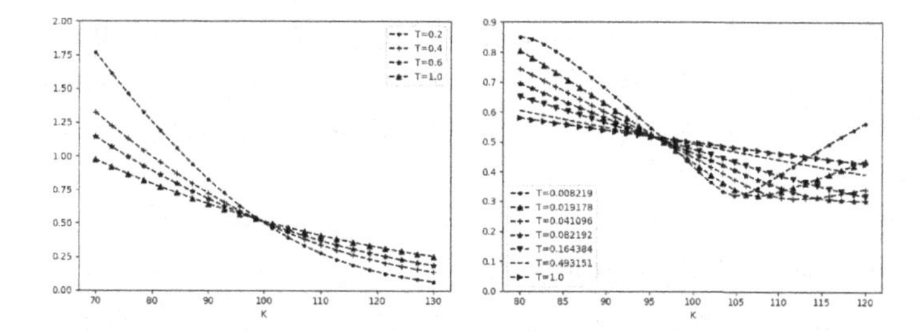

Figure 2.13 Implied volatility skews for a Heston model with $S_0 = 100$, $\nu = 0.75, k = 0.6, \rho = -0.5, v_0 = 0.25, \theta = 0.5$ (left), and for a rough Bergomi model with $\nu = 0.8, \rho = \frac{d\langle B,W\rangle_t}{dt} = -0.2$, $\sigma_0^2 = 0.05$, and $H = 0.05$ (right).

where ${}_2F_1$ denotes the hypergeometric function. Moreover, for all $s \neq t$

$$E(Z_t W_s) = \frac{\rho\sqrt{2H}}{H+\frac{1}{2}} \left(t^{H+\frac{1}{2}} - (t - \min(t,s))^{H+\frac{1}{2}}\right). \tag{2.5.4}$$

This joint covariance matrix allows us to apply classical simulation methods for multivariate normal distributions, and then to compute Monte Carlo option prices. In Figure 2.13 we can see the corresponding implied volatility skews for $S_0 = 100, \nu = 0.8, \rho = \frac{d\langle B,W\rangle_t}{dt} = -0.2$, $\sigma_0^2 = 0.05$, and $H = 0.05$, where the processes Z and B have been simulated via the Cholesky transform[3]; and we compare them with the skews of a Heston model with $S_0 = 100$, $\nu = 0.75, k = 0.6, \rho = -0.5, v_0 = 0.25$, and $\theta = 0.5$. We can observe that, for short maturities, the skew slope flattens very fast in the rBergomi model, according to market data (see Lee (2005)). Moreover, we can see that for ATM, for short maturities, the smile effect is more pronounced than the skew effect, a property that fits empirical observations and that we will study in detail in Chapter 8.

2.6 VOLATILITY DERIVATIVES

As the volatility of the market is a random process, it has become an asset in its own right (see Gangahar (2006)). Several instruments have appeared to hedge the risk associated with this randomness. Classical examples include variance and volatility swaps, defined in Chapter 1. In

[3]This simulation method is detailed in Chapter 5.

recent years, several other tools have emerged, such as gamma swaps or derivatives on the VIX (the Chicago Board of Exchange Volatility Index). We briefly introduce these concepts, which will be used throughout the book.

2.6.1 Variance swaps and the VIX

Variance swaps are forward contracts on the future realised variance, computed as the sum of the squares of daily log-returns. That is, the payoff of a variance swap is given by

$$\sum_{i=0}^{n} \ln^2\left(\frac{S_i}{S_{i-1}}\right) - K, \tag{2.6.1}$$

where n is the number of trading days for the maturity of the variance swap, S_i are the daily closing quotes of the underlying, and K is the corresponding forward price. Taking $X = \ln S$ we can write this payoff as

$$\sum_{i=0}^{n}(X_i - X_{i-1})^2 - K. \tag{2.6.2}$$

Now, observe that the above sum can be seen (if n is big enough) as an approximation of the quadratic variation of X. If we assume a risk-neutral stochastic volatility model of the form (2.3.1) (where we can take $r = 0$ without losing generality), a direct application of Itô's formula gives us that

$$dX_t = -\frac{1}{2}\sigma_t^2 dt + \sigma_t dW_t, \tag{2.6.3}$$

which implies that the first term in (2.6.2) is nothing more than an estimator of $\int_0^T \sigma_s^2 ds$. Then, the discrete variance swap payoff in (2.6.1) is approximately equal to the continuous variance swap payoff given by

$$\int_0^T \sigma_s^2 ds - K, \tag{2.6.4}$$

that is easier to handle from the mathematical point of view. Both payoffs are usually identified in the literature. An analytical study of the corresponding discretisation error can be found in Bernard and Cui (2014) and Alòs and Fukasawa (2020).

Assume now the payoff (2.6.4). Then, the fair strike K is given by $K = E_t \int_0^T \sigma_s^2 ds$. Then, from (2.6.3),

$$E(X_T - X_0) = -\frac{1}{2}E_t \int_0^T \sigma_t^2 dt, \tag{2.6.5}$$

that can be written as

$$E_t \int_0^T \sigma_t^2 dt = 2E_t \left(\ln \left(\frac{S_T}{S_0} \right) \right)$$

that is the classical relationship between the variance swap and the log-contract. This equality is relevant since it is a model-free identity that allows us to compute the fair price of a variance swap via the value of a European option on the asset, with payoff $h(S_T) = \ln S_T$. By the works of Breeden and Litzenberger (1978) and Carr-Madan (1998), a payoff of the form $G(S_T)$ can be written in terms of puts and calls. More precisely, for any $k > 0$ the following representation holds

$$\begin{aligned} G(x) &= G(k) + G'(k)(x-k) \\ &+ \int_{K>k} G''(K)(x-K)_+ dK + \int_{0<K<k} G''(K)(K-x)_+ dK. \end{aligned} \tag{2.6.6}$$

Even when in calls and puts do not trade at all strikes, in liquid markets, they may trade at enough strikes to get adequate approximations to the integrals in (2.6.6) (see Leung and Lorig (2016)).

A particular example of a variance swap is given by the *CBOE volatility index* (VIX), computed as the square root of the fair price of a variance swap on the S&P index, over a 30-days time interval. More precisely,

$$VIX_t = \sqrt{E_t \int_t^{t+30days} \sigma_s^2 ds}.$$

Being the VIX a financial index, we can also consider futures and options on it. VIX futures and options (introduced in 2004 and 2006, respectively) are exchange-traded instruments that allow the investors to manage the volatility risk.

2.6.2 Volatility swaps

A volatility swap is a forward contract on future realised volatility. That is, its payoff is of the form

$$\sqrt{\sum_{i=0}^{n} \ln^2 \left(\frac{S_i}{S_{i-1}} \right)} - K. \tag{2.6.7}$$

As before, it is common to identify this payoff with its continuous limit

$$\sqrt{\int_0^T \sigma_s^2 ds} - K. \tag{2.6.8}$$

Then, in a similar way as before, the fair price K is given by

$$E\sqrt{\int_0^T \sigma_s^2 ds}.$$

Although it seems to be similar to the variance swap, the volatility swap is much more complex from the mathematical point of view. As quoted in Gangahar (2006), 'Variance is easier to hedge. Volatility can be a nightmare.' Variance swaps can be replicated by vanilla options, but there is not a similar model-free exact replication for volatility swaps. More precisely, this exact replication can be constructed only in the uncorrelated case $\rho = 0$ (see Carr and Lee (2008)). Classical approaches to this problem include:

- **The convexity adjustment**: Notice that, because of Jensen's inequality, $E\sqrt{\int_0^T \sigma_s^2 ds} \leq \sqrt{E\int_0^T \sigma_s^2 ds}$. More precisely, a Taylor expansion up to second order allows us to write, after taking expectations,

$$E\sqrt{\int_0^T \sigma_s^2 ds} \approx \sqrt{E\int_0^T \sigma_s^2 ds} - \frac{E\left(\int_0^T \sigma_s^2 ds - E\int_0^T \sigma_s^2 ds\right)^2}{8(\sqrt{E\int_0^T \sigma_s^2 ds})^3} \tag{2.6.9}$$

  (see Brockhaus and Long (2000)). We remark that this approximation will not work very well if the higher-order terms in the Taylor expansion are not negligible, as in the Heston case (see Broadie and Jain (2008)).

- **Laplace transforms**: An alternative methodology, introduced by Broadie and Jain (2008), is based on the following identity (Schürger 2002):

$$\sqrt{x} = \frac{1}{2\sqrt{\pi}} \int_0^\infty \frac{1 - e^{ux}}{u^{\frac{3}{2}}} du,$$

  and then the fair price of a volatility swap admits the representation

$$E\sqrt{\int_0^T \sigma_s^2 ds} = \frac{1}{2\sqrt{\pi}} \int_0^\infty \frac{1 - E(e^{u\int_0^T \sigma_s^2 ds})}{u^{\frac{3}{2}}} du. \tag{2.6.10}$$

  Then, if we have an explicit expression for the Laplace transform of $\int_0^T \sigma_s^2 ds$, the fair continuous volatility strike can be evaluated using (2.6.10). This explicit expression does not exist in general, but it is possible to construct an approximated replication of the payoffs $e^{u\int_0^T \sigma_s^2 ds}$ (see Carr and Lee (2009)).

2.6.3 Weighted variance swaps and gamma swaps

Weighted average swaps are forward contracts on a weighted realised variance of the type

$$\int_0^T a(u)\sigma_u^2 du.$$

Taking into account Itô's formula, we know that, for every function f, (assume again $r = 0$ for the sake of simplicity)

$$f(S_T/S_0) = f(1) + \int_0^T f'(S_u/S_0)dS_u + \frac{1}{2S_0^2}\int_0^T f''(S_u/S_0)S_u^2\sigma_u^2 du.$$

Then, a weighed variance swap can be replicated by a European option with payoff $f(S_T/S_0)$, where $f(1) = 0$ and $f''(x)x^2 = a(x)$.

Let us consider the particular case of a **gamma swap**, where $a(u) = \frac{S_u}{S_0}$. While a standard variance swap is insensitive to changes in the asset price, a gamma swap is designed to hedge the volatility exposure when the underlying price drops. Then, taking $f(x) = x\ln(x)$ we get

$$E\left[\frac{S_T}{S_0}\ln\left(\frac{S_T}{S_0}\right)\right] = E\int_0^T \frac{S_u}{S_0}\sigma_u^2 du,$$

which tell us that, as in the case of a variance swap, a gamma swap can be replicated by vanilla options.

2.7 CHAPTER'S DIGEST

The volatility $\sigma_t^2 := \frac{d\langle X,X\rangle_t}{dt}$ is a stochastic process that cannot be directly observed in the market, but it has to be estimated. We have seen how to do this estimation both for integrated and spot volatilities via realised variances and Fourier-Malliavin estimators. The statistical analysis of the volatility path allows us to study some relevant properties as the existence of jumps, the volatility clustering, short- and long-memory properties, and the roughness of the corresponding paths.

Volatility modelling takes into account not only the statistical properties of the spot volatility, but its main objective is to find a model fitting option prices observed in the market. The simplest class of models with random volatilities are the so-called local volatility models, where the volatility process is of the form $\sigma_t = \sigma(t, S_t)$, being σ the corresponding local volatility function. Under some conditions, Gyongy's lemma (see Theorem 2.2.1) guarantees that there exists a local volatility model reproducing vanilla option prices. Moreover, the local volatility function can be computed explicitly through the Dupire formula (2.2.12). Nevertheless, local volatilities mimic only the marginal distributions of asset prices. Then, the applications of local volatility are very restricted to European options. Moreover, this class of models presents some dynamic inconsistencies of the corresponding skews and smiles.

The second class of models is given by stochastic volatilities, where the spot volatility is assumed to follow a diffusion process. Stochastic volatility models as Heston or SABR are more consistent from the dynamical point of view and are more adequate for non-vanilla options. But, even when they can reproduce skews and smiles, they cannot reproduce the term-structure of the implied volatility. Then, they fail in reproducing the whole implied volatility surface. A way to overcome this difficulty is given by stochastic-local volatility models, where local and stochastic volatilities are combined. Another recent approach assumes the volatility to be a fractional Brownian motion. Fractional volatility models can better explain the long-term and short-term behaviour of the implied volatility surface.

Volatility has become an asset in its own right and several instruments have appeared to hedge the risk associated with this randomness, such as variance and volatility swaps. Variance swaps can be replicated using a static portfolio of European vanilla options, while there is not a similar replicating strategy for volatility swaps. Other derivatives on the volatility include gamma swaps (designed to hedge the volatility exposure when the asset price falls) or derivatives on the VIX (the Chicago Board of Exchange Volatility Index).

II

Mathematical tools

CHAPTER 3

A primer on Malliavin Calculus

This chapter is a first introduction to the main tools of Malliavin calculus. We define and study the main properties of the Malliavin derivative operator and its adjoint, the divergence operator. We see how to compute Malliavin derivatives for different types of processes. In particular, we see that the Malliavin derivative of a diffusion process can be computed as the solution of a stochastic differential equation (SDE) and that, alternatively, it can be represented as the solution of an ordinary differential equation (ODE). We detail the explicit computations in several classical asset models, like the Black-Scholes and the CEV. We also study the case of stochastic volatilities as the SABR, the Heston, or the 3/2 model. Moreover, we compare numerically the efficiency of the different proposed computational methods.

3.1 DEFINITIONS AND BASIC PROPERTIES

The basic tools in Malliavin calculus are the Malliavin derivative operator and its adjoint, the divergence operator. This section is devoted to briefly present the definitions and basic properties of these operators. For a deeper approach to this topic, we refer to Da Prato (2014), Di Nunno, Øksendal, and Proske (2009), Malliavin and Thalmaier (2006), Nualart (2006), and Sanz-Solé (2005).

3.1.1 The Malliavin derivative operator

Let us consider a standard Brownian motion $W = \{W_t, t \in [0,T]\}$ defined in a probability space $(\Omega, \mathcal{F}, P)$. We denote by $(\mathcal{F}_t)_{t\in[0,T]}$ the filtration generated by W. Set $H = L^2([0,T])$, and denote $W(h) = \int_0^T h(s)dW_s$ the Wiener integral of a deterministic function $h \in H$. Notice that

$$E(W(h)W(g)) = \int_0^T h(s)g(s)ds,$$

for all $h, g \in H$.

One of the objectives of Malliavin calculus is to define, in an adequate sense, the concept of 'derivative' of a random variable with respect to the white noise (see for example Hairer (2016)). Consider first a random variable of the form $W(h) = \int_0^T h(s)dW_s$. Intuitively, we should define a derivative operator D in such a way that $DW(h) = h$. Moreover, we would like this operator to satisfy the chain rule. That is, given a random variable A and a differentiable function f

$$Df(A) = f'(A)DA.$$

All these observations lead us to the following definition (see for example Nualart (2006)).

Definition 3.1.1 *Consider a random variable F of the form*

$$F = f(W(h_1), ..., W(h_n)), \tag{3.1.1}$$

where $f : \mathbb{R}^n \to \mathbb{R}$ is a function such that f and all its derivatives have polynomial growth, $h_1, ..., h_n \in H$ and $n \geq 1$. Then, the Malliavin derivative of F, which we will denote by DF, is the random variable with values in H defined by

$$DF = \sum_{i=1}^{n} \frac{\partial f}{\partial x_i}(W(h_1), ..., W(h_n))h_i.$$

The domain of D (that we denote by $\mathbb{D}^{1,2}$) is given by the closure of the class of random variables of the form (3.1.1) with respect to the norm

$$\|F\|_{1,2} = \left(E|F|^2 + E\|DF\|_H^2\right)^2.$$

We also denote by $\mathbb{D}^{n,2}$ the domain of the iterated operator D^n. The Malliavin derivative of some classical processes can be computed directly from this definition, as we see in the following examples.

Example 3.1.2 (The Malliavin derivative of a Brownian motion) Consider the random variable $W_t = W(\mathbf{1}_{[0,t]})$. Then, definition 3.1.1 gives us that $DW_t = \mathbf{1}_{[0,t]}$. That is, $D_s W_t = \mathbf{1}_{[0,t]}(s)$.

Example 3.1.3 (The Malliavin derivative of a Black-Scholes price) In the Black-Scholes model (1.3.1), asset prices $S = \{S_t, t \in [0,T]\}$ can be written (under the risk-neutral probability) as

$$S_t = S_0 \exp\left(\left(r - \frac{\sigma^2}{2}\right)t + \sigma W_t\right),$$

where r denotes the interest rate and σ is the volatility. Notice that the above equality gives us that S_t is a function of W_t. Then, it follows that the Malliavin derivative of S_t is given by[1]

$$D_r S_t = \sigma S_t \mathbf{1}_{[0,t]}(r).$$

It is easy to prove the following properties of the Malliavin derivative operator, which we will use throughout the book.

3.1.1.1 Basic properties

Proposition 3.1.4 (Integration by parts) *Consider a random variable F of the form (3.1.1). Then, for all $h \in H$,*

$$E\langle DF, h\rangle_H = E(FW(h)).$$

Proof. For the sake of simplicity, we only proof the result in the case $F = f(W(g))$, for some $f : \mathbb{R} \to \mathbb{R}$ such that f and all its derivatives have polynomial growth, and some $g \in H$. From the definition of the Malliavin derivative operator we get

$$E\langle DF, h\rangle = E(f'(W(g)))\langle g, h\rangle_H = \langle g, h\rangle_H \int_{\mathbb{R}} f'(x)\phi(x)dx,$$

where $\phi(x)$ denotes the Gaussian density with zero mean and standard deviation equal to $\|g\|_H$. On the other hand,

$$E(FW(h)) = E(f(W(g))W(h)) = \int_{\mathbb{R}^2} f(x)y\hat{\phi}(x,y)dxdy,$$

where $\hat{\phi}$ denotes the bivariate normal distribution of the random vector $(W(g), W(h))$. Then, the properties of the bivariate normal distribution allow us to complete the proof. ■

[1] Even when the exponential function has not polynomial growth, the result follows from an approximation argument.

Proposition 3.1.5 (The chain rule) *Consider $n \geq 1$. Assume that $\psi \in C_b^1(\mathbb{R}^n)$ and $F_1, ..., F_n \in \mathbb{D}^{1,2}$. Then, $\psi(F_1, ..., F_n) \in \mathbb{D}^{1,2}$ and*

$$D\psi(F_1, ..., F_n) = \sum_{i=1}^{n} \frac{\partial \psi}{\partial x_i}(F_1, ..., F_n) DF_i.$$

Proof. If F is a random variable of the form (3.1.1) the result is obvious. In the general case, a limit argument gives us the result. ■

Remark 3.1.6 *As an application Proposition 3.1.5 we get the following product rule: Consider two random variables $F, G \in \mathbb{D}^{1,2}$ such that F and $\|DF\|_H$ are bounded. Then, $FG \in \mathbb{D}^{1,2}$ and $D(FG) = (DF)G + F(DG)$. This, jointly with Proposition 3.1.4 gives us that*

$$E\langle (DF)G + F(DG), h\rangle_H = E(FGW(h)).$$

That is,

$$E(G\langle DF, h\rangle_H) + E(F\langle DG, h\rangle_H) = E(FGW(h)). \tag{3.1.2}$$

Proposition 3.1.7 (The future Malliavin derivative of an adapted process) Consider a random variable $A \in \mathbb{D}^{1,2}$ adapted to $\mathcal{F}_t$, for some $t < T$. Then $D_r F_t = 0$ a.s., for all $r > t$.

Proof. For the sake of simplicity, assume $A = f(W_t)$, for some $f \in C_b^1$. Then, the Malliavin derivative of A is given by $D_s(f(W_t)) = f'(W_t)\mathbf{1}_{[0,t]}(s) = 0$. In the general case, the result follows from a limit argument. ■

Remark 3.1.8 *In a similar way, we can prove that if A is $\mathcal{F}_B$-measurable, for some $B \in [0, T]$, then $D_s A = 0$ for all $s \in B^c$. In particular, if A and the Brownian motion W are independent, $DA = 0$.*

The above proposition can be generalised as follows (see Proposition 1.2.8 in Nualart (2006)).

Proposition 3.1.9 (The Malliavin derivative of a conditional expectation) Assume $A \in \mathbb{D}^{1,2}$. Then, for all $t \in (0, T)$, $E_t(A) \in \mathbb{D}^{1,2}$, and we have

$$D_s(E_t(A)) = E_t(D_s A)\mathbf{1}_{[0,t]}(s), a.s.$$

3.1.2 The divergence operator

Definition 3.1.10 *The divergence operator δ is the adjoint of the Malliavin derivative operator. That is, for every $u \in Dom\delta$, $\delta(u)$ is defined by the equality*

$$E(\langle DF, u\rangle_H) = E(F\delta(u)). \tag{3.1.3}$$

We will denote $\delta(u) = \int_0^T u(s)dW_s$. We notice that the domain of δ (Domδ) is given by the set of processes $u \in L^2(\Omega \times [0,T])$ such that $|E\langle DF, u\rangle_H| \leq c\|F\|_2$, for all $F \in \mathbb{D}^{1,2}$, and for some $c > 0$ depending on u. Notice that, taking $F = 1$ in (3.1.3) we get $E(\delta(u)) = 0$. The divergence operator with respect to a Brownian motion W coincides with the Skorohod integral, an extension of the Itô integral to the case of non-adapted processes introduced independently by Hitsuda (1972, 1978) and Skorohod (1975).

Let us compute this integral for different kind of process. Let us consider first a deterministic processes of the form $u = h$, for some $h \in H$. In this case, Proposition 3.1.4 gives us that $\delta(u) = W(h)$. As a next step, assume $u = Ah$, where A is a random variable in $\mathbb{D}^{1,2}$ and $h \in H$. Then, a direct application of (3.1.2) gives us

$$E(A\langle DF, h\rangle_H) = E(FAW(h) - F\langle DA, h\rangle_H),$$

from where it follows that

$$\delta(Ah) = AW(h) - \langle DA, h\rangle_H. \tag{3.1.4}$$

The above observations prove the following result.

Proposition 3.1.11 *Consider the class of simple processes u of the form*

$$u(s) = \sum_i^n A_i h_i(s), \tag{3.1.5}$$

where $A_i \in \mathbb{D}^{1,2}$ and $h_i \in H$, for all $i = 1, ..., n$, and where $n \geq 1$. Then, the Skorohod integral of u is given by

$$\delta(u) = \sum_i^n \left(A_i W(h_i) - \langle DA_i, h_i\rangle_H\right). \tag{3.1.6}$$

The Skorohod integral of a general process u in Domδ can then be computed using a limit procedure.

Remark 3.1.12 *If u is a process of the form*

$$u(s) = \sum_i^n A_i \mathbf{1}_{(t_i, t_{i+1}]}(s), \tag{3.1.7}$$

where $A_i \in \mathbb{D}^{1,2}$ and $t_i = iT/n$, we have

$$\delta(u) = \sum_i^n A_i (W_{t_{i+1}} - W_{t_i}) - \sum_i^n \int_{t_i}^{t_{i+1}} D_s A_i ds. \tag{3.1.8}$$

If u is adapted (that is, A_i is $\mathcal{F}_{t_i}$-adapted), then $D_s A_i = 0$ for all $s > t_i$ (see Proposition 3.1.7), and then the above equality reduces to

$$\delta(u) = \sum_i^n A_i (W_{t_{i+1}} - W_{t_i}). \tag{3.1.9}$$

That is, the divergence operator coincides with the Itô integral in the case of adapted and square integrable processes, where it is defined as a $L^2(\Omega)$ limit of Riemann sums. When the integrand u is not adapted, the Skorohod integral can still be defined, but it is not the limit of Riemann sums.

Remark 3.1.13 *Notice finally that 3.1.6 implies that the Skorohod integral is not only well defined for adapted and square integrable processes but also, for example, for processes in the space $\mathbb{L}^{1,2}$, where $\mathbb{L}^{n,2} = \mathbb{D}^{n,2}(L^2([0,T])$.*

The following result, that can be seen as an extension of (3.1.4), gives us the relationship between $\delta(u)$ and $\delta(Au)$, for a random variable A and a random process u.

Proposition 3.1.14 *Consider a random variable $A \in \mathbb{D}^{1,2}$ and a square integrable process u such that u and $Au \in Dom\delta$. Then*

$$\delta(Au) = A\delta(u) - \langle DA, u \rangle_H. \tag{3.1.10}$$

3.2 COMPUTATION OF MALLIAVIN DERIVATIVES

This section is devoted to learn how to compute the Malliavin derivatives of some of the most common processes in stochastic volatility modelling.

3.2.1 The Malliavin derivative of an Itô process

Consider an Itô process of the form

$$X_t = X_0 + \int_0^t \mu_s ds + \int_0^t \sigma_s dW_s, t \in [0,T], \tag{3.2.1}$$

where μ, σ are adapted and square integrable processes in $\mathbb{L}^{1,2}$. Given $n > 1$, let us consider the processes μ^n, σ^n defined by

$$\mu_s^n = \sum_{i=1}^n \mu_{t_i} \mathbf{1}_{(t_i,t_{i+1}]}(s),$$

and

$$\sigma_s^n = \sum_{i=1}^n \sigma_{t_i} \mathbf{1}_{(t_i,t_{i+1}]}(s),$$

where $t_i = iT/n$, for all $i = 1, ..., n$. Consider now the process

$$\begin{aligned} X_t^n &= X_0 + \int_0^t \mu_s^n ds + \int_0^t \sigma_s^n dW_s \\ &= \sum_{i=1}^n \mu_{t_i}(t_{i+1} - t_i) + \sum_1^n \sigma_{t_i}(W_{t_{i+1}} - W_{t_i}). \end{aligned}$$

It is easy to see that

$$\begin{aligned} D_r X_t^n &= \sum_{i=1}^n D_r \mu_{t_i}(t_{i+1} - t_i) + \sum_{i=1}^n \left(\sigma_{t_i} \mathbf{1}_{(t_i,t_{i+1}]}(r) + (D_r \sigma_{t_i})(W_{t_{i+1}} - W_{t_i})\right) \\ &= \int_0^t D_r \mu_s^n ds + \sigma_r^n \mathbf{1}_{[0,t]}(r) + \int_0^t D_r \sigma_s^n dW_s. \end{aligned} \tag{3.2.2}$$

Then, a limit argument gives us that $X \in \mathbb{L}^{1,2}$ and

$$DX_t = \int_0^t D\mu_s ds + \sigma \mathbf{1}_{[0,t]} + \int_0^t D\sigma_s dW_s. \tag{3.2.3}$$

This result allows us to compute the Malliavin derivative of a diffusion process, as we see in the following subsection.

3.2.2 The Malliavin derivative of a diffusion process

Our goal in this section is to evaluate the Malliavin derivative of the solution of a stochastic differential equation (SDE) of the form

$$dX_t = a(t, X_t)dt + b(t, X_t)dW_t, \tag{3.2.4}$$

where $a, b \in \mathcal{C}_b^1$. We can write this equation in integral form

$$X_t = X_0 + \int_0^t a(s, X_s)ds + \int_0^t b(s, X_s)dW_s. \tag{3.2.5}$$

Then, the Malliavin derivative DX can be computed by making use of two different techniques, as we see in the following sections.

3.2.2.1 *The Malliavin derivative of a diffusion process as a solution of a linear SDE*

Equation (3.2.3) and the fact that $D_r X_s = 0$ for $r > s$ give us that $X \in \mathbb{L}^{1,2}$ and

$$D_r X_t = \int_r^t \frac{\partial a}{\partial x}(s, X_s) D_r X_s ds + b(r, X_r)\mathbf{1}_{[0,t]}(r) + \int_r^t \frac{\partial b}{\partial x}(s, X_s) D_r X_s dW_s. \tag{3.2.6}$$

That is, $D_r X$ is the solution of a *linear* stochastic differential equation.[2]

Example 3.2.1 *In Example 3.1.3 we computed the Malliavin derivative of the Black-Scholes asset price S_t using the fact that this price is a function of W_t. Now, let us compute this Malliavin derivative from the Black-Scholes SDE*

$$S_t = S_0 + r\int_0^t S_u du + \sigma \int_0^t S_u dW_u,$$

where r denotes the interest rate and sigma is the volatility. A direct application of the above argument gives us that

$$D_\theta S_t = r\int_\theta^t D_\theta S_u du + \sigma S_\theta \mathbf{1}_{[0,t]}(\theta) + \sigma \int_\theta^t D_\theta S_u dW_u.$$

That is, $D_\theta S$ follows a geometric Brownian motion. Then, we get

$$\begin{aligned} D_\theta S_t &= \sigma S_\theta \mathbf{1}_{[0,t]}(\theta) \exp\left(\left(r - \frac{\sigma^2}{2}\right)(t-\theta) + \sigma(W_t - W_\theta)\right) \\ &= \sigma S_t \mathbf{1}_{[0,t]}(\theta), \end{aligned}$$

according to the result in Example 3.1.3.

[2]The process X is also differentiable if a, b are globally Lipschitz functions with polynomial growth (see Theorem 2.2.1 in Nualart (2006)). Then (3.2.6) still holds, replacing $\frac{\partial a}{\partial x}(s, X_s)$ and $\frac{\partial b}{\partial x}(s, X_s)$ by adequate processes.

Example 3.2.2 *Consider the CEV model for asset prices introduced in Example 2.2.6*

$$dS_t = rS_t dt + \sigma S_t^\gamma dW_t, \tag{3.2.7}$$

for some positive r, σ, *and* γ. *Then, (3.2.6) takes the form*

$$D_\theta S_t = r \int_\theta^t D_\theta S_u du + \sigma S_\theta^\gamma \mathbf{1}_{[0,t]}(\theta) + \sigma\gamma \int_\theta^t S_u^{\gamma-1} D_\theta S_u dW_u.$$

In the case $\gamma \in (1/2, 1)$ *classical results (see for example Lemmas 4.1 and 4.2 in Bossy and Diop (2004)) give us some upper-bounds on inverse moments and exponential moments, which imply that the above SDE has a solution given by*

$$D_\theta S_t \quad = \quad \sigma S_\theta^\gamma \mathbf{1}_{[0,t]}(\theta) \exp\left(\int_\theta^t \left(r - \frac{\gamma^2\sigma^2 S_u^{2\gamma-2}}{2} \right) du + \sigma\gamma \int_\theta^t S_u^{\gamma-1} dW_u \right).$$

3.2.2.2 Representation formulas for the Malliavin derivative of a diffusion process

Here we present the method proposed in Detemple, García, and Rindisbacher (2005) that allows us to compute the Malliavin derivative of a diffusion process not as a solution of an SDE, but as a solution of an ordinary differential equation (ODE). One of the advantages of this approach is that it improves the rate of convergence in the numerical estimation of Malliavin derivatives. As we will see later, this method was the one used in Alòs and Ewald (2008) to explicitly compute the Malliavin derivative of the Heston volatility in terms of the volatility process.

Let us describe informally the main idea of this approach. We refer to Detemple, García, and Rindisbacher (2005) for a more rigorous explanation. Given (3.2.5), we apply the following change-of-variable: we consider the new process

$$Z_t := F(t, X_t),$$

where F is a deterministic function such that

$$\frac{\partial F}{\partial x}(t, x) = \frac{1}{b(t, x)}. \tag{3.2.8}$$

Notice that this implies

$$\frac{\partial^2 F}{\partial x^2}(t, x) = -\frac{1}{b^2(t, x)} \frac{\partial b}{\partial x}(t, x). \tag{3.2.9}$$

Then, from a direct application of the classical Itô's formula we get

$$\begin{aligned} dZ_t &= \frac{\partial F}{\partial t}(t, X_t)dt + \frac{\partial F}{\partial x}(t, X_t)(a(t, X_t)dt + b(t, X_t)dW_t) \\ &+ \frac{1}{2}\frac{\partial^2 F}{\partial x^2}(t, X_t)b^2(t, X_t)dt. \end{aligned}$$

Now, using (3.2.8) and (3.2.9)

$$dZ_t = \left(\frac{\partial F}{\partial t}(t, X_t)dt + \frac{a(t, X_t)}{b(t, X_t)} - \frac{1}{2}\frac{\partial b}{\partial x}(t, X_t)\right) dt + dW_t,$$

that can be written as

$$\begin{aligned} dZ_t &= \left(\frac{\partial F}{\partial t}(t, F^{-1}(t, Z_t)) + \frac{a(t, F^{-1}(t, Z_t))}{b(t, F^{-1}(t, Z_t))} - \frac{1}{2}\frac{\partial b}{\partial x}(t, F^{-1}(t, Z_t))\right) dt + dW_t \\ &:= f(t, Z_t)dt + dW_t. \end{aligned}$$

The key point of this trick is that now, the Malliavin derivative is expressed as the solution of an ODE

$$D_r Z_t = 1 + \int_r^t \frac{\partial f}{\partial z}(s, Z_s) D_r Z_s ds, \tag{3.2.10}$$

whose solution is given by

$$D_r Z_t = \exp\left(\int_r^t \frac{\partial f}{\partial z}(s, Z_s)ds\right).$$

Finally, notice that $F(t, X_t) = Z_t$ implies

$$DZ_t = \frac{\partial F}{\partial x}(t, X_t)DX_t = \frac{1}{b(t, X_t)}DX_t,$$

which gives us that

$$D_r X_t = b(t, X_t)\exp\left(\int_r^t \frac{\partial f}{\partial z}(s, F_t(s, X_s))ds\right). \tag{3.2.11}$$

Example 3.2.3 *Consider again the Black-Scholes model as in Example 3.1.3. Take $Z_t := F(S_t)$, where $F(x) := \frac{1}{\sigma}\ln x$. Then, a direct application of Itô's formula gives us that*

$$Z_t = \int_0^t \frac{1}{\sigma}\left(r - \frac{\sigma^2}{2}\right) ds + W_t.$$

Then, $D_r Z_t = \mathbf{1}_{[0,t]}(r)$ *and, as* $D_r Z_t = F'(S_t) D_r S_t$*, it follows that*

$$D_r S_t = \frac{D_r Z_t}{F'(S_t)} = \sigma S_t \mathbf{1}_{[0,t]}(r),$$

which coincides with the results in Examples 3.1.3 and 3.2.1.

Example 3.2.4 *(Adapted from Section 4.2 in Detemple, García, and Rindisbacher (2005)) Consider a CEV model as in Example 2.2.6. Take* $Z_t := F(S_t)$*, where* $F(x) = \frac{x^{1-\gamma}}{\sigma(1-\gamma)}$*. Then, Itô's formula allows us to write*

$$dZ_t = \left(\frac{\mu}{\sigma} S_t^{1-\gamma} - \frac{\gamma\sigma}{2} S_t^{\gamma-1} \right) dt + dW_t.$$

Now, taking into account that $Z_t = \frac{S_t^{1-\gamma}}{\sigma(1-\gamma)}$ *we get*

$$dZ_t = \left(\mu(1-\gamma) Z_t - \frac{\gamma}{2(1-\gamma)} Z_t^{-1} \right) dt + dW_t.$$

Now, we compute the Malliavin derivative of Z

$$D_r Z_t = \mathbf{1}_{[0,t]}(r) + \int_r^t \left(\mu(1-\gamma) + \frac{\gamma}{2(1-\gamma)} Z_u^{-2} \right) D_r Z_u du,$$

which gives us that

$$D_r Z_t = \mathbf{1}_{[0,t]}(r) \exp\left(\int_r^t \left(\mu(1-\gamma) + \frac{\gamma}{2(1-\gamma)} Z_u^{-2} \right) du. \right.$$

Finally, as $D_r Z_t := F'(S_t) D_r S_t$ *we get*

$$D_r S_t = \sigma S_t^{\gamma} D_r Z_t = \sigma S_t^{\gamma} \mathbf{1}_{[0,t]}(r) \exp\left((1-\gamma) \int_r^t \left(\mu + \frac{\gamma\sigma^2}{2} S_u^{2\gamma-2} \right) du \right).$$

Notice that the Malliavin derivative in the Black-Scholes case $\gamma = 1$ *is consistent with the above expression. In the case* $\gamma \neq 1$*, this representation is different from the one in Example 2.2.6, which involves an Itô integral. The representation of a Malliavin derivative as a solution of an ODE has several advantages. One is the improvement of the numerical simulation of Malliavin derivatives. To observe this phenomenon, we compute the mean error* $|D_0^N S_1 - D_0 S_1|$ *for* $N = 1, .., 12$*, where* $D_0 S_1$ *denotes the benchmark true value for the Malliavin derivative (computed from a* 2^{14} *points discretisation), and* $D_0^N S_1$ *denotes the approximation from a* 2^N *points discretisation and where the number of Monte Carlo*

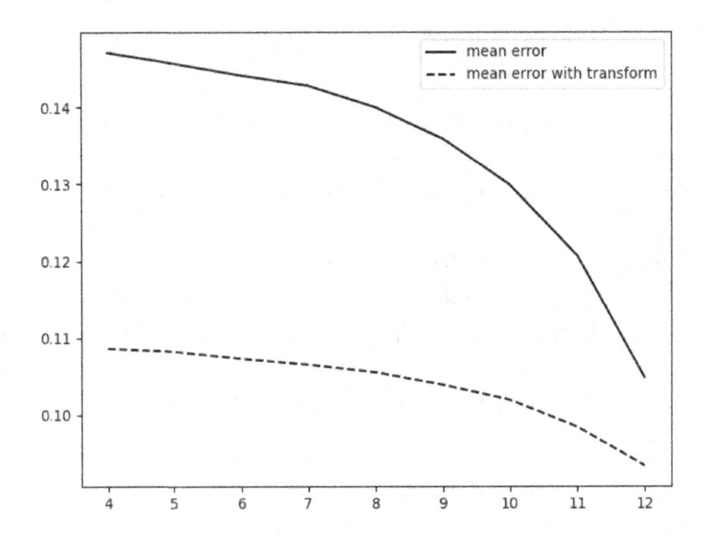

Figure 3.1 Mean error $|D_0^N S_1 - D_0 S_1|$ for the CEV model with $S_0 = 10, \sigma = 0.2$, and $\gamma = 0.6$ as a function of N, computed as the solutions of an SDE and an ODE.

simulations is taken to be equal to $M = 50000$. *In Figures 3.1 and 3.2 we can see this mean error when* $D_0^N S_1$ *is computed using a classical Euler scheme (solid line) and when we consider the transformation (dotted line). We can clearly see that the method based on the ODE representation converges faster to the true value.*

3.3 MALLIAVIN DERIVATIVES FOR GENERAL SV MODELS

In this subsection, we compute the Malliavin derivative of some classical processes used to describe the spot volatility.

3.3.1 The SABR volatility

As introduced in Section 2.3.2, the SABR model for a forward price F is given by

$$\begin{cases} dF_t = \sigma_t F_t^\beta dZ_t \\ d\sigma_t = \alpha \sigma_t dW_t, \quad t \in [0, T], \end{cases} \tag{3.3.1}$$

where we take $\beta \in [0, 1]$, $\alpha > 0$ and where W, Z are two Brownian motions that can be correlated. Let us compute the Malliavin derivative

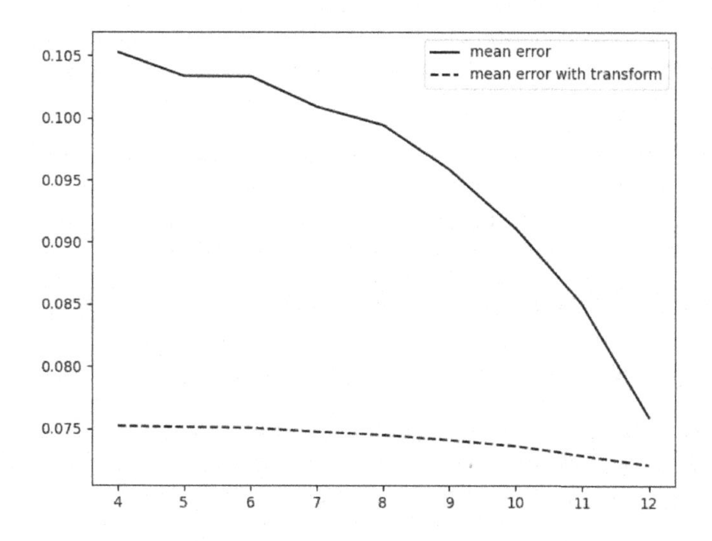

Figure 3.2 Mean error $|D_0^N S_1 - D_0 S_1|$ for the CEV model with $S_0 = 3, \sigma = 0.2$, and $\gamma = 1.2$ as a function of N, computed as the solution of a SDE and as the solution of a ODE.

of σ with respect to the Brownian motion W. As the volatility is a geometric Brownian motion we have

$$\sigma_t = \sigma_0 \exp\left(-\frac{\alpha^2}{2}t + \alpha W_t\right).$$

Then, the chain rule gives us that the Malliavin derivative with respect to W is given by

$$D_r^W \sigma_t = \alpha \sigma_t \mathbf{1}_{[0,t]}(r).$$

3.3.2 The Heston volatility

In the Heston model (see Section 2.3.1), the volatility σ is given by the square root of a Cox-Ingersoll-Ross (CIR) process. That is, $\sigma_t = \sqrt{v_t}$, where

$$dv_t = k(\theta - v_t)dt + \nu\sqrt{v_t}dW_t,$$

for some positive constants k, θ, ν. Taking (formally)[3] Malliavin derivatives in the above equation, we get the following linear SDE for $D_r v_t$

$$D_r v_t = -k \int_r^t D_r v_s ds + \nu \sqrt{v_r} \mathbf{1}_{[0,t]}(r) + \frac{\nu}{2} \int_r^t \frac{1}{\sqrt{v_s}} D_r v_s dW_s.$$

Notice that, taking conditional expectations, it follows that

$$E_r(D_r v_t) = -k \int_r^t E_r(D_r v_s ds) + \nu \sqrt{v_r} \mathbf{1}_{[0,t]}(r),$$

which implies that $E_r(D_r \sigma_t^2) = \nu \sqrt{v_r} \mathbf{1}_{[0,t]}(r) \exp(-k(t-r))$.

Let us now apply the method in Detemple, García, and Rindisbacher (2005) as in Section 3.2.2.2. Consider the process $\sigma_t = \sqrt{v_t}$. Then, a direct application of Itô's formula gives us that

$$\begin{aligned} d\sigma_t &= \frac{1}{2\sigma_t} \left(k(\theta - \sigma_t^2) dt + \nu \sigma_t dW_t \right) - \frac{\nu^2}{8\sigma_t} dt \\ &= \left(\left(\frac{k\theta}{2} - \frac{\nu^2}{8} \right) \frac{1}{\sigma_t} - \frac{k}{2} \sigma_t \right) dt + \frac{\nu}{2} dW_t. \end{aligned} \tag{3.3.2}$$

This allows us to find the ODE for the Malliavin derivative of σ

$$D_r \sigma_t = \int_r^t \left(- \left(\frac{k\theta}{2} - \frac{\nu^2}{8} \right) \frac{1}{\sigma_s^2} - \frac{k}{2} \right) D_r \sigma_s ds + \frac{\nu}{2} \mathbf{1}_{[0,t]}(r),$$

whose solution is given by

$$D_r \sigma_t = \frac{\nu}{2} \mathbf{1}_{[0,t]}(r) \exp \left(\int_r^t \left(- \left(\frac{k\theta}{2} - \frac{\nu^2}{8} \right) \frac{1}{\sigma_s^2} - \frac{k}{2} \right) ds \right).$$

Finally, $D_r v_t = D_r \sigma_t^2$, and by the chain rule $D_r v_t = 2\sigma_t D\sigma_t$, which allows us to write

$$D_r v_t = \frac{\nu \sqrt{v_t}}{2} \mathbf{1}_{[0,t]}(r) \exp \left(\int_r^t \left(- \left(\frac{k\theta}{2} - \frac{\nu^2}{8} \right) \frac{1}{v_s} - \frac{k}{2} \right) ds \right),$$

that gives us an explicit expression for $D_r v_t$ in terms of v. We refer to Alòs and Ewald (2008) for a deeper analysis on the regularity of the Heston volatility in the Malliavin calculus sense.

[3] The coefficients of the Heston model do not satisfy the standard assumptions since the square root is not differentiable nor globally Lipschitz. We refer to Alòs and Ewald (2008) for a rigorous approach.

3.3.3 The 3/2 Heston volatility

In this model, the volatility is given by $\sqrt{v_t}$, where v follows an equation of the form

$$dv_t = v_t k(\theta - v_t)dt + \nu v_t^{3/2} dW_t,$$

for some positive k, θ, ν. It is well known that the inverse of the 3/2 model follows a Heston dynamics. In fact, a direct application of Itô's formula for $u = 1/v$ (which implies $u' = -v^{-2}$ and $u'' = 2v^{-3}$) gives us that

$$du_t = -k(u_t\theta - 1)dt - \nu u_t^{1/2} dW_t + \nu^2 dt,$$

from where it follows that

$$du_t = -k\theta\left(u_t - \frac{1}{\theta} - \frac{\nu^2}{k\theta}\right)dt - \nu u_t^{1/2} dW_t.$$

That is, u follows a Heston dynamics with different parameters. Then, the results in the previous section allow us to write

$$\begin{aligned} D_r v_t &= -\frac{1}{u_t^2} D_r u_t \\ &= -\frac{\nu' v_t^{3/2}}{2} \mathbf{1}_{[0,t]}(r) \exp\left(\int_r^t \left(-\left(\frac{k'\theta'}{2} - \frac{(\nu')^2}{8}\right) v_s - \frac{k'}{2}\right) ds\right), \end{aligned}$$

where $k' := k\theta$, $v' = -v$, and $\theta' = \frac{1}{\theta} + \frac{\nu^2}{k\theta}$. Finally, a direct computation gives us the following expression for $D_r v_t$

$$D_r v_t = \frac{\nu v_t^{3/2}}{2} \mathbf{1}_{[0,t]}(r) \exp\left(\int_r^t \left(-\left(\frac{k}{2} + 3\frac{\nu^2}{8}\right) v_s - \frac{k\theta}{2}\right) ds\right). \quad (3.3.3)$$

We can also try to use the approach in Section 3.2.2.2. Let us define $Z_t := F(v_t)$, where $F(x) = x^{-1/2}$ (this implies $f'(x) = -\frac{1}{2}x^{-3/2}$ and $f''(x) = \frac{3}{2}x^{-5/2}$). Then, Itô's formula gives us that

$$dZ_t = -\frac{1}{2} v_t^{-1/2} k(\theta - v_t)dt - \frac{\nu}{2} dW_t + \frac{3\nu^2}{8} v_t^{1/2} dt.$$

That is,

$$dZ_t = -\frac{1}{2} k Z_t \left(\theta - Z_t^{-2}\left(1 + 3\frac{\nu^2}{4k}\right)\right) - \frac{\nu}{2} dW_t.$$

This implies that for all $r < t$

$$D_r Z_t = -\frac{\nu}{2} \exp\left(-\frac{1}{2} k \int_r^t \left(\theta + Z_u^{-2} \left(1 + 3\frac{\nu^2}{4k} \right) \right) ds \right).$$

Now, as $D_r Z_t = -\frac{1}{2} v^{-3/2} D_r Z_t$, we get

$$D_r v_t = \frac{\nu v_t^{3/2}}{2} \exp\left(-\frac{1}{2} k \int_r^t \left(\theta + v_s \left(1 + 3\frac{\nu^2}{4k} \right) \right) ds \right),$$

an expression that coincides with (3.3.3).

3.4 CHAPTER'S DIGEST

One of the objectives of Malliavin calculus is the adequate definition of a 'derivative' with respect to white noise. The Malliavin derivative operator D is defined in such a way that $DW(h) = h$ and it satisfies the chain rule. From its definition, we can prove some basic properties of this operator, as the integration-by-parts formula, or the fact that the future Malliavin derivative of an adapted process is equal to zero. Its adjoint, the divergence operator, coincides with the Skorohod integral, an extension of the Itô integral to the case of non-adapted processes.

The Malliavin derivative of a diffusion process can be computed as the solution of an SDE. An adequate change of variable leads to a representation of this Malliavin derivative as the solution of an ODE, a characterisation that improves its numerical simulation.

Malliavin derivatives are not just abstract random variables but they can be computed explicitly for most models in finance. As examples, we have studied the Black-Scholes and the CEV model, as well as some common volatility models such as Heston, SABR, and the 3/2 model.

CHAPTER 4

Key tools in Malliavin Calculus

This chapter is devoted to present some key Malliavin calculus tools that we will use throughout the book, together with some direct financial applications. More precisely, we introduce the Clark-Ocone-Haussmann formula, the general integration-by-parts formula, and the anticipating Itô's formula.

4.1 THE CLARK-OCONE-HAUSSMAN FORMULA

One of the first applications of Malliavin calculus is to get an explicit martingale representation for any random variable F measurable with respect to the $\sigma - algebra$ generated by a Brownian motion W. This result is a powerful tool in several scenarios, as we will see in this section.

4.1.1 The Clark-Ocone-Haussman formula and the martingale representation theorem

Assume that we have a square integrable and $\mathcal{F}_T$-measurable random variable F, where $\mathcal{F}$ denotes the σ-algebra generated by a Brownian motion W. Then, the martingale representation theorem (see, for instance, Proposition 3.2 in Chapter V in Revusz-Yor (1999)) states that there exists a $\mathcal{F}_t$-adapted process $m(T, \cdot)$ such that

$$F = E(F) + \int_0^T m(T, s) dW_s. \tag{4.1.1}$$

The intuition behind this formula is as follows. Consider first the set of random variables of the form

$$F = \mathcal{E}\left(X_T^h\right), \tag{4.1.2}$$

where $X_t^h := \int_0^t h(s)dW_s$ for some $h \in H$, and $\mathcal{E}$ denotes the Dóleans-Dade exponential

$$\mathcal{E}(X_t^h) := \exp\left(\int_0^t h(s)dW_s - \frac{1}{2}\int_0^t h^2(s)ds\right).$$

A direct application of Itô's formula gives us that

$$F = 1 + \int_0^T \mathcal{E}\left(X_s^h\right) h(s)dW_s,$$

which proves (4.1.1) with $m(T,s) = \mathcal{E}\left(X_s^h\right) h(s)$. As the set of random variables of the form (4.1.2) is total in $L^2(\Omega, \mathcal{F}_T, P)$, the result follows from a limit argument.

Even when this result gives a powerful tool with a wide range of applications, it remains to get an explicit expression for the process $m(T,\cdot)$. The Clark-Ocone-Haussman formula is the key result that allows us to find this explicit representation. This result makes use of Malliavin calculus to identify the process $m(T,\cdot)$ in the martingale representation theorem

Proposition 4.1.1 (Clark-Ocone-Haussman formula) *Consider a random variable $F \in \mathbb{D}^{1,2}$. Then*

$$F = E(F) + \int_0^T E_r(D_r F)dW_r.$$

Proof. For the sake of simplicity, we can assume $E(F) = 0$. Now, given a random variable G of the form $G = \int_0^T h(s)dW_s$, for some $h \in H$, we can write

$$\begin{aligned} E(FG) &= E\left(\int_0^T D_s F h(s)ds\right) \\ &= E\left(\int_0^T E_s(D_s F)h(s)ds\right). \end{aligned} \tag{4.1.3}$$

On the other side, the martingale representation theorem gives us that there exists a unique process $m(T,\cdot) \in L_a^2$ such that

$$F = \int_0^T m(T,s)dW_s,$$

which implies that

$$E(FG) = \int_0^T m(T,s)h(s)ds. \tag{4.1.4}$$

Now, comparing (4.1.3) and (4.1.4) the result follows. ■

Let us see some examples of the application of this formula.

Example 4.1.2 (The martingale representation of a CEV model) *Consider a CEV model as in Example (2.2.6). That is,*

$$dS_t = \mu S_t dt + \sigma S_t^{\gamma} dW_t, \tag{4.1.5}$$

for some positive constants μ and γ. Then

$$D_\theta S_t = \mu \int_\theta^t D_\theta S_u du + \sigma S_\theta^\gamma \mathbf{1}_{[0,t]}(\theta) + \sigma\gamma \int_\theta^t S_u^{\gamma-1} D_\theta S_u dW_u,$$

from where we deduce

$$E_\theta(D_\theta S_t) = \mu \int_\theta^t E_\theta(D_\theta S_u) du + \sigma S_\theta^\gamma \mathbf{1}_{[0,t]}(\theta),$$

which implies that $E_\theta(D_\theta S_t) = \sigma S_\theta^\gamma \mathbf{1}_{[0,t]}(\theta) \exp\left(\mu(t-\theta)\right)$. Then, we get the representation

$$\begin{aligned} S_t &= E(S_t) + \sigma \int_0^t S_\theta^\gamma \exp(\mu(t-\theta)) dW_\theta \\ &= S_0 \exp(\mu t) + \sigma \int_0^t S_\theta^\gamma \exp(\mu(t-\theta)) dW_\theta. \end{aligned} \tag{4.1.6}$$

In particular, in the Black-Scholes case $\gamma = 1$

$$S_t = S_0 \exp(\mu t) + \sigma \int_0^t S_\theta \exp(\mu(t-\theta)) dW_\theta.$$

Example 4.1.3 (The variance of a random variable) *Consider the random variable $F = W_T^2$. We can make use of the Clark-Ocone-Haussmann representation formula to compute its variance. More precisely, we can write*

$$Var(F) = E(F - E(F))^2 = E\left(\int_0^T E_r(D_r F) dW_r\right)^2 = E \int_0^T (E_r(D_r F))^2 dr.$$

Now, using the chain rule and taking into account the expression for the Malliavin derivative of a Brownian motion in Example 3.1.2 we get

$$D_r W_T^2 = 2W_T D_r W_T = 2W_T \mathbf{1}_{[0,T]}(r).$$

This implies that, for all $r < T$,

$$E_r(D_r W_T^2) = 2E_r(W_T) = 2W_r,$$

which allows us to write

$$Var(F) = E\int_0^T 4W_r^2 dr = 2T^2.$$

Example 4.1.4 (The covariance between two random variables) *Consider a Black-Scholes model as in (1.3.1). Let us compute the covariance between the asset price S_T and the Brownian motion W_T. Towards this end, we apply the Clark-Ocone formula to get*

$$S_T = E(S_T) + \int_0^T E_u(D_u S_T) dW_u.$$

Now, as $D_u S_T = \sigma S_T \mathbf{1}_{[0,T]}(u)$ and $E_u(S_T) = S_u e^{r(T-u)}$ we can write

$$S_T = E(S_T) + \sigma \int_0^T S_u e^{r(T-u)} dW_u.$$

Then

$$\begin{aligned} Cov(S_T, W_T) &= E\left[\left(\sigma \int_0^T S_u e^{r(T-u)} dW_u\right)\left(\int_0^T dW_u\right)\right] \\ &= \sigma E \int_0^T S_u e^{r(T-u)} du \\ &= \sigma S_0 e^{rT} T. \end{aligned} \tag{4.1.7}$$

4.1.2 Hedging in the Black-Scholes model

The most popular application in finance of the Clark-Ocone-Haussman formula is hedging. Consider for simplicity a market with one bond and one stock (S). Assume that the bond price β is given by $\beta_t = e^{rt}$, for some real parameter, and that the asset price S follows a Black-Scholes model under a risk-neutral probability

$$dS_t = rS_t dt + \sigma S_t dW_t,$$

for some positive constant σ. In this context, a trading strategy is defined by the number of assets (α_t) and the number of non-risky assets (β_t) that the investor has at every time t, and the value of this portfolio at time t is given by

$$V_t = \alpha_t S_t + \beta_t e^{rt}.$$

In a self-financing strategy (see Chapter 1), these quantities satisfy that

$$\begin{aligned} V_t &= V_0 + \int_0^t \alpha_u dS_u + \int_0^t \beta_u d\beta_u \\ &= V_0 + \int_0^t \alpha_u dS_u + r\int_0^t \beta_u e^{ru} du. \end{aligned} \tag{4.1.8}$$

Then it follows that

$$\begin{aligned} e^{-rt}V_t &= V_0 + \int_0^t \alpha_u e^{-ru} dS_u + r\int_0^t \beta_u du - r\int_0^t e^{-ru} V_u du \\ &= V_0 + \int_0^t \alpha_u e^{-ru} dS_u + r\int_0^t \beta_u du - r\int_0^t e^{-ru}(\alpha_u S_u + \beta_u e^{ru}) du \\ &= V_0 + \int_0^t \alpha_u e^{-ru}(dS_u - rS_u du) \\ &= V_0 + \sigma\int_0^t \alpha_u S_u dW_u. \end{aligned} \tag{4.1.9}$$

In particular, $e^{-rt}V_t$ is a martingale. On the other hand, a direct application of Clark-Ocone-Haussmann formula to the random variable V_T gives us that

$$V_T = E(V_T) + \int_0^T E_u(D_u V_T) dW_u.$$

Now, as $e^{-rt}V_t$ is a martingale it follows that $E_t(e^{-rT}V_T) = e^{-rt}V_t$, and then we can write

$$\begin{aligned} e^{-rt}V_t &= e^{-rT}E_t(V_T) \\ &= e^{-rT}\left(E(V_T) + \int_0^t E_u(D_u V_T) dW_u\right) \\ &= V_0 + \int_0^t e^{-rT} E_u(D_u V_T) dW_u. \end{aligned} \tag{4.1.10}$$

Then, a direct comparison of (4.1.9) and (4.1.10) gives us the number of assets α_u in a self-financing portfolio with final value V_T. That is,

$$\alpha_u \sigma S_u = e^{-rT} E_u(D_u V_T),$$

which gives us that

$$\alpha_u = \frac{e^{-rT} E_u(D_u V_T)}{\sigma S_u}. \tag{4.1.11}$$

Now, assume $V_T = h(S_T)$, the payoff of a European option. The chain rule gives us that $D_u(h(S_T)) = h'(S_T) D_u S_T$ and then, the results in Example (3.1.3) allow us to write, for all $u \in (0, T)$

$$E_u(D_u V_T) = \sigma E_u((h'(S_T)) S_T). \tag{4.1.12}$$

Now, as S_T follows, under the Black-Scholes model, a log-normal distribution, we get

$$E_u(h'(S_T) S_T) = \int_{\mathbb{R}} h'(e^x) e^x \phi(x, S_u) dx,$$

where $\phi(\cdot, S_u)$ denotes the density of Gaussian random variable with mean $\ln(S_u) + \left(r - \frac{\sigma^2}{2}\right)(T-u)$ and standard deviation $\sigma\sqrt{T-u}$. Then, using the integration by parts formula is easy to see that

$$\begin{aligned} E_u(h'(S_T) S_T) &= \int_{\mathbb{R}} h(e^x) \frac{\partial \phi}{\partial x}(x, S_u) dx \\ &= S_u \frac{\partial}{\partial S_u} \int_{\mathbb{R}} h(e^x) \phi(x, S_u) dx \\ &= S_u \frac{\partial}{\partial S_u} \left(e^{-rT} E_u(h(S_T)) \right). \end{aligned} \tag{4.1.13}$$

This, jointly with (4.1.11) and (4.1.12), gives us that

$$\alpha_u = \frac{\partial}{\partial S_u} \left(e^{-rT} E_u(h(S_T)) \right) = \frac{\partial V_T}{\partial S_u}, \tag{4.1.14}$$

which is the classical Delta.

Remark 4.1.5 *The above arguments can be applied to other complete markets. In the case of incomplete markets, S is not adapted to the filtration generated by one Brownian motion W and it, not possible to replicate the option just with bonds and stocks. Nevertheless, Clark-Ocone-Haussman formula still allows us to construct minimal variance hedging strategies in this framework (see for example Benth, Di Nunno, Lokka, Oksendal, and Proske (2003)).*

4.1.3 A martingale representation for spot and integrated volatilities

The Clark-Ocone-Haussman formula allows us to obtain a martingale representation for spot and integrated volatilities, a useful tool in several applications, as we see in the following chapters. Consider a volatility process σ, adapted to the filtration $\mathcal{F}$ generated by a Brownian motion W. Under some general conditions, the Clark-Ocone-Haussman formula gives us that

$$\sigma_t^2 = E(\sigma_t^2) + \int_0^t E_u(D_u\sigma_t^2)dW_u.$$

Then, the realised variance can be represented as

$$\int_0^T \sigma_t^2 dt = \int_0^T E(\sigma_t^2)dt + \int_0^T \left(\int_0^t E_u(D_u\sigma_t^2)dW_u\right)dt.$$

Now, the stochastic Fubini's theorem allows us to write

$$\int_0^T \sigma_t^2 dt = E\int_0^T (\sigma_t^2)dt + \int_0^T \left(\int_u^T E_u(D_u\sigma_t^2)dt\right)dW_u, \quad (4.1.15)$$

which gives us a martingale representation for the integrated volatility $\int_0^T \sigma_t^2 dt$. Notice that, in particular, the above equality proves that

$$dM_u = \left(\int_u^T E_u(D_u\sigma_t^2)dt\right)dW_u, \quad (4.1.16)$$

where $M_u := E_u\int_0^T \sigma_t^2 dt$. More generally, we have the following result.

Lemma 4.1.6 *Consider a process of the form* $u_t := E_t\int_0^T \mu_s ds$, *for* $t \in [0,T]$ *and where* $\mu \in \mathbb{L}^{1,2}$. *Then*

$$du_t = \left(\int_t^T E_t(D_t\mu_s)ds\right)dW_t.$$

In the following examples, we find this explicit representation formula for different volatility models.

4.1.3.1 The SABR volatility

Consider a SABR model as in Section 2.3.2, where $d\sigma_t = \alpha\sigma_t dW_t$. As computed in Section 3.3.1, $D_u\sigma_t = \alpha\sigma_t$, for all $u < t$. Then

$$D_u\sigma_t^2 = 2\sigma_t D_u(\sigma_t) = 2\alpha\sigma_t^2$$

and this leads to the representation

$$\begin{aligned}\sigma_t^2 &= E(\sigma_t^2) + \int_0^t E_u(D_u \sigma_t^2) dW_u \\ &= E(\sigma_t^2) + 2\alpha \int_0^t E_u(\sigma_t^2) dW_u. \end{aligned} \tag{4.1.17}$$

Now, as σ_t is, conditional to $\mathcal{F}_u$, a log-normal random variable with mean $\sigma_u - \frac{\alpha^2(t-u)}{2}$ and standard deviation $\alpha\sqrt{t-u}$, we get $E_u(\sigma_t^2) = \sigma_u^2 \exp(2\alpha^2(t-u))$. This gives us that

$$\sigma_t^2 = \sigma_0^2 \exp(2\alpha^2 t) + 2\alpha \int_0^t \sigma_u^2 \exp(2\alpha^2(t-u)) dW_u. \tag{4.1.18}$$

This, jointly with (4.1.15), allows us to write

$$\begin{aligned}\int_0^T \sigma_t^2 dt &= \sigma_0^2 \int_0^T \exp(2\alpha^2 t) dt + 2\alpha \int_0^T \sigma_u^2 \left(\int_u^T \exp(2\alpha^2(t-u)) dt \right) dW_u, \\ &= \sigma_0^2 \int_0^T \exp(2\alpha^2 t) dt + 2\alpha \int_0^T \sigma_u^2 \left(\int_u^T \exp(2\alpha^2(t-u)) dt \right) dW_u. \end{aligned} \tag{4.1.19}$$

As a direct consequence, we observe that, under the SABR model, the variance of the squared integrated volatility is given by

$$\begin{aligned} Var\left(\int_0^T \sigma_t^2 dt \right) &= 4\alpha^2 E \int_0^T \sigma_u^4 \exp(8\alpha^2 u) \left(\int_u^T \exp(2\alpha^2(t-u)) dt \right)^2 du \\ &= 4\alpha^2 \sigma_0^4 \int_0^T \left(\int_u^T \exp(2\alpha^2(t+u)) dt \right)^2 du. \end{aligned} \tag{4.1.20}$$

4.1.3.2 The Heston volatility

Consider a Heston model as in Section 2.3.1. That is,

$$v_t = v_0 + \int_0^t k(\theta - v_t) dt + \nu \int_0^t \sqrt{v_s} dW_s.$$

A direct computation gives us that

$$E(v_t) = v_0 + \int_0^t k(\theta - E(v_t)) dt,$$

which implies that

$$E(v_t) = \theta + (v_0 - \theta)e^{-kt}.$$

On the other hand, from Section 3.3.2

$$E_r(D_r v_t) = \nu\sqrt{v_r}\mathbf{1}_{[0,t]}(r)\exp(-k(t-r)).$$

Then, we get the following martingale representation for the Heston model

$$v_t = \theta + (v_0 - \theta)e^{-kt} + \nu\int_0^t e^{-k(t-u)}\sqrt{v_u}dW_u.$$

In particular,

$$dM_u = \nu\sqrt{v_u}\left(\int_u^T e^{-k(t-u)}dt\right)dW_u,$$

where $M_u := E_u \int_0^T v_t dt$.

This martingale representation of the integrated volatility can be applied to the study of variance and volatility swaps, as we see in the following example.

Example 4.1.7 (Volatility and variance swaps) *The classical convexity adjustment (2.6.9) gives us that*

$$E\sqrt{\int_0^T \sigma_s^2 ds} \approx \sqrt{E\int_0^T \sigma_s^2 ds} - \frac{E\left(\int_0^T \sigma_s^2 ds - E\int_0^T \sigma_s^2 ds\right)^2}{8\left(\sqrt{E\int_0^T \sigma_s^2 ds}\right)^3}.$$

Let us obtain a version of this adjustment from Malliavin calculus techniques. If the volatility process is adapted with respect to the filtration generated by a Brownian motion W, Lemma 4.1.16 allows us to write

$$\int_0^T \sigma_s^2 ds - E\int_0^T \sigma_s^2 ds = \int_0^T \left(\int_s^T D_s^W \sigma_u^2 du\right) dW_s.$$

Now, a direct application of Itô's formula gives us that

$$\begin{aligned}\sqrt{\int_0^T \sigma_s^2 ds} &= \sqrt{E\int_0^T \sigma_s^2 ds} + \int_0^T \frac{1}{2\sqrt{E_r \int_0^T \sigma_s^2 ds}} \left(E_r \int_r^T D_r^W \sigma_s^2 ds \right) dW_s \\ &- \int_0^T \frac{1}{8\left(\sqrt{E_r \int_t^T \sigma_s^2 ds}\right)^3} \left(E_r \int_r^T D_r^W \sigma_s^2 ds \right)^2 ds. \qquad (4.1.21)\end{aligned}$$

Then, taking expectations

$$E\sqrt{\int_0^T \sigma_s^2 ds} = \sqrt{E\int_0^T \sigma_s^2 ds} - \frac{1}{8} E \int_0^T \frac{\left(E_r \int_r^T D_r^W \sigma_s^2 ds\right)^2}{\left(\sqrt{E_r \int_0^T \sigma_s^2 ds}\right)^3} ds. \qquad (4.1.22)$$

Notice that this is a closed-form expression for the difference between the square root of the variance swap and the volatility swap (see Alòs and Muguruza (2021) for further details).

4.1.4 A martingale representation for non-log-normal assets

Several options are defined by a payoff of the form

$$(A - K)_+,$$

where A does not follow classical log-normal dynamics. For example, A can be a volatility index (as the VIX), a realised variance, an average of asset prices (as in the case of Asian options), a maximum or minimum of asset prices (as in lookback options), etc.

In this context, the Clark-Ocone-Haussman formula allows us to find a log-normal type dynamics for the underlying A. More precisely, assume that A is a square integrable random variable adapted to $\mathcal{F}_T$, being $\mathcal{F}$ the filtration generated by a Brownian motion W. Then, define $N_t := E_t(A)$, for $t \in [0, T]$ (notice that $N_T = A$). If $A \in \mathbb{D}^{1,2}$

$$N_t = E(N_t) + \int_0^t E_s(D_s N_T) dW_s. \qquad (4.1.23)$$

Then, N_t admits the following Black-Scholes type dynamics

$$N_t = E(N_t) + \int_0^t \phi_t N_s dW_s, \tag{4.1.24}$$

where

$$\phi_s := \frac{E_s(D_s N_T)}{N_s}. \tag{4.1.25}$$

That is, (4.1.24) can be seen as a stochastic volatility model for N, where the volatility is given by the process ϕ.

Remark 4.1.8 *The above result allows us to adapt previous results on vanilla options in stochastic volatility models, but with one technical difference: notice that the volatility ϕ_s depends on T. This will have relevant consequences in the behaviour of the corresponding implied volatility surface, as we will see in Chapter 10.*

As an example, let us compute the volatility ϕ for the VIX volatility index.

Example 4.1.9 (VIX options) *Consider as underlying the random variable VIX_T given by*

$$VIX_T = \sqrt{\frac{1}{\Delta} E_T \int_T^{T+\Delta} \sigma_s^2 ds}, \tag{4.1.26}$$

where σ denotes the volatility process, adapted to the filtration generated by a Brownian motion W, and where $(T, T+\Delta)$ denotes a 30-days interval. Then, under some general hypotheses, the chain rule for the Malliavin derivative operator gives us that

$$D_t(VIX_T) = \frac{1}{2VIX_T\Delta} \int_T^{T+\Delta} D_t(E_T(\sigma_s^2))ds. \tag{4.1.27}$$

Now, Proposition 3.1.9 allows us to write, for all $t \in (0,T)$

$$D_t(VIX_T) = \frac{1}{2VIX_T\Delta} \int_T^{T+\Delta} E_T(D_t(\sigma_s^2))ds. \tag{4.1.28}$$

Then, the VIX volatility ϕ is given by

$$\phi_t = \frac{1}{2\Delta N_t} E_t\left(\frac{1}{VIX_T} \int_T^{T+\Delta} (D_t\sigma_s^2)ds\right), \tag{4.1.29}$$

where $N_t := E_t(VIX_T)$ (see Alòs, García-Lorite, and Muguruza (2018)). This representation gives us a tool to study the VIX implied volatility, as we will see in Chapter 10.

4.2 THE INTEGRATION BY PARTS FORMULA

One of the most well-known applications of Malliavin calculus in finance is the device of efficient Monte Carlo methods for the computation of the Greeks, the derivatives of options prices with respect to the parameters of the market. A natural approach in the computation of these derivatives is the Monte Carlo simulation of the finite approximation of the differentials, but this method performs very poorly for non-smooth payoffs (see Glyn (1989), Glasserman and Yao (1992), and L'Ecuyer and Perron (1994)). Another technique, introduced by Broadie and Glasserman (1996), is based on the differentiation of the payoff inside the expectation operator. Nevertheless, this method can be applied only in the case of very simple payoffs. Malliavin calculus techniques allow us to avoid the computation of the derivatives. More precisely, it gives us a way to represent these derivatives as the expectation of the product of the payoff times some random weight. This methodology was introduced by Fournié, Lasry, Lebuchoux, Lions, and Touzi (1999) and it is based on one of the main results in Malliavin calculus: the integration by parts formula.

4.2.1 The integration-by-parts formula for the Malliavin derivative and the Skorohod integral operators

The duality relationship between the Malliavin derivative and the Skorohod integral allows us to prove the following integration by parts formula (see Proposition 6.2.1 in Nualart (2006)).

Proposition 4.2.1 *Consider two random variables F, G and a random process u such that $F \in \mathbb{D}^{1,2}$, $\langle DF, u\rangle_H \neq 0$, and $\frac{Gu}{\langle DF,u\rangle_H} \in Dom\delta$. Then, for all $f \in \mathcal{C}_b^1$,*

$$E(f'(F)G) = E\left(f(F)\delta\left(\frac{Gu}{\langle DF, u\rangle_H}\right)\right).$$

Proof. The chain rule for the Malliavin derivative operator gives us that

$$\langle D(f(F)), u\rangle = \langle f'(F)DF, u\rangle_H,$$

from where we get

$$f'(F) = \frac{\langle D(f(F)), u\rangle_H}{\langle DF, u\rangle_H}.$$

Then

$$E(f'(F)G) = E\left(\frac{\langle D(f(F)), u\rangle_H}{\langle DF, u\rangle_H} G\right) = E\left\langle D(f(F)), \frac{Gu}{\langle DF, u\rangle_H}\right\rangle_H.$$

Then, by the duality relationship (3.1.3) it follows that

$$E(f'(F)G) = E\left(f(F)\delta\left(\frac{Gu}{\langle DF, u\rangle_H}\right)\right),$$

as we wanted to prove. ■

Remark 4.2.2 *Notice that Proposition 3.1.14 gives us that:*

$$\delta\left(\frac{uG}{\langle DF, h\rangle_H}\right) = \frac{G}{\langle DF, u\rangle_H}\delta(u) - \left\langle D\left(\frac{G}{\langle DF, h\rangle_H}\right), u\right\rangle_H. \tag{4.2.1}$$

Proposition 4.2.1 allows us to estimate the expectation of a derivative by Monte Carlo methods without directly simulating it, as we see in the following example.

Example 4.2.3 *Let $F = W_T$, and take $G = 1$, $u = \mathbf{1}_{[0,T]}$. Then, $DW_T = \mathbf{1}_{[0,T]}$, and for every $f \in C_b^1$, Proposition 4.2.1 gives us that*

$$E(f'(W_T)) = E\left(f(W_T)\delta\left(\frac{1}{T}\right)\right) = \frac{1}{T}E(f(W_T)W_T).$$

4.2.2 Delta, Vega, and Gamma in the Black-Scholes model

Let us now show how to apply the integration by parts formula for the computation of the Greeks in the Black-Scholes model, as defined in Example 3.1.3. Further analysis in the framework of stochastic volatility models can be found in Fournié, Lasry, Lebuchoux, and Lions (2001) and Benhamou (2000) and is discussed in Section 4.2.4.

4.2.2.1 The delta

The delta (Δ) is the derivative of the option price with respect to the initial price of the stock S_0. Consider a payoff of the type $f(S_T)$, where $f \in C_b^1$. The option price is given by

$$V = e^{-rT}E(f(S_T)),$$

where E denotes the expectation under the risk-neutral probability. Taking derivatives we obtain that the delta is given by

$$\Delta = \frac{\partial V}{\partial S_0} = e^{-rT} E\left(f'(S_T)\frac{\partial S_T}{\partial S_0}\right) = \frac{e^{-rT}}{S_0} E\left(f'(S_T)S_T\right).$$

A direct application of Proposition 4.2.1 with $G = S_T$, $u = \mathbf{1}_{[0,T]}$, and the fact that $\langle DS_T, u\rangle = \sigma T S_T$ allows us to write

$$\frac{\partial V}{\partial S_0} = \frac{e^{-rT}}{S_0} E\left(f(S_T)\delta\left(\frac{1}{\sigma T}\right)\right) = \frac{e^{-rT}}{\sigma T S_0} E\left(f(S_T)W_T\right). \tag{4.2.2}$$

An approximation argument allows us to see that the above equality also holds when f is the payoff function of a European call with strike K ($f(x) = (x-K)_+$) or a European put ($f(x) = (K-x)_+$). Notice that the estimation of the term in the right-hand side of (4.2.2) does not requiere the computation of the derivative $f'(S_T)$.

4.2.2.2 The vega

The vega ($\mathcal{V}$) is the derivative of the option price with respect to the volatility. If $f \in \mathcal{C}_b^1$, we have

$$\mathcal{V} = \frac{\partial V}{\partial \sigma} = e^{-rT} E\left(f'(S_T)\frac{\partial S_T}{\partial \sigma}\right) = e^{-rT} E\left(f'(S_T)S_T(W_T - \sigma T)\right).$$

Then, taking $G = S_T(W_T - \sigma T)$ and $u = \mathbf{1}_{[0,T]}$ we can write

$$\frac{\partial V}{\partial \sigma} = e^{-rT} E\left(f(S_T)\delta\left(\frac{W_T - \sigma T}{\sigma T}\right)\right). \tag{4.2.3}$$

Proposition 3.1.14 gives us that

$$\delta\left(\frac{W_T - \sigma T}{\sigma T}\right) = \frac{W_T^2 - \sigma T W_T}{\sigma T} - \frac{1}{\sigma}. \tag{4.2.4}$$

Then

$$\frac{\partial V}{\partial \sigma} = \frac{e^{-rT}}{\sigma T} E\left(f(S_T)\left(W_T^2 - \sigma T W_T - T\right)\right). \tag{4.2.5}$$

Notice that (4.2.5) also holds when f is the payoff function of a European call or a European put.

4.2.2.3 The gamma

The gamma (Γ) is the second derivative of the option price with respect to the initial price of the stock S_0. That is, given a payoff of the type $f(S_T)$, for some $f \in \mathcal{C}_b^1$

$$\begin{aligned}\Gamma = \frac{\partial^2 V}{\partial S_0^2} &= e^{-rT}\frac{\partial^2}{\partial S_0^2}E\left(f(S_T)\right) \\ &= e^{-rT}\frac{\partial}{\partial S_0}\left(E\left(f'(S_T)\frac{S_T}{S_0}\right)\right). \end{aligned} \tag{4.2.6}$$

As S_T/S_0 does not depend on S_0, we get

$$\frac{\partial^2 V}{\partial S_0^2} = e^{-rT}\left(E\left(f''(S_T)\left(\frac{S_T}{S_0}\right)^2\right)\right) = \frac{e^{-rT}}{S_0^2}\left(E\left(f''(S_T)S_T^2\right)\right).$$

Then, taking $G = S_T^2$ and $u = \mathbf{1}_{[0,T]}$ it follows that

$$\frac{\partial^2 V}{\partial S_0^2} = \frac{e^{-rT}}{S_0^2}\left(E\left(f'(S_T)\delta\left(\frac{S_T^2}{\sigma T S_T}\right)\right)\right) = \frac{e^{-rT}}{\sigma T S_0^2}\left(E\left(f'(S_T)\delta\left(S_T\right)\right)\right).$$

Now, Equation (3.1.4) gives us that

$$\delta(S_T) = S_T W_T - T\sigma S_T$$

and this allows us to write

$$\frac{\partial^2 V}{\partial S_0^2} = \frac{e^{-rT}}{\sigma T S_0^2}\left(E\left(f'(S_T)(S_T W_T - T\sigma S_T)\right)\right). \tag{4.2.7}$$

Now, take $G = (S_T W_T - T\sigma S_T)$ and $u = \mathbf{1}_{[0,T]}$. Then, by Propositions 4.2.1 and 3.1.14

$$\begin{aligned}\frac{\partial^2 V}{\partial S_0^2} &= \frac{e^{-rT}}{\sigma T S_0^2}\left(E\left(f(S_T)\delta\left(\frac{W_T - T\sigma}{\sigma T}\right)\right)\right) \\ &= \frac{e^{-rT}}{\sigma^2 T^2 S_0^2}\left(E\left(f(S_T)\left(W_T^2 - \sigma T W_T - T\right)\right)\right). \end{aligned} \tag{4.2.8}$$

As expected, the obtained formulas for Γ and $\mathcal{V}$ satisfy the *gamma-vega relationship*

$$\Gamma = \frac{1}{\sigma T S_0^2}\mathcal{V}. \tag{4.2.9}$$

4.2.3 The Delta of an Asian option in the Black-Scholes model

Now consider the case of an arithmetic Asian option, with a payoff given by $(A_T - K)_+$, where

$$A_T := \frac{1}{T}\int_0^T S_\tau d\tau$$

is the arithmetic mean of the asset during the life of the option. In this context, the delta is given by

$$\begin{aligned}\Delta = \frac{\partial V}{\partial S_0} &= e^{-rT} E\left(f'(A_T)\frac{\partial A_T}{\partial S_0}\right)\\ &= \frac{e^{-rT}}{TS_0} E\left(f'(A_T)\int_0^T S_\tau d\tau\right)\\ &= \frac{e^{-rT}}{S_0} E\left(f'(A_T)A_T\right). \qquad (4.2.10)\end{aligned}$$

A direct application of Proposition 4.2.1 with $G = A_T$, $u_t = S_t$, and the fact that

$$\langle DA_T, S\rangle = \frac{\sigma}{T}\int_0^T \left(\int_\theta^T S_\tau d\tau\right) S_\theta d\theta = \frac{\sigma T}{2} A_T^2$$

allows us to write

$$\frac{\partial V}{\partial S_0} = \frac{2e^{-rT}}{\sigma T S_0} E\left(f(A_T)\delta\left(\frac{2S.}{A_T}\right)\right). \qquad (4.2.11)$$

Then, by Proposition 3.1.14

$$\begin{aligned}\delta\left(\frac{S.}{A_T}\right) &= \frac{1}{A_T}\int_0^T S_\tau dW_\tau - \left\langle D\left(\frac{1}{A_T}\right), S\right\rangle_H\\ &= \frac{1}{A_T}\int_0^T S_\tau dW_\tau - \left\langle \frac{-DA_T}{A_T^2}, S\right\rangle_H\\ &= \frac{1}{A_T}\int_0^T S_\tau dW_\tau + \frac{1}{A_T^2}\left\langle DA_T, S\right\rangle_H\\ &= \frac{1}{A_T}\int_0^T S_\tau dW_\tau + \frac{\sigma T}{2}\\ &= \frac{1}{A_T}\left(S_T - S_0 - \frac{r}{\sigma}TA_T\right) + \frac{\sigma T}{2}. \qquad (4.2.12)\end{aligned}$$

This, jointly with (4.2.11), gives us that

$$\frac{\partial V}{\partial S_0} = \frac{e^{-rT}}{S_0} E\left(f(A_T)\left(2\left(\frac{S_T - S_0}{\sigma T A_T} - \frac{r}{\sigma^2}\right) + 1\right)\right). \qquad (4.2.13)$$

4.2.4 The Stochastic volatility case

Consider a stochastic volatility model (under a risk-neutral probability) of the form

$$S_t = r\int_0^t S_\theta d\theta + \int_0^t \sigma_\theta S_\theta d(\rho W_\theta + \sqrt{1-\rho^2}B_t), \tag{4.2.14}$$

where W, B are two independent Brownian motions, $\rho \in (-1,1)$, and σ is a positive and square integrable process adapted to the filtration generated by another Brownian motion W. A direct application of Itô's formula gives us that, under some regularity conditions on σ,

$$\ln S_t = \ln S_0 + \int_0^t \left(r - \frac{\sigma_\theta^2}{2}\right) d\theta + \int_0^t \sigma_\theta d(\rho W_\theta + \sqrt{1-\rho^2}B_t).$$

That is,

$$S_t = S_0 \exp\left(\int_0^t \left(r - \frac{\sigma_\theta^2}{2}\right) d\theta + \int_0^t \sigma_\theta d(\rho W_\theta + \sqrt{1-\rho^2}B_t)\right). \tag{4.2.15}$$

Under this framework, the price of a European option defined by a payoff function f is given by

$$V = e^{-rT}E(f(S_T)).$$

Now, if $f \in C_b^1$ and given some parameter α of the model (that is, α can denote S_0, σ_0, ρ, etc.)

$$\frac{\partial V}{\partial \alpha} = e^{-rT}E\left(f'(S_T)\frac{\partial S_T}{\partial \alpha}\right).$$

Let $\mathbb{D}_B^{1,2}$ be the domain of the Malliavin derivative operator D^B with respect to the Brownian motion B and denote by δ^B the Skorohod integral with respect to B. The chain rule, Remark 3.1.8, and (3.2.3) give us that $S_T \in \mathbb{D}_B^{1,2}$ and that $D_\tau^B S_T = \sigma_\tau\sqrt{1-\rho^2}S_T$. Then, Proposition 4.2.1 with $G = \frac{\partial S_T}{\partial \alpha}$ and $u = \frac{1}{\sigma}$ gives us

$$\frac{\partial V}{\partial \alpha} = e^{-rT}E\left(f(S_T)\delta^B\left(\frac{1}{\sigma T\sqrt{1-\rho^2}S_T}\frac{\partial S_T}{\partial \alpha}\right)\right). \tag{4.2.16}$$

4.2.4.1 *The delta in stochastic volatility models*

In particular, if the parameter α is taken to be the initial stock price S_0, Equation (4.2.16) gives us the following expression for the Delta

$$\begin{aligned}\Delta = \frac{\partial V}{\partial S_0} &= \frac{e^{-rT}}{S_0} E\left(f(S_T)\delta^B \left(\frac{1}{\sqrt{1-\rho^2}T\sigma} \right) \right) \\ &= \frac{e^{-rT}}{T\sqrt{1-\rho^2}S_0} E\left(f(S_T) \int_0^T \frac{1}{\sigma_s} dB_s \right). \quad (4.2.17)\end{aligned}$$

Notice that this equality reduces to (4.2.2) in the constant volatility case.

4.2.4.2 *The gamma in stochastic volatility models*

A direct computation gives us that Γ is given by

$$\Gamma = \frac{\partial^2 V}{\partial S_0^2} = \frac{e^{-rT}}{S_0^2} E\left(f''(S_T) S_T^2 \right).$$

Now, applying Proposition 4.2.1 with $G = S_T^2$ and $u = \frac{1}{\sigma}$ we get

$$\begin{aligned}\frac{\partial^2 V}{\partial S_0^2} &= \frac{e^{-rT}}{T\sqrt{1-\rho^2}S_0^2} E\left(f'(S_T)\delta^B \left(S_T \frac{1}{\sigma} \right) \right) \\ &= \frac{e^{-rT}}{T\sqrt{1-\rho^2}S_0^2} E\left(f'(S_T) S_T \left(\delta^B \left(\frac{1}{\sigma} \right) - \sqrt{1-\rho^2}T \right) \right).\end{aligned}$$

Then, taking $G = S_T \left(\delta^B \left(\frac{1}{\sigma} \right) \right)$ and $u = \frac{1}{\sigma}$ it follows that

$$\frac{\partial^2 V}{\partial S_0^2} = \frac{e^{-rT}}{\sqrt{1-\rho^2}TS_0^2} \left(E\left(f'(S_T) S_T \left(\delta^B \left(\frac{1}{\sigma} \right) - \sqrt{1-\rho^2}T \right) \right) \right).$$

Now, we apply again Proposition 4.2.1 with $G = S_T \left(\delta^B \left(\frac{1}{\sigma} \right) - \sqrt{1-\rho^2}T \right)$ and $u = \frac{1}{\sigma}$ and we get

$$\frac{\partial^2 V}{\partial S_0^2} = \frac{e^{-rT}}{(1-\rho^2)T^2S_0^2} \left(E\left(f(S_T)\delta^B \left(\frac{\left(\delta^B \left(\frac{1}{\sigma} \right) - \sqrt{1-\rho^2}T \right)}{\sigma} \right) \right) \right).$$

Now, as σ and B are independent

$$\begin{aligned}
&\delta^B\left(\frac{\left(\delta^B\left(\frac{1}{\sigma}\right)-\sqrt{1-\rho^2}T\right)}{\sigma}\right)\\
&= \delta^B\left(\frac{1}{\sigma}\right)\delta^B\left(\frac{1}{\sigma}\right)-\left\langle D^B\left(\delta^B\left(\frac{1}{\sigma}\right)\right),\frac{1}{\sigma}\right\rangle\\
&- \sqrt{1-\rho^2}T\delta^B\left(\frac{1}{\sigma}\right)\\
&= \left(\int_0^T\frac{1}{\sigma_s^2}dBs\right)^2-\int_0^T\frac{1}{\sigma_s^2}ds-\sqrt{1-\rho^2}T\int_0^T\frac{1}{\sigma_s}dB_s,
\end{aligned}$$

which allows us to write

$$\begin{aligned}
\Gamma &= \frac{e^{-rT}}{(1-\rho^2)T^2S_0^2}\\
&\times E\left(f(S_T)\left(\left(\int_0^T\frac{1}{\sigma_s}dBs\right)^2-\int_0^T\frac{1}{\sigma_s^2}ds-\sqrt{1-\rho^2}T\int_0^T\frac{1}{\sigma_s}dB_s\right)\right).
\end{aligned}\tag{4.2.18}$$

Remark 4.2.4 *Notice that the weights in (4.2.17) and (4.2.18) can be computed at the same time we carry out the Monte Carlo simulation. This results in a reduction of the computing time, as we see in the following examples.*

Example 4.2.5 (Delta and gamma in the Heston case) *In this example, we estimate the delta and gamma for a European call option under a Heston model, by three different methodologies:*

- *Finite differences.*
- *Malliavin calculus.*
- *Quasi-analytic Fourier representation.*

Each methodology has its advantages. For example, finite differences are widely used in the industry because of its universality. They are easy to implement and we can use the same architecture in the library for computing the sensitivities of a broad class of derivatives. However, the weaknesses of this method rely on the lack of stability when trying to compute high-order Greeks, and on its high computational cost. On the

Numerical Method	Price	Delta	Gamma	CPU Time (sg)
Numerical integration	32.429	0.9631	0.0012493	0.0439941
MC and Malliavin	32.379	0.96234	0.0013125	5.449218
MC and Finite differences	32.379	0.96234	0.0012024	10.882471

Table 4.1 $K = 70$. The amount between parenthesis is the MC error.

other hand, Fourier methods depend on the characteristic function of the log-asset price, which is unknown even for standard models like the SABR. Nevertheless, when this function is known, the method is fast and robust. In particular, in case of the Heston model, it allows us to compute the sensitivities via a deterministic numerical integration. As pointed out in Remark 4.2.4, the main advantage of Malliavin calculus is a reduction of the computational time in comparison with methods based on finite differences. Let us see these properties in a numerical experiment. Consider a Heston model of the form

$$\begin{aligned} dS_t &= \sqrt{v_t} S_t dW_t \\ dv_t &= k(\theta - v_t)dt + \nu\sqrt{v_t} dB_t, \end{aligned}$$

where we assume for the sake of simplicity that $r = 0$ *and* $\langle W, B \rangle_t = \rho dt$, *for some* $\rho \in (-1, 1)$. *In this example, the Monte Carlo simulations of the Heston model have been carried out via a quadratic exponential (QE) scheme (see Andersen (2008)), and the Malliavin computation of the Greeks has been optimised via the localisation methods introduced in Fournié, Lasry, Lebuchoux, Lions, and Touzi (1999). We have taken* $n = 500000$ *Monte Carlo antithetic simulations and a time step* $\Delta = 2^{-5}$. *In Tables 4.1, 4.2, and 4.3 we compare the three methods for the set of parameters* $k = 0.5, \theta = 0.05, \nu = 0.9, \rho = -0.8, v_0 = 0.05, S_0 = 100$, *and* $T = 2$, *and for the strikes* $K = 70$, $K = 100$, *and* $K = 150$. *Taking as a benchmark the estimation obtained via the quasi-analytic Fourier representation, we can see that both Malliavin calculus and finite differences give accurate results, but that the computational cost is less for the Malliavin estimators.*

Example 4.2.6 (Delta and gamma in the SABR case) *In this example, we repeat the same experiment as before but for the SABR model (see Section 2.3.2). We assume again that the interest rate is zero and*

Numerical Method	Price	Delta	Gamma	CPU Time (sg)
Numerical integration	7.0843	0.78323	0.018702	0.0439941
MC and Malliavin	7.0623	0.78088	0.01975	5.449218
MC and Finite differences	7.0623	0.78097	0.019981	10.882471

Table 4.2 $K = 100$.

Numerical Method	Price	Delta	Gamma	CPU Time (sg)
Numerical integration	0.21122	0.044323	0.010082	0.0439941
MC and Malliavin	0.21342	0.044919	0.010049	5.449218
MC and Finite differences	0.2134	0.044971	0.0099692	10.882471

Table 4.3 $K = 150$.

Numerical Method	Price	Delta	Gamma	CPU Time (sg)
MC and Malliavin	38.592	0.96234	0.0021333	2.72711
MC and Finite differences	38.592	0.92298	0.002153	5.432223

Table 4.4 $K = 70$.

we take the SABR parameter $\beta = 1$. That is, we assume the following dynamics

$$
\begin{aligned}
dF_t &= \sigma_t F_t dW_t \\
d\sigma_t &= \nu \sigma_t dB_t \\
d\langle W_t, B\rangle_t &= \rho dt,
\end{aligned}
$$

for some positive ν and for some $\rho \in (-1, 1)$. Notice that the SABR models does not have an explicit expression for the characteristic function and then we do not have a closed-form Fourier representation. Then, this example is restricted to the study of the finite differences method and the Malliavin calculus approach. The applied simulation scheme for the SABR follows the steps in Cai, Song, and Chen (2017) and it is based on the previous simulation of the integrated volatility. This integrated volatility has been estimated via the probabilistic approach introduced in Kennedy, Mitra, and Pham (2012). As before, the number of antithetic simulations and the time step are $n = 500000$ and $\Delta = 2^{-5}$, respectively. SABR parameters are $\alpha = 0.4, \rho = -0.8$, and $\nu = 1.1$. The results are summarised in Tables 4.4, 4.5, and 4.6. As before, we can see a clear reduction of the computing time when using the Malliavin approach.

Numerical Method	Price	Delta	Gamma	CPU Time (sg)
MC and Malliavin	18.059	0.7614	0.008501	2.72711
MC and Finite differences	18.059	0.761823	0.00869	5.432223

Table 4.5 $K = 100$.

Numerical Method	Price	Delta	Gamma	CPU Time (sg)
MC and Malliavin	2.1226	0.15766	0.010501	2.72711
MC and Finite differences	2.1226	0.158204	0.010780	5.432223

Table 4.6 $K = 150$.

4.3 THE ANTICIPATING ITÔ'S FORMULA

One of the main tools of Malliavin calculus is the anticipating Itô's formula, which allows us to work with non-adapted processes. A first extension of this type was introduced by Hitsuda (1972). Different anticipating formulas can also be found in Sveljakov (1981), Sekiguchi and Shiota (1985), Üstünel (1986), and Nualart and Pardoux (1988).

4.3.1 The anticipating Itô's formula as an extension of Itô's formula

In this section, we consider two independent Brownian motions W and B and denote by D^W the Malliavin derivative operator with respect to W. Similarly, $\mathbb{D}_W^{1,2}$ is the domain of D^W and $\mathbb{L}_W^{1,2} := L^2([0,T]; \mathbb{D}_W^{1,2})$. Moreover, given a process Y of the form $Y_t = \int_t^T a_s ds$, where a is an adapted process in $\mathbb{L}^{1,2}$, we denote $D^- Y_t := \int_t^T D_t^W a_s ds$.

The following version of the anticipating Itô's formula is an adaptation of the result by Nualart and Pardoux (1988).

Proposition 4.3.1 *Consider a process of the form* $X_t = X_0 + \int_0^t u_s dW_s + \int_0^t u'_s dB_s + \int_0^t v_s ds$, *where* X_0 *is a constant and* $u, v \in L_a^2([0,T] \times \Omega)$. *Consider also a process* $Y_t = \int_t^T \theta_s ds$, *for some* $\theta \in \mathbb{L}_W^{1,2}$ *adapted to the filtration generated by* W. *Let* $F : [0,T] \times \mathbb{R}^2 \to \mathbb{R}$ *be a function in* $C^{1,2}([0,T] \times \mathbb{R}^2)$ *such that there exists a positive constant* C *such that, for all* $t \in [0,T]$, F *and its partial derivatives evaluated in*

(t, X_t, Y_t) *are bounded by* C*. Then, it follows that*

$$\begin{aligned}
F(t, X_t, Y_t) &= F(0, X_0, Y_0) + \int_0^t \frac{\partial F}{\partial s}(s, X_s, Y_s)ds \\
&+ \int_0^t \frac{\partial F}{\partial x}(s, X_s, Y_s)v_s ds \\
&+ \int_0^t \frac{\partial F}{\partial x}(s, X_s, Y_s)(u_s dW_s + u'_s dB_s) \\
&- \int_0^t \frac{\partial F}{\partial y}(s, X_s, Y_s)\theta_s ds + \int_0^t \frac{\partial^2 F}{\partial x \partial y}(s, X_s, Y_s)D^- Y_s u_s ds \\
&+ \frac{1}{2}\int_0^t \frac{\partial^2 F}{\partial x^2}(s, X_s, Y_s)(u_s^2 + (u'_s)^2)ds. \qquad (4.3.1)
\end{aligned}$$

Proof. The proof follows the same steps as in the proof of the classical Itô's formula. For the sake of simplicity, we assume $u' = v = 0$. Then, for every $n > 0$ we can write

$$F(t, X_t, Y_t) - F(0, X_0, Y_0) = \sum_{i=1}^n \left[F(t_{t_{i+1}}, X_{t_{i+1}}, Y_{t_{i+1}}) - F(t_{t_i}, X_{t_i}, Y_{t_i})\right],$$

where $t_i := ti/n$. Then, a Taylor expansion gives us that

$$\begin{aligned}
&\sum_{i=1}^n \left[F(t_{t_{i+1}}, X_{t_{i+1}}, Y_{t_{i+1}}) - F(t_{t_i}, X_{t_i}, Y_{t_i})\right], \\
&= \sum_{i=1}^n \frac{\partial F}{\partial t}(t_{t_i}, X_{t_i}, Y_{t_i})(t_{i+1} - t_i) \\
&+ \sum_{i=1}^n \frac{\partial F}{\partial x}(t_{t_i}, X_{t_i}, Y_{t_i})\left(\int_{t_i}^{t_{i+1}} u_s dW_s\right) \\
&+ \sum_{i=1}^n \frac{\partial F}{\partial y}(t_{t_i}, X_{t_i}, Y_{t_i})(Y_{t_{i+1}} - Y_{t_i}) \\
&+ \frac{1}{2}\sum_{i=1}^n \frac{\partial^2 F}{\partial x^2}(t_{t_i}, X_{t_i}, Y_{t_i})\left(\int_{t_i}^{t_{i+1}} u_s dW_s\right)^2 \\
&+ A_n \\
&=: T_1 + T_2 + T_3 + T_4 + A_n, \qquad (4.3.2)
\end{aligned}$$

where $A_n \to 0$ as $n \to \infty$. Now, straightforward computations and the fact that $dY_t = -\theta_t$ allows us to write

$$\lim_{n\to\infty}(T_1+T_3+T_4) = \int_0^t \frac{\partial F}{\partial s}(s,X_s,Y_s)ds - \int_0^t \frac{\partial F}{\partial y}(s,X_s,Y_s)\theta_s ds + \frac{1}{2}\int_0^t \frac{\partial^2 F}{\partial x^2}(s,X_s,Y_s)u_s^2 ds. \tag{4.3.3}$$

The only essential difference with the proof of the classical Itô's formula relies on the limit of T_2. As $\frac{\partial F}{\partial x}(t_{t_i}, X_{t_i}, Y_{t_i})$ is not adapted (because Y_t is not adapted), T_2 does not tend to an Itô integral. To study this term, we make use of Proposition 3.1.14. This result allows us to write

$$\begin{aligned} T_2 &= \sum_{i=1}^n \int_{t_i}^{t_{i+1}} \frac{\partial F}{\partial x}(t_{t_i}, X_{t_i}, Y_{t_i}) u_s dW_s \\ &+ \sum_{i=1}^n \int_{t_i}^{t_{i+1}} D_s^W \left(\frac{\partial F}{\partial x}(t_{t_i}, X_{t_i}, Y_{t_i})\right) u_s ds \\ &= T_2^1 + T_2^2. \end{aligned} \tag{4.3.4}$$

It is easy to see that T_2^1 tends to the Skorohod integral $\int_0^t \frac{\partial F}{\partial x}(s,X_s,Y_s)u_s dW_s$. On the other hand,

$$\begin{aligned} T_2^2 &= \sum_{i=1}^n \int_{t_i}^{t_{i+1}} \left(\frac{\partial^2 F}{\partial x \partial y}(t_{t_i}, X_{t_i}, Y_{t_i})\right) D_s^W Y_{t_i} u_s ds \\ &= \sum_{i=1}^n \int_{t_i}^{t_{i+1}} \left(\frac{\partial^2 F}{\partial x \partial y}(t_{t_i}, X_{t_i}, Y_{t_i})\right) \left(\int_s^T D_s^W \theta_r dr\right) u_s ds, \end{aligned} \tag{4.3.5}$$

which tends to

$$\int_0^t \partial_{xy}^2 F(s,X_s,Y_s) D^- Y_s u_s ds,$$

and this allows us to complete the proof. ■

Remark 4.3.2 *Notice that the only difference with the classical Itô's formula is the term*

$$\int_0^t \frac{\partial^2 F}{\partial x \partial y}(s,X_s,Y_s) D^- Y_s u_s ds,$$

that comes from the non-adaptedness of the process Y.

Remark 4.3.3 *The anticipating Itô's formula also holds for processes of the form $F(s, X_s, Y_s^1, ..., Y_s^n)$, where $Y^i = \int_t^T \theta_s^i ds$, $i = i, ..., n$. One just needs to replace the term*

$$\int_0^t \frac{\partial^2 F}{\partial x \partial y}(s, X_s, Y_s) D^- Y_s u_s ds, \tag{4.3.6}$$

by

$$\sum_i^n \int_0^t \frac{\partial^2 F}{\partial x \partial y}(s, X_s, Y_s^1, ..., Y_s^n) D^- Y_s^i ds.$$

Remark 4.3.4 *The anticipating Itô's formula can also be extended to the case of processes of the from $F(s, X_s, Y_s^1, ..., Y_s^n)$, where Y^i, $i = i, ..., n$ are non-adapted processes satisfying some regularity conditions in the Malliavin calculus sense. Then, the additional term (4.3.6) is substituted by*

$$\sum_i^n \int_0^t \frac{\partial^2 F}{\partial x \partial y^i}(s, X_s, Y_s^1, ..., Y_s^n) D^- Y_s^i ds,$$

where now $D^- Y_s^i := \lim_{r \uparrow s} D_r Y_s^i$.

4.3.2 The law of an asset price as a perturbation of a mixed log-normal distribution

Consider the following stochastic volatility model for the log-price of a stock under a risk-neutral probability measure P

$$X_t = X_0 + \frac{1}{2} \int_0^t \left(r - \sigma_s^2 \right) ds + \int_0^t \sigma_s \left(\rho dW_s + \sqrt{1 - \rho^2} dB_s \right), \quad t \in [0, T]. \tag{4.3.7}$$

Here, X_0 is the current log-price, W and B are standard and independent Brownian motions, σ is a square-integrable and right-continuous stochastic process adapted to the filtration generated by W, and $\rho \in (-1, 1)$ is the correlation parameter. We also denote $\mathcal{F}^W$ and $\mathcal{F}^B$ the filtrations generated by W and B. Moreover, we define $\mathcal{F} := \mathcal{F}^W \vee \mathcal{F}^B$.

Notice that the model (4.3.7) does not assume a concrete structure for the volatility process σ, which can be a diffusion (like in the Heston or the SABR models), but also a non-Markovian process as in the case of rough volatilities (see Chapter 2).

Our main objective in this section is to study the law of the asset price X_t, for $t \in [0,T]$. Consider first the uncorrelated case. If $\rho = 0$, X, conditioned to $\mathcal{F}_T^W$, is a normal process such that

$$E(X_t) = X_0 + \frac{1}{2}\int_0^t \left(r - \sigma_s^2\right) ds$$

and

$$Var(X_t) = \int_0^t \sigma_s^2 ds.$$

That is, the asset price e^{X_t}, conditioned to $\mathcal{F}_T^W$, follows a Black-Scholes dynamics with volatility equal to $\sqrt{\frac{1}{t}\int_0^t \sigma_s^2 ds}$. We will say that the asset prices e^{X_t} follow a **mixed log-normal distribution**.

In order to analyse the correlated case $\rho \neq 0$, we study the characteristic function of X_t. Now our objective is to see that the characteristic function of X_t

$$\Psi_t(s) = E(e^{isX_t}) \tag{4.3.8}$$

can be decomposed as the sum of a Gaussian characteristic function and a correction term due by the correlation. Towards this end, we introduce the following notation:

- $F_t(s,u,x,\sigma)$ denotes the characteristic function of a normal random variable with mean $\mu_u := x + \left(r - \frac{1}{2}\sigma^2\right)(t-u)$ and standard deviation $\sigma_u := \sigma\sqrt{t-u}$. That is, $F_t(s,u,x,\sigma) = e^{is\mu_u - \frac{1}{2}s^2\sigma_u^2}$.
- $v_u^t := \sqrt{\frac{1}{t-u}Y_u}$, where $Y_u := \int_u^t \sigma_\theta^2 d\theta$.

Notice that, fixed u and σ, F_t satisfies the Black-Scholes equation for log-prices. That is

$$\frac{\partial F_t}{\partial u} + \frac{\partial F_t}{\partial x}\left(r - \frac{1}{2}\sigma^2\right) + \frac{1}{2}\frac{\partial^2 F_t}{\partial x^2}\sigma^2 = 0. \tag{4.3.9}$$

Moreover, it satisfies the following *Delta-Gamma-Vega relationship*

$$\frac{\partial F_t}{\partial y}\frac{1}{\sigma(t-u)} = \left(\frac{\partial^2 F_t}{\partial x^2} - \frac{\partial F_t}{\partial x}\right). \tag{4.3.10}$$

Now we are in a position to prove the following theorem, which is an adaptation of Alòs (2006).

Theorem 4.3.5 *Assume the model (4.3.7). Then*

$$\begin{aligned}\Psi_t(s) &= E(F_t(s,0,X_0,v_0^t)) \\ &+ \frac{\rho}{2}E\int_0^t e^{-r(u-t)}H_t(s,u,X_u,v_u^t)\sigma_u\left(\int_u^t D_u^W\sigma_\theta^2 d\theta\right)du,\end{aligned} \tag{4.3.11}$$

where $H_t(s,u,X_u,v_u^t) := (-is^3+s^2)\,F_t(s,u,X_u,v_u^t)$.

Proof. We apply the anticipating Itô's formula of Proposition 4.3.1 to the process $e^{-ru}F_t(s,u,X_u,v_u^t) = F_t\left(s,u,X_u,\sqrt{\frac{Y_u}{t-u}}\right)$. This allows us to write

$$\begin{aligned}e^{-rt}F_t(s,t,X_t,v_t) &= F_t(s,0,X_0,v_0^t) - r\int_0^t e^{-ru}F_t(s,u,X_u,v_u)du \\ &+ \int_0^t e^{-ru}\frac{\partial F_t}{\partial u}(s,u,X_u,v_u)du \\ &+ \int_0^t e^{-ru}\frac{\partial F_t}{\partial x}(s,u,X_u,v_u^t)\left(r-\frac{1}{2}\sigma_u^2\right)du \\ &+ \int_0^t e^{-ru}\frac{\partial F_t}{\partial x}(s,u,X_u,v_u^t)\sigma_u(\rho X_u dW_u+\sqrt{1-\rho^2}X_u dB_u) \\ &- \int_0^t e^{-ru}\frac{\partial F_t}{\partial y}(s,u,X_u,v_u^t)\frac{1}{2v_u^t(t-u)}\sigma_u^2 du \\ &+ \rho\int_0^t e^{-ru}\frac{\partial^2 F_t}{\partial x\partial y}(s,u,X_u,v_u^t)\frac{1}{2v_u^t(t-u)}\sigma_u\left(\int_u^t D_u^W\sigma_\theta^2 d\theta\right)du \\ &+ \frac{1}{2}\int_0^t e^{-ru}\frac{\partial^2 F_t}{\partial x^2}(s,u,X_u,v_u^t)\sigma_u^2 du.\end{aligned} \tag{4.3.12}$$

Now, observe that $F_t(s,t,X_t,v_t^t) = e^{isX_t}$ and

$$E\left(\int_0^t e^{-ru}\frac{\partial F_t}{\partial x}(s,u,X_u,v_u^t)\sigma_u(\rho X_u dW_u+\sqrt{1-\rho^2}X_u dB_u)\right) = 0.$$

These equalities give us, after taking expectations and multiplying by

e^{rt}, that

$$\begin{aligned}
\Psi_t(s) &= E\Big\{F_t(s,0,X_0,v_0^t) - r\int_0^t e^{-r(u-t)}F_t(s,u,X_u,v_u)du \\
&+ \int_0^t e^{-r(u-t)}\frac{\partial F_t}{\partial u}(s,u,X_u,v_u^t)du \\
&+ \int_0^t e^{-r(u-t)}\frac{\partial F_t}{\partial x}(s,u,X_u,v_u^t)\left(r-\frac{1}{2}\sigma_u^2\right)du \\
&+ \int_0^t e^{-r(u-t)}\frac{\partial F_t}{\partial y}(s,u,X_u,v_u^t)\frac{1}{2v_u^t(t-u)}((v_u^t)^2-\sigma_u^2)du \\
&+ \rho\int_0^t e^{-r(u-t)}\frac{\partial^2 F_t}{\partial x\partial y}(s,u,X_u,v_u^t)\frac{1}{2v_u^t(t-u)}\sigma_u\left(\int_u^t D_u^W\sigma_\theta^2 d\theta\right)du \\
&+ \frac{1}{2}\int_0^t e^{-r(u-t)}\frac{\partial^2 F_t}{\partial x^2}(s,u,X_u,v_u^t)\sigma_u^2 du\Big\}. \qquad (4.3.13)
\end{aligned}$$

Now, notice that

$$\begin{aligned}
&\left(\frac{\partial}{\partial u}+\left(r-\frac{1}{2}\sigma_u^2\right)\frac{\partial}{\partial x}+\frac{1}{2}\sigma_u^2\frac{\partial^2}{\partial x^2}\right)F_t(s,u,X_u,v_u^t) \\
&= \left(\frac{\partial}{\partial u}+\left(r-\frac{1}{2}(v_u^t)^2\right)\frac{\partial}{\partial x}+\frac{1}{2}(v_u^t)^2\frac{\partial^2}{\partial x^2}\right)F_t(s,u,X_u,v_u^t) \\
&+\frac{1}{2}\left(\frac{\partial^2 F_t}{\partial x^2}(s,u,X_u,v_u^t)-\frac{\partial F_t}{\partial x}(s,u,X_u,v_u^t)\right)(\sigma_u^2-(v_u^t)^2). \qquad (4.3.14)
\end{aligned}$$

Then, (4.3.14) and (4.3.9) allow us to rewrite (4.3.13) as

$$\begin{aligned}
\Psi_t(s) &= E(F_t(s,0,X_0,v_0^t)) \\
&+ \rho E\int_0^t e^{-r(u-t)}\frac{\partial^2 F_t}{\partial x\partial y}(s,u,X_u,v_u^t)\frac{1}{2v_u^t\sigma(t-u)}\sigma_u \\
&\left(\int_u^t D_u^W\sigma_\theta^2 d\theta\right)du. \qquad (4.3.15)
\end{aligned}$$

Finally, due to (4.3.10)

$$\begin{aligned}
\frac{\partial^2 F_t}{\partial x\partial y}(s,u,X_u,v_u^t)\frac{1}{2v_u^t\sigma(t-u)} &= \frac{1}{2}\left(\frac{\partial^3}{\partial x^3}-\frac{\partial^2}{\partial x^2}\right)F_t(s,u,X_u,v_u^t) \\
&= \frac{1}{2}\left(-is^3+s^2\right)F_t(s,u,X_u,v_u^t). \qquad (4.3.16)
\end{aligned}$$

Now the proof is complete. ■

Remark 4.3.6 *Notice that the above formula decomposes the characteristic function of X_t into two parts:*

- *The term $E(F_t(s,0,X_0,v_0^t))$ does not depend on ρ. It coincides with the characteristic function of the uncorrelated case, where X_t follows a mixed normal distribution.*
- *The term*

$$\frac{\rho}{2}E\int_0^t e^{-r(u-t)}H_t(s,u,X_u,v_u^t)\sigma_u\left(\int_u^t D_u^W\sigma_\theta^2 d\theta\right)du$$

is **exactly** *the contribution of the correlation to the characteristic function of X_t.*

In Chapter 6, similar arguments will allow us get adequate decomposition formulas that will become our basic tool in the analysis of the implied volatility.

4.3.3 The moments of log-prices in stochastic volatility models

We can apply the above results in the computation of the moments of the log-prices X_t. For the sake of simplicity, we take $r=0$. As $E(X_t^n) = i^{-n}\Psi_t^{n)}(0)$, it follows that

$$\begin{aligned}E(X_t^n) &= i^{-n}E\left(\frac{\partial^n F_t}{\partial s^n}(0,0,X_0,v_0^t)\right)\\ &+ i^{-n}\frac{\rho}{2}E\int_0^t e^{-r(u-t)}\frac{\partial^n H_t}{\partial s^n}(0,u,X_u,v_u^t)\sigma_u\left(\int_u^t D_u^W\sigma_\theta^2 d\theta\right)du\\ &=: T_n^1+T_n^2. \end{aligned} \quad (4.3.17)$$

This gives us a decomposition of the moment $E(X_t^n)$ as the sum of two terms: T_n^1 is the moment of order n of a mixed normal distribution with mean μ_0 and variance $(v_0^t)^2 t$, and the impact of the correlation is given by T_n^2. Now, as $H_t(s,u,X_u,v_u^t) := (-is^3+s^2)\,F_t(s,u,X_u,v_u^t)$, we get

$$\frac{\partial^n H_t}{\partial s^n}(0,u,X_u,v_u^t) = \sum_{i,j\geq 0}^{n}\binom{n}{i}\frac{\partial^i}{\partial s^i}\left(-is^3+s^2\right)\frac{\partial^{n-i}F_t}{\partial s^{n-i}}(s,u,X_u,v_u^t)\Big|_{s=0}.$$

That is,

- $\frac{\partial H_t}{\partial s}(0, u, X_u, v_u^t) = 0$
- $\frac{\partial^2 H_t}{\partial s^2}(0, u, X_u, v_u^t) = 2F_t(0, u, X_u, v_u^t) = 2$
- $\frac{\partial^3 H_t}{\partial s^3}(0, u, X_u, v_u^t) = \binom{3}{2}2\left(\frac{\partial F_t}{\partial s} - 6iF\right)(0, u, X_u, v_u^t) = 6i(\mu_u - 1)$, and
- $\frac{\partial^n H_t}{\partial s^n}(0, u, X_u, v_u^t) = \left(\binom{n}{2}2\frac{\partial^{n-2} F_t}{\partial s^{n-2}} - \binom{n}{3}6i\frac{\partial^{n-3} F_t}{\partial s^{n-3}}\right)(0, u, X_u, v_u^t)$, for $n > 3$.

This implies that $E(X_t) = T_1^1 = E(\mu_0)$. For the second moment, we have

$$\begin{aligned} E(X_t^2) &= E(T_2^1) - \rho E\int_0^t \sigma_u \left(\int_u^t D_u^W \sigma_\theta^2 d\theta\right) du \\ &= E(\mu_0^2 + (v_0^t)^2 t) - \rho E\int_0^t \sigma_u E_u \left(\int_u^t D_u^W \sigma_\theta^2 d\theta\right) du \\ &= E(\mu_0^2 + (v_0^t)^2 t) - E\langle X, M\rangle_t, \end{aligned} \tag{4.3.18}$$

where $M_u = E_u \int_0^T \sigma_s^2 ds$ (see Lemma 4.1.6). The above equation decomposes the second moment of X as the second moment of a mixed normal distribution minus the covariation between X and M. Now, the variance is given by

$$\begin{aligned} E(X_t - E(\mu_0))^2 &= E(\mu_0^2 + (v_0^t)^2 t) - E\langle X, M\rangle_t - (E(\mu_0))^2 \\ &= Var(\mu_0) + E((v_0^t)^2 t) - E\langle X, M\rangle_t \\ &= \frac{1}{4}Var((v_0^t)^2 t) + E((v_0^t)^2 t) - E\langle X, M\rangle_t. \end{aligned} \tag{4.3.19}$$

That is, the variance depends on the expectation and the variance of the realised variance $(v_0^t)^2$, and the correlation between X and this realised variance.

Let us consider now the third moment

$$\begin{aligned} E(X_t^3) &= E(T_3^1) - 3\rho E\left(\int_0^t (\mu_u - 1)\sigma_u \left(\int_u^t D_u^W \sigma_\theta^2 d\theta\right) du\right) \\ &= E(\mu_0^3 + 3\mu_0(v_0^t)^2 t) \\ &- 3\rho E\left(\int_0^t (\mu_u - 1)\sigma_u \left(\int_u^t D_u^W \sigma_\theta^2 d\theta\right) du\right) \\ &= E(\mu_0^3 + 3\mu_0(v_0^t)^2 t) + 3\rho E\left(\int_0^t (\mu_u - 1) dR_u\right), \end{aligned} \tag{4.3.20}$$

where $R_u := \int_u^t \sigma_s \left(\int_s^t D_s \sigma_\theta^2 d\theta \right) ds$. To study the second term on the right-hand side, we apply again the anticipating Itô formula to the process

$$\lambda_u := (\mu_u - 1) R_u.$$

This gives us that

$$\begin{aligned} \lambda_t &= \lambda_0 + E\left(\int_0^t (\mu_u - 1) dR_u \right) \\ &+ \int_0^t R_u \left(dX_u + \frac{1}{2}\sigma_u^2 du \right) + \rho \int_0^t \sigma_u D^- R_u du. \end{aligned} \quad (4.3.21)$$

Then, as $\lambda_t = 0$ and taking into account Lemma 4.1.6 we get

$$\begin{aligned} E\left(\int_0^t (\mu_u - 1) \, dR_u \right) &= E\left((\mu_0 - 1) R_0 \right) + \rho E \int_0^t \sigma_u (D^- R_u) du \\ &= E\left((\mu_0 - 1) R_0 \right) + \rho E \int_0^t \sigma_u E_u (D^- R_u) du \\ &= E((\mu_0 - 1) R_0) + E\langle X, r \rangle, \end{aligned} \quad (4.3.22)$$

where $r_u := E_u(R_0)$. This, jointly with (4.3.20), implies that

$$E(X_t^3) = E(\mu_0^3 + 3\mu_0 (v_0^t)^2 t) - 3\rho E\left((\mu_0 - 1) R_0 + \langle X, r \rangle \right). \quad (4.3.23)$$

Then, as $\rho E(R_0) = E\langle X, M \rangle$, the third central moment is given by

$$\begin{aligned} & E(X_t - (E(\mu_0)))^3 \\ &= E(X_t^3) - 3E(X_t^2) E(\mu_0) + 3E(X_t)(E(\mu_0))^2 - (E(\mu_0))^3 \\ &= E(\mu_0^3 + 3\mu_0 (v_0^t)^2) - 3\rho E\left((\mu_0 - 1) R_0 + \langle X, r \rangle \right) \\ & -3E(\mu_0) \left(E(\mu_0^2 + v_0^2) - E\langle X, M \rangle \right) \\ & +3E(\mu_0)(E(\mu_0)))^2 - (E(\mu_0))^3 \\ &= E(\mu_0^3 + 3\mu_0 v_0^2) - 3E(\mu_0) \left(E(\mu_0^2 + v_0^2) \right) + 2(E(\mu_0))^3 \\ & -3\rho E\left((\mu_0 - E(\mu_0) - 1) R_0 + \langle X, r \rangle \right). \end{aligned} \quad (4.3.24)$$

That is, the effect of the correlation on the central third moment is given by

$$\begin{aligned} & - \quad 3\rho E\left((\mu_0 - E(\mu_0) - 1) R_0 + \langle X, r \rangle \right) \\ &= -3\rho E\left((\mu_0 - E(\mu_0) - 1) R_0 \right) + O(\rho^2). \\ &= \frac{3}{2} \rho E\left(((v_0^t)^2 t - E(v_0^t)^2 t + 2) R_0 \right) + O(\rho^2). \end{aligned} \quad (4.3.25)$$

Example 4.3.7 (The variance and the skewness in the SABR model) *Consider the model (4.3.7) where the volatility assumed to be as in Section 2.3.2. As*

$$E_u(D_u\sigma_\theta^2) = 2\alpha\sigma_u^2 \exp(\alpha^2(\theta - u)),$$

the effect of the correlation on the variance of X_t is given by

$$\begin{aligned} -E\langle X, M\rangle &= -\rho E \int_0^T \sigma_s E_s \left(\int_s^T D_s^W \sigma_u^2 du \right) ds \\ &= -2\alpha\rho E \int_0^T \sigma_s^3 \left(\int_s^T e^{2\alpha^2(u-s)} du \right) ds \\ &= -2\alpha\rho\sigma_0^3 \int_0^T e^{\frac{9}{2}\alpha^2 s} \left(\int_s^T e^{2\alpha^2(u-s)} du \right) ds. \quad (4.3.26) \end{aligned}$$

Notice that, as t tends to zero, as $\rho E(R_0) = E\langle X, M\rangle$, the skewness of X_t behaves as

$$\frac{E(X_t - E(X_t))^3}{(E(X_t - E(X_t))^2)^{\frac{3}{2}}} \approx \frac{3E\langle X, M\rangle}{(v_0^t)^3 t^{\frac{3}{2}}},$$

according to the results in pg. 314 of Bergomi (2016).

4.3.4 Some applications to volatility derivatives

4.3.4.1 *Leverage swaps and gamma swaps*

We define a leverage swap as the difference between a gamma swap and a variance swap. Its fair price $\mathcal{L}(T)$ is given by

$$\mathcal{L}(T) = \mathcal{G}(T) - w(T),$$

where $\mathcal{G}(T) := 2E[(X_T - X_0)e^{X_T - X_0}]$ and $w(T) = E((v_0^T)^2 T)$ are the fair prices of a gamma swap and a variance swap, respectively (see

Section 2.6). Notice that

$$\begin{aligned}
E(X_T e^{X_T}) &= -i\Psi'_T(-i) \\
&= -iE\left(\frac{\partial F_T}{\partial s}(-i,0,X_0,v_0^T)\right) \\
&- \frac{i\rho}{2}E\int_0^T \frac{\partial H_T}{\partial s}(-i,u,X_u,v_u^T)\sigma_u\left(\int_u^T D_u^W\sigma_\theta^2 d\theta\right)du \\
&= E\left(\frac{T(v_0^T)^2}{2}+X_0\right)\exp(X_0) \\
&+ \frac{\rho}{2}E\int_0^T e^{X_u}\sigma_u\left(\int_u^T D_u^W\sigma_\theta^2 d\theta\right)du. \qquad (4.3.27)
\end{aligned}$$

Then, straightforward computations give us that

$$\begin{aligned}
2E[(X_T-X_0)e^{X_T-X_0}] &= w(T)+\rho E\int_0^T e^{X_u-X_0}\sigma_u\left(\int_u^T D_u^W\sigma_\theta^2 d\theta\right)du \\
&= w(T)+\rho E\int_0^T e^{X_u-X_0}\sigma_u E_u\left(\int_u^T D_u^W\sigma_\theta^2 d\theta\right)du,
\end{aligned}$$

which gives us the fair price of a gamma swap. As a direct consequence,

$$\begin{aligned}
\mathcal{L}(T) &= \rho E\int_0^T e^{X_u-X_0}\sigma_u E_u\left(\int_u^T D_u^W\sigma_\theta^2 d\theta\right)du \\
&= e^{-X_0}E\langle e^X, M\rangle_t, \qquad (4.3.28)
\end{aligned}$$

where M is the future expected variance defined as in Section 4.3.3. This equality fits the previous results by Fukasawa (2014).

In the next example, we compute explicitly the leverage swap in the case of the Heston model.

Example 4.3.8 (The leverage swap in the Heston model) *Consider the model (4.3.7) where the volatility is given by a CIR model as in Section 2.3.1. Assume, for the sake of simplicity, that the interest rate $r=0$. The results in Section 3.3.2 give us that*

$$E_u(D_u\sigma_\theta^2)=\nu\sigma_u\exp(-k(\theta-u)).$$

Then

$$\mathcal{L}(T)=\nu\rho e^{-X_0}E\int_0^T e^{X_u}\sigma_u^2\lambda(T,u)du,$$

where $\lambda(T,u) := \int_u^T e^{-k(\theta-u)} d\theta$. *Now, as*

$$e^{X_u} = e^{X_0} + \int_0^u e^{X_s}\sigma_s\left(\rho dW_s + \sqrt{1-\rho^2}dB_s\right)$$

and

$$\sigma_u^2 = E(\sigma_u^2) + \int_0^u E_\theta(D_\theta\sigma_u^2)dW_\theta = E(u_s^2) + \nu\int_0^u \sigma_\theta \exp(-k(u-\theta))dW_\theta$$

we get

$$\begin{aligned}\mathcal{L}(T) &= \nu\rho E\int_0^T \sigma_u^2\lambda(T,u)du \\ &\quad + \nu^2\rho^2 e^{-X_0} E\int_0^T \left(\int_0^u e^{X_s}\sigma_s dW_s\right)\left(\int_0^u \sigma_s \exp(-k(u-s))dW_s\right) \\ &\quad \times \lambda(T,u)du \\ &= \nu\rho E\int_0^T \sigma_u^2\lambda(T,u)du \\ &\quad + \nu^2\rho^2 e^{-X_0} E\int_0^T e^{X_s}\sigma_s^2\left(\int_s^T \exp(-k(u-s))\lambda(T,u)du\right)ds.\end{aligned} \tag{4.3.29}$$

Then, a recursive argument and the fact that $E(\sigma_u^2) = \theta + e^{-ku}(\sigma_0^2 - \theta)$ *allow us to write*

$$\begin{aligned}\mathcal{L}_T &= \sum_{n=1}^{\infty}\frac{(\nu\rho)^n}{n!}\int_0^T E(\sigma_u^2)\lambda^n(T,u)du \\ &= \int_0^T [\theta + e^{-ku}(\sigma_0^2-\theta)](e^{\nu\rho\lambda(T,u)} - 1)du,\end{aligned} \tag{4.3.30}$$

according to Corollary 5.2 in Alòs, Gatheral, and Radoičić (2019).

4.3.4.2 Arithmetic variance swaps

An arithmetic variance swap is a forward contract on the quadratic variation of the asset price

$$\int_0^T S_u^2\sigma_u^2 du = \int_0^T e^{2X_u}\sigma_u^2 du.$$

These variance swaps were introduced in Leontsinis and Alexander (2017). This definition avoids the technical errors and biases that arise from the computation of the log-returns. Moreover, arithmetic variance swaps can be defined for underlying assets that can be negative, such as spreads, energy prices, or even interest rates. A direct application of Itô's formula gives us that the fair strike of an arithmetic variance swap is given by

$$E(e^{2X_t}) - e^{2X_0} = \Psi_t(-2i) - e^{2X_0}.$$

Now, Theorem 4.3.5 implies that

$$\begin{aligned}
\Psi_t(-2i) &= E(e^{2\mu_0+2t(v_0^t)^2}) + 2\rho E\int_0^t e^{2\mu_u+2(t-u)(v_u^t)^2}\sigma_u\left(\int_u^t D_u^W\sigma_\theta^2 d\theta\right)du \\
&= E(e^{2X_0+t(v_0^t)^2}) + 2\rho E\int_0^t e^{2X_u+(t-u)(v_u^t)^2}\sigma_u\left(\int_u^t D_u^W\sigma_\theta^2 d\theta\right)du \\
&= E(e^{2X_0+t(v_0^t)^2}) - 2\rho E\int_0^T e^{2X_u+(t-u)(v_u^t)^2}dR_u, \qquad (4.3.31)
\end{aligned}$$

where R is defined as in Section 4.3.3.

In order to study the second term in the right-hand side, we apply again the anticipating Itô formula to the process

$$\lambda_u := e^{2X_u+t(v_u^t)^2}R_u.$$

This gives us that

$$\begin{aligned}
\lambda_t &= \lambda_0 + \left(\int_0^t e^{2X_u+(t-u)(v_u^t)^2}dR_u\right) \\
&+ \int_0^t e^{2X_u+(t-u)(v_u^t)^2}R_u\left(2dX_u - \sigma_u^2 du\right) \\
&+ \rho\int_0^t e^{2X_u+(v_u^t)^2}\sigma_u\left(D^-R_u + (2X_u + (v_u^t)^2)\left(\int_u^t D_u^W\sigma_s^2\right)\right)du.
\end{aligned}$$

Then, as $\lambda_t = 0$, we get

$$E\left(\int_0^t e^{2X_u+(t-u)(v_u^t)^2}dR_u\right) = E\left(e^{2X_0+t(v_0^t)^2}R_0\right) + O(\rho), \quad (4.3.32)$$

and then the impact of correlation on the arithmetic variance swap admits the representation

$$-2e^{2X_0}\rho E\left(e^{t(v_0^t)^2}R_0\right) + O(\rho^2).$$

In the following example, we analyse the concrete case of the SABR model.

Example 4.3.9 (The arithmetic variance swap in the SABR model) *The same arguments as in Example 4.3.7 and the fact that* $\rho E(R_0) = E\langle X, M\rangle$, *where* M *is the future expected variance (defined as in 4.3.3), give us that the effect of the correlation in an arithmetic variance swap under the SABR model admits the representation*

$$-2E\left(e^{2X_0+TE(v_0^T)^2}\right)E\langle X, M\rangle + O(\rho\alpha^3 + \rho^2).$$

4.4 CHAPTER'S DIGEST

In this chapter, we have discussed some key tools of Malliavin calculus in finance. More precisely, we have studied:

- **The Clark-Ocone-Haussman formula**, which leads to an explicit martingale representation for random variables adapted to the σ-algebra generated by a Brownian motion. The most classical application of this result is the construction of hedging strategies. Other applications include the martingale representation of spot and integrated volatilities (that lead to an expression of the convexity adjustment for volatility swaps in terms of the Malliavin derivative operator of the volatility process), as well as the construction of log-normal dynamics for an arbitrary underlying asset, as the VIX volatility index.

- **The integration by parts formula**, which allows us to avoid the computation of the derivatives in the estimation of the Greeks. More precisely, it gives us a way to represent these derivatives as the expectation of the product of the payoff times some random weight. As particular examples, we have studied the estimation of the delta, gamma, and vega under different models, as well as the delta for an Asian option under the Black-Scholes model. Our numerical analysis shows that this approach leads to a reduction in the computational time.

- **The anticipating Itô formula**, which extends the classical Itô's formula to the case of non-adapted processes, as the future average volatility. As an application of it, we can see that, in stochastic volatility models, the characteristic function of an asset price can be decomposed as the sum of the characteristic function of a mixed log-normal distribution, plus a second term due to the impact of correlation. This representation gives us an easy-to-apply tool in the analysis of the properties of the asset distribution, as the computation of the moments. In consequence, we recover some well-known results in the literature on volatility derivatives.

CHAPTER 5

Fractional Brownian motion and rough volatilities

The fractional Brownian motion (fBm) $B = \{B_t, t \in [0, T]\}$ is an extension of the Brownian motion, where we allow the increments to be correlated. This process was introduced in Kolmogorov (1940) and the name 'fractional Brownian motion' was coined in Mandelbrot and Van Ness (1968). The fBm is of interest in several applications in fields such as physics, biology, and finance, due to its capability of modelling short-range and long-range dependent phenomena. From the mathematical point of view, it differs from the classical Brownian motion in many ways: it is not Markovian, nor a semimartingale. In particular, finance models based on the fBm are mathematically more complex than models based on diffusion processes.

In this chapter, we study the main properties of the fBm and related processes and briefly discuss its role in volatility modelling. For a complete approach to this field, we refer to Biagini, Hu, Oksendal, and Zhang (2008), Mishura (2008), Nourdin (2012), and Nualart (2006). A brief introduction including practical issues of the fBm as simulation or identification can be found in Shevchenko (2015).

5.1 THE FRACTIONAL BROWNIAN MOTION

This section is devoted to define the fractional Brownian motion and to study its basic properties. In particular, we see that it is not a

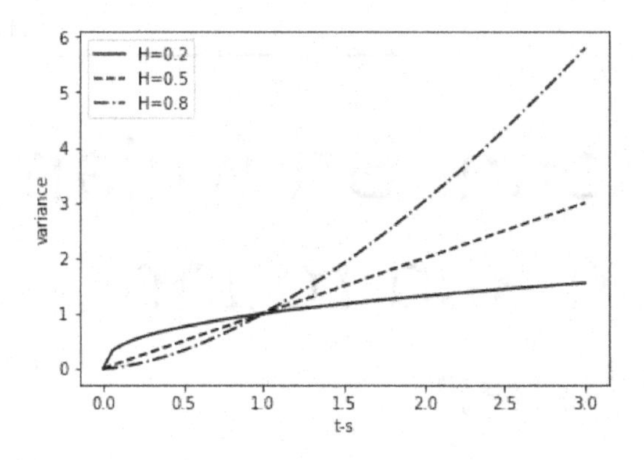

Figure 5.1 Variance function of $B_t - B_s$ as a function of $s - t$.

semimartingale, nor a Markovian process. These facts have relevant consequences in calibration, simulation, and option prices computation, as we discuss in the following chapters.

The fractional Brownian motion (fBm) with Hurst parameter H is defined as a centred Gaussian process $B^H = \{B_t^H, t \in [0,T]\}$ with covariance function given by

$$E(B_t^H B_s^H) = \frac{1}{2}\left(t^{2H} + s^{2H} - |t-s|^{2H}\right), \tag{5.1.1}$$

for some $H \in (0,1)$ that we call the **Hurst parameter**. If $H = \frac{1}{2}$, the fBm B^H is a classical Brownian motion. Notice that (5.1.1) implies that every increment of the form $B_t^H - B_s^H$ is a centred Gaussian random variable with variance equal to $(t-s)^{2H}$.

From this definition, we can prove directly some basic properties of the fBm.

5.1.1 Correlated increments

Consider $t > s \geq u > r$. Then (5.1.1) allows us to write

$$\begin{aligned}
&E[(B_t - B_s)(B_u - B_r)] \\
&= E(B_t^H B_u^H) - E(B_t^H B_r^H) - B_s^H B_u^H + E(B_s^H B_r^H) \\
&= \frac{1}{2}\left(t^{2H} + u^{2H} - (t-u)^{2H}\right) \\
&-\frac{1}{2}\left(t^{2H} + r^{2H} - (t-r)^{2H}\right) \\
&-\frac{1}{2}\left(s^{2H} + u^{2H} - (s-u)^{2H}\right) \\
&+\frac{1}{2}\left(s^{2H} + r^{2H} - (s-r)^{2H}\right) \\
&= \frac{1}{2}(-(t-u)^{2H} + (t-r)^{2H} + (s-u)^{2H} - (s-r)^{2H}).
\end{aligned}$$

If $H = \frac{1}{2}$, the above quantity is equal to zero, according to the fact that the Brownian motion has independent increments. If $H \neq \frac{1}{2}$

$$\begin{aligned}
&E[(B_t^H - B_s^H)(B_u^H - B_r^H)] \\
&= H(2H-1)\int_r^u \left(\int_s^t (\tau - \theta)^{2H-2} d\tau\right) d\theta. \qquad (5.1.2)
\end{aligned}$$

Notice that this quantity is positive if $H > \frac{1}{2}$, while it is negative otherwise. That is, the fBm has negatively correlated increments if $H < \frac{1}{2}$ (then the fBm is **counterpersistent**), and positively correlated increments if $H > \frac{1}{2}$ (then the fBm is **persistent**). In practice, this means that if the last increment has been negative (positive), it is more likely the next one to be positive (negative) if $H < \frac{1}{2}$ and negative (positive) if $H > \frac{1}{2}$.

5.1.2 Long and short memory

Denote $X_n = B_n^H - B_{n-1}^H$ and define

$$\rho_H(n) := Cov(X_1^H, X_{n+1}^H) = Corr(X_1^H, X_{n+1}^H).$$

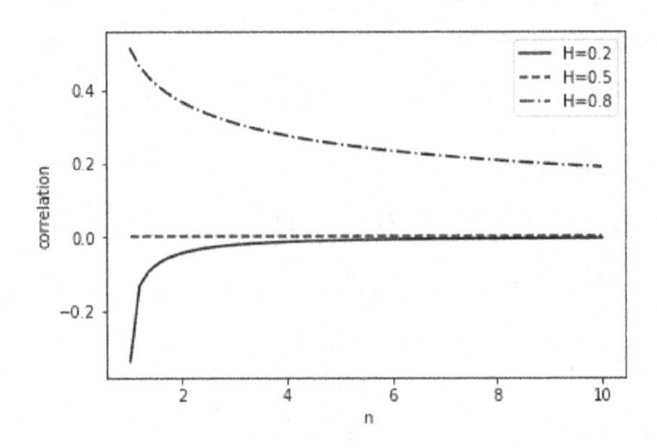

Figure 5.2 Covariance function for the increments of the fBm, as a function of n.

Then, we have

$$\begin{aligned}\rho_H(n) &= E(B_1^H(B_{n+1}^H - B_n^H)) \\ &= \frac{1}{2}(1^{2H} + (n+1)^{2H} - n^{2H}) \\ &- \frac{1}{2}(1^{2H} + n^{2H} - (n-1)^{2H}) \\ &= \frac{1}{2}((n+1)^{2H} + (n-1)^{2H} - 2n^{2H}). \end{aligned} \tag{5.1.3}$$

We can observe the behaviour of ρ_H for different values of H in Figure 5.2. Moreover, we can see that

$$\rho_H(n) \approx H(2H-1)n^{2H-2}.$$

If $H > \frac{1}{2}$, the power $2H - 2 > -1$, and then the sum $\sum_{n=1}^{\infty} \rho_H(n) = \infty$. Then, we say the fBm is **long memory**. If $H < \frac{1}{2}$, this sum is finite, and we say the fBm is **short memory**.

5.1.3 Stationary increments and self-similarity

Take $t > s \geq 0$. Then for every $\tau \in \mathbb{R}$ we have that $E[(B_{t+\tau}^H - B_{s+\tau}^H)^2] = (t-s)^{2H}$. That is, this variance does not depend on τ. As all the increments are Gaussian and centred, this result implies that the law of an increment $B_t^H - B_s^H$ depends only on $t - s$. Then we say the increments are **stationary**.

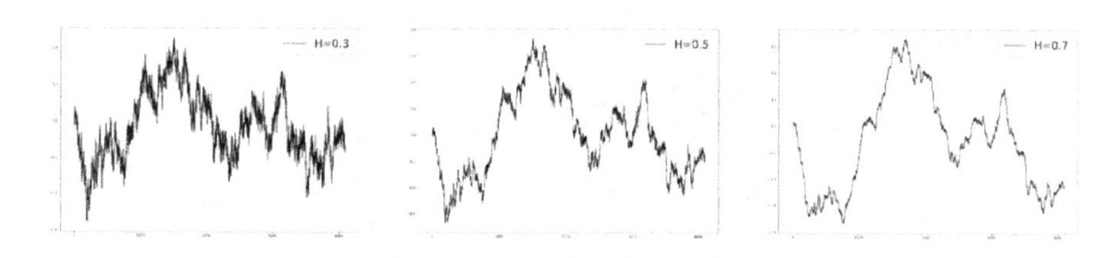

Figure 5.3 fBm simulations for $H = 0.3, H = 0.5$, and $H = 0.7$.

On the other hand, a direct computation gives us that, for all $a > 0$, the distribution of the random process $\{B^H_{at}, t \geq 0\}$ is the same as the distribution of $\{a^H B^H_t, t \geq 0\}$. Then we say that B^H is **statistically self-similar**.

5.1.4 Hölder continuity

Take $t > s$. Notice that, as $B_t - B_s$ is a centred Gaussian random variable with variance equal to $(t-s)^{2H}$, if follows that

$$E|B_t - B_s|^p = \frac{2^{\frac{p}{2}}\Gamma\left(\frac{p+1}{2}\right)}{\sqrt{\pi}}(t-s)^{pH}.$$

Then, the Kolmorogov's continuity criteria gives us that there exists a continuous modification of the fBm such that this continuous modification is λ-Hölder continuous for every $\lambda \in \left(0, \frac{pH-1}{p}\right)$. Now, letting $p \to \infty$, we get that it is Hölder continuous for every $\lambda < H$. For the sake of simplicity, B^H denotes, in the sequel, this continuous modification.

The above computations show that the Hölder continuity of B^H depends on H and that the paths are smoother for bigger values of H. In Figure 5.3 we can observe several simulations of fBm paths corresponding to different values of the Hurst parameters H.

5.1.5 The p-variation and the semimartingale property

It is well known that the fBm is not a semimartingale. This result appears early in the literature (see Liptser and Shiryayev (1986)). This property can be studied from the p-variation of the fBm. Given a stochastic process X with continuous trajectories, and a fixed $p > 0$, we define the p-variation of X on an interval $[0, T]$ as the following limit in probability

$$\lim_{n\to\infty}\sum_{j=1}^{n}\left|X_{\frac{jT}{n}}-X_{\frac{(j-1)T}{n}}\right|^{p}. \tag{5.1.4}$$

In order to study this limit for the fBm, we define the process

$$Y_{n,p}=\left(\frac{T}{n}\right)^{1-pH}\sum_{j=1}^{n}\left|B^H_{\frac{jT}{n}}-B^H_{\frac{(j-1)T}{n}}\right|^{p}.$$

As B^H is self-similar, the law of $Y_{n,p}$ is the same as the law of

$$\tilde{Y}_{n,p}=\frac{T}{n}\sum_{j=1}^{n}\left|B^H_j-B^H_{j-1}\right|^{p}.$$

As the increments $B^H_j - B^H_{j-1}$ are stationary and ergodic, the Ergodic theorem gives us that $\tilde{Y}_{n,p}$ tends in $L^1(\Omega)$ (and then in probability) to $TE(|B_1|^p)$. Then, (5.1.4) tends (as $n \to \infty$) to zero if $pH > 1$, and to infinity if $pH < 1$.
These limit results imply that, if $H > \frac{1}{2}$, the fBm has zero quadratic variation but infinite total variation (see for example Nualart (2006)). In a similar way, if $H < \frac{1}{2}$, the fBm has infinite quadratic variation. Therefore, the fBm cannot be a semimartingale for $H \neq \frac{1}{2}$.

5.1.6 Representations of the fBm

Maldelbrot and Van Ness (1968) proved that the fBm admits a representation of the form

$$B^H_t=\frac{1}{\Gamma(H+\frac{1}{2})}\left(\int_{-\infty}^{0}((t-s)^{H-\frac{1}{2}}-(-s)^{H-\frac{1}{2}})dB_s+\int_0^t (t-s)^{H-\frac{1}{2}}dB_s\right),$$

for all $t > 0$, and where B denotes a Brownian motion in the real line. That is, the fBm can be represented as a Wiener integral with respect to some Brownian motion B. Another classical representation (see Molchan and Golosov (1969), Decreusefond and Üstünel (1999), Norros, Valkeila, and Virtamo (1999), or Alòs, Mazet, and Nualart (2001)) allows us to write B^H_t in terms of a Brownian motion in the time interval $[0, t]$. More precisely,

$$B^H_t=\int_0^t K_H(t,s)dB_s, \tag{5.1.5}$$

where B is a Brownian motion and

$$K_H(t,s) \\ := c_H(t-s)^{H-\frac{1}{2}} + c_H\left(\frac{1}{2}-H\right)\int_s^t (u-s)^{H-\frac{3}{2}}\left(1-\left(\frac{s}{u}\right)^{\frac{1}{2}-H}\right)du. \tag{5.1.6}$$

These representation formulas allow us to write a fBm in terms of a Brownian motion. Then, they become a useful tool in several applications.

5.2 THE RIEMANN-LIOUVILLE FRACTIONAL BROWNIAN MOTION

The following truncated version of the fBm

$$\hat{B}_t^H := \frac{1}{\Gamma(H+\frac{1}{2})}\int_0^t (t-s)^{H-\frac{1}{2}} dB_s \tag{5.2.1}$$

is a Riemann-Liouville fractional process (see Picard (2011)), and it is also called a type II fractional Brownian motion (see Marinucci and Robinson (1999)). In this book, we follow the notation in Lim and Sithi (1995) and we refer to this process as the Riemann-Liouville fractional Brownian motion (RLfBm).

It is easy to see that $\hat{B}^H$ is also a centred Gaussian self-similar process but its increments are not stationary (see Lim and Sithi (1995)). In particular, $E(\hat{B}^H)^2 = t^{2H}$ and its covariance function (see Example 2.5.1) is given by

$$E(\hat{B}_t^H \hat{B}_s^H) = s^{H+\frac{1}{2}} t^{H-\frac{1}{2}} \frac{1}{(H+\frac{1}{2})\Gamma(H+\frac{1}{2})^2}\, {}_2F_1\left(\frac{1}{2}-H, 1, H+\frac{3}{2}, \frac{s}{t}\right).$$

As in the fBm case, if $H=\frac{1}{2}$, the process $\hat{B}^H$ is a classical Brownian motion. We can observe the behaviour of the RLfBm in Figures 5.4 and 5.5. As for the fBm, the paths are rougher for small values of the parameter H. Moreover, the autocorrelation function reveals a negative correlation between consecutive increments in the case $H<\frac{1}{2}$, with a fast decay (in absolute value) to zero. In the case $H>\frac{1}{2}$, this correlation is positive, with a slower decay.

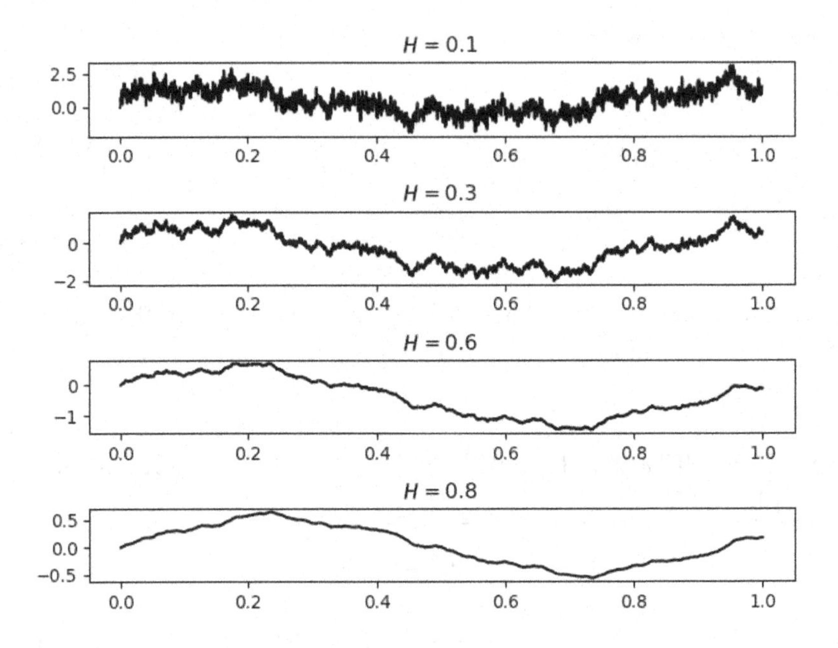

Figure 5.4 RLfBm simulations for $H = 0.1, H = 0.3$, $H = 0.6$, and $H = 0.8$.

Because of the simplicity of its representation, the RLfBm has been widely considered in the construction of fractional volatility models (see for example Comte and Renault (1998), Alòs, León, and Vives (2007), or Bayer, Friz, and Gatheral (2016)). In particular, in the rBergomi model introduced in Example 2.5.1 the volatility is given by

$$\sigma_t^2 = \sigma_0^2 \exp\left(\nu\sqrt{2H} Z_t - \frac{1}{2}\nu^2 t^{2H}\right), t \in [0, T], \tag{5.2.2}$$

for some positive real values σ_0^2 and ν, and where

$$Z_t = \int_0^t (t-s)^{H-\frac{1}{2}} dB_s,$$

for some $H \in (0, 1)$ and some Brownian motion B. That is, the rBergomi volatility is a function of a RLfBm.

5.3 STOCHASTIC INTEGRATION WITH RESPECT TO THE FBM

The definition of a stochastic integral with respect to the fractional Brownian motion is not a trivial problem. We have seen that, for $H \neq \frac{1}{2}$,

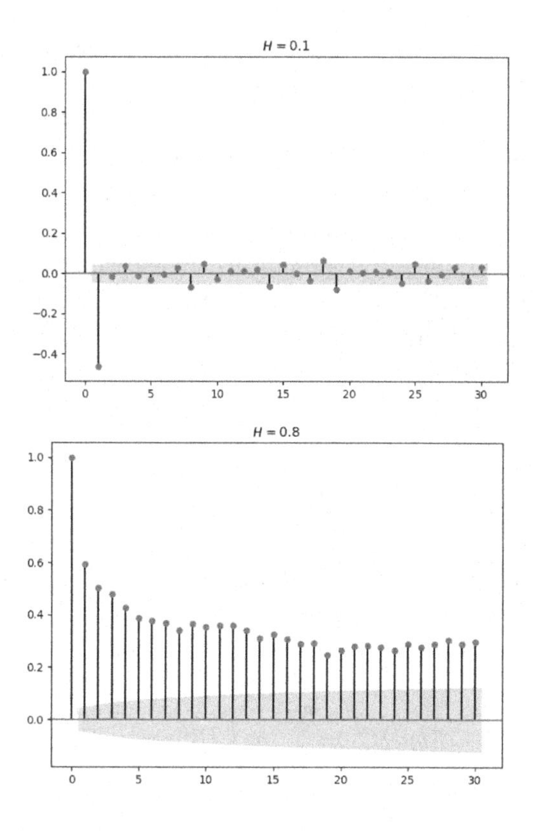

Figure 5.5 Autocorrelation function of the increments of simulated RLfBm, with $H = 0.1$ (left) and $H = 0.8$ (right).

the fBm is not a semimartingale and then we cannot apply the classical Itô's calculus.

Several alternatives have been proposed in the literature. The pathwise approach allows us to construct this integral in the Riemann-Stieljes sense, for integrands whose paths satisfy some regularity conditions, as finite p-variation with $\frac{1}{p} + H > 1$ (see for example Lin (1995) or Day and Heyde (1996)), or λ-Hölder continuity, for all $\lambda < 1 - H$, as in Zähle (1998, 1999). If $H > \frac{1}{2}$, this approach allows us to integrate a function of the fBm $f(B^H)$, but if $H < \frac{1}{2}$ we cannot, in general, integrate such a process $f(B^H)$.

Another approach is based on Malliavin calculus. We have seen in Chapter 3 that the Skorohod integral is an extension of the classical Itô integral, in the sense that both integrals coincide in the case of square-integrable and adapted processes. As Malliavin calculus is valid for an arbitrary Gaussian process (see Malliavin (1997) and Nualart (2006)), the divergence operator with respect to the fBm can be interpreted as a stochastic integral (see Decreusefond and Üstünel (1998, 1999), Duncan, Hu, and Pasik-Duncan (2000), Alòs, Mazet, and Nualart (2001), and Hu and Øksendal (2003)).

In order to exemplify the difficulties that arise in the definition of a stochastic integral with respect to the fBm, let us consider the simple case of a deterministic integrand, and let us try to construct its stochastic integral as the limit of Riemann sums. Consider the class $\mathcal{S}$ of deterministic processes $\phi = \{\phi(t), t \in [0,T]\}$ of the form

$$\phi(t) = \sum_{i=1}^{n} a_i \mathbf{1}_{[\frac{iT}{n}, \frac{(i+1)T}{n}]}(t), \tag{5.3.1}$$

for some $n > 1$ and for some real values $a_i, i = 1, ...n$. For all $\phi \in \mathcal{S}$, we define the stochastic integral of ϕ with respect to a fBm $B^H = \int_0^t K_H(t,s)dB_s$ as the sum

$$\int_0^T \phi(t)dB_t^H = \sum_{i=0}^{n-1} a_i \left(B^H_{\frac{(i+1)T}{n}} - B^H_{\frac{iT}{n}} \right). \tag{5.3.2}$$

Then, the representation formula (5.1.5) allow, us to write, for all $\phi \in \mathcal{S}$, 79

$$\begin{aligned}
\int_0^T \phi(t)dB_t^H &= \sum_{i=1}^{n-1} a_i \left(\int_0^{\frac{iT}{n}} \left(K_H\left(\frac{(i+1)T}{n}, s\right) - K_H\left(\frac{iT}{n}, s\right)\right) dB_s \right. \\
&+ \left. \int_{\frac{iT}{n}}^{\frac{(i+1)T}{n}} K_H\left(\frac{(i+1)T}{n}, s\right) dB_s \right) \\
&= \sum_{i=0}^{n-1} a_i \sum_{j=0}^{i-1} \left(\int_{\frac{jT}{n}}^{\frac{(j+1)T}{n}} \left(\int_{\frac{iT}{n}}^{\frac{(i+1)T}{n}} \frac{\partial K_H}{\partial u}(u,s)du \right) dB_s \right. \\
&+ \left. \int_{\frac{iT}{n}}^{\frac{(i+1)T}{n}} K_H\left(\frac{(i+1)T}{n}, s\right) dB_s \right) \\
&= \sum_{j=0}^{n-1} \int_{\frac{jT}{n}}^{\frac{(j+1)T}{n}} \left[a_j K_H\left(\frac{(j+1)T}{n}, s\right) + \sum_{i=j+1}^{n-1} a_i \left(\int_{\frac{iT}{n}}^{\frac{(i+1)T}{n}} \frac{\partial K_H}{\partial u}(u,s)du \right) \right] dB_s \\
&= \sum_{j=0}^{n-1} \int_{\frac{jT}{n}}^{\frac{(j+1)T}{n}} \left[a_j K_H(T,s) + \sum_{i=j+1}^{n-1} (a_i - a_j) \left(\int_{\frac{iT}{n}}^{\frac{(i+1)T}{n}} \frac{\partial K_H}{\partial u}(u,s)du \right) \right] dB_s.
\end{aligned} \tag{5.3.3}$$

That is,

$$\int_0^T \phi(t)dB_t^H = \int_0^T (K_H^*\phi)(s)dB_s, \tag{5.3.4}$$

where

$$(K_H^*\phi)(s) := \phi(s)K_H(T,s) + \int_s^T (\phi(u)-\phi(s))\frac{\partial K_H}{\partial u}(u,s)du. \tag{5.3.5}$$

If we define $\mathcal{H}$ as the closure of $\mathcal{S}$ by the norm $\|\phi\|_{\mathcal{H}} = \|(K_H^*\phi)\|_{L^2([0,T])}$, a limit argument allows us to define the integral $\int_0^T \phi(t)dB_t^H$ as

$$\int_0^T \phi(t)dB_t^H = \int_0^T (K_H^*\phi)(s)dB_s, \tag{5.3.6}$$

for every $\phi \in \mathcal{H}$. Notice that we have two cases:

- If $H > \frac{1}{2}$, $K_H(s,s) = 0$ and then

$$\begin{aligned}(K_H^*\phi)(s) &= \phi(s)K_H(s,s) + \int_s^T \phi(u)\frac{\partial K_H}{\partial u}(u,s)du \\ &= \int_s^T \phi(u)\frac{\partial K_H}{\partial u}(u,s)du, \end{aligned} \tag{5.3.7}$$

  from where we can see, in particular, that $L^2([0,T]) \in \mathcal{H}$.

- If $H < \frac{1}{2}$, $\mathcal{H}$ is a smaller set of functions, and $L^2([0,T])$ is not included in this completion. The integrability of a function ϕ requires some conditions on the increments $\phi(u) - \phi(s)$. For example, this integral is well defined if ϕ is γ-Hölder continuous for all $\gamma < \frac{1}{2} - H$.

The last point gives us a clue about the main problem in stochastic integration with respect to the fBm. If now we want to extend the above definition to the case of random processes of the form $f(B_t^H)$, one of the main difficulties is that this process is, under general conditions, γ-Hölder continuous for $\gamma < H$. As $H < \frac{1}{2} - H$ for $H < \frac{1}{4}$, this integral would be well defined only in the case $H > \frac{1}{4}$.

Remark 5.3.1 *The arguments in the definition of the integral with respect to the fBm can be extended to the case of other Gaussian processes defined for other kernels different from $K_H(t,s)$. For example, in the case of a RLfBm $\hat{B}^H$, the same arguments apply by simply substituting*

$K_H(t,s)$ by $\frac{(t-s)^{H-\frac{1}{2}}}{\Gamma(H+\frac{1}{2})}$. Then we get that, for all deterministic ϕ in the corresponding closure of $\mathcal{S}$

$$\int_0^T \phi(t)d\hat{B}_t^H = \int_0^T (\hat{K}_H^*\phi)(s)dB_s, \tag{5.3.8}$$

where

$$\begin{aligned}(\hat{K}_H^*\phi)(s) &= \frac{1}{\Gamma(H+\frac{1}{2})}\phi(s)(T-s)^{H-\frac{1}{2}} \qquad (5.3.9)\\ &+ \frac{H-\frac{1}{2}}{\Gamma(H+\frac{1}{2})}\left(\int_s^T (\phi(u)-\phi(s))(u-s)^{H-\frac{3}{2}}du\right).\end{aligned}$$

Again, if $H > \frac{1}{2}$, the above expression reduces to

$$\begin{aligned}(\hat{K}_H^*\phi)(s) &= \frac{H-\frac{1}{2}}{\Gamma(H+\frac{1}{2})}\left(\int_s^T \phi(u)(u-s)^{H-\frac{3}{2}}du\right)\\ &= \frac{1}{\Gamma(H-\frac{1}{2})}\left(\int_s^T \phi(u)(u-s)^{H-\frac{3}{2}}du\right). \qquad (5.3.10)\end{aligned}$$

We notice that in classical diffusion models, like Heston or SABR, the volatility process is the solution of a stochastic differential equation. In the case of fractional volatilities, the situation is different, because for $H \neq \frac{1}{2}$, the fBm and the RLfBm are not semimartingales and then we cannot apply the classical Itô's calculus. Due to the difficulties in the definition of a stochastic integral with an adequate economical meaning (and, in consequence, in the construction of stochastic differential equations with respect to the fBm), most rough volatilities are defined, in general, not as solutions of an SDE driven by a fBm/RLfBm, but simply as functions of the integral of some deterministic function with respect to the fBm/RLfBm.

5.4 SIMULATION METHODS FOR THE FBM AND THE RLFBM

This section is devoted to present a simple method for the simulation of the fBm and the RLfBm. As the increments of the fBm are not independent, the simulation of the fBm is not straightforward. Several approaches have been proposed and analysed in the literature (see for example Dieker (2004), Bennedsen, Lunde, and Pakkanen (2017), or Banna, Mishura, Ralchenko, and Shklyar (2019)). All these methods can be classified into two groups: the so-called *exact methods*, which directly simulate a discretised fBm as a Gaussian vector, and the *approximate methods*, where the fBm is approximated by some other processes.

In this section we present one of the simplest exact methods, based on the Cholesky decomposition of the covariance matrix. Assume we want to simulate a Gaussian-centred random vector of the form

$$\left(X_{\frac{1}{n}}, X_{\frac{2}{n}}, X_{\frac{3}{n}}, ..., X_{\frac{n-1}{n}}, X_1\right) \tag{5.4.1}$$

whose $n \times n$ variance-covariance matrix given by

$$\Sigma = (\Sigma_{i,j}) = \Sigma_{i,j} = E\left(X_{\frac{i}{n}} X_{\frac{j}{n}}\right).$$

It can be proved that there exists a lower triangular matrix Γ such that $\Gamma\Gamma' = \Sigma$. Then, because of the properties of the multivariate normal distribution, the law of the vector (5.4.1) is the same as the law of the random vector

$$\Gamma u,$$

where u denotes a n-dimensional standard multivariate normal vector. The matrix Γ can be computed by means of the Cholesky decomposition (see for example Dieker (2002)). Then our simulation method can be summarised as follows:

- We simulate a sequence of n-independent standard Gaussian numbers $u = u_1, ..., u_n$.
- We use Cholesky's decomposition method to compute the matrix Γ.
- We compute the product Γu.

Even when the Cholesky method is easy to implement, its computation complexity is $O(n^3)$ which renders this method very slow in practice. Nevertheless, in the simulation of multiple paths, the matrix Γ has to be computed only once, and this allows us to reduce the computing time. Some alternative ways to compute a squared root of Σ include the Hosting and the Davis and Harte methods (see for example Dieker (2002) and Banna, Mishura, Ralchenko, and Shklyar (2019)). Some simulated paths of the fBm and the RLfBm obtained using the Cholesky method can be seen in Figures 5.3 and 5.4, respectively.

Notice that the Cholesky method can be applied to volatility models based both on the fBm and on the RLfBm. For models based on the RLfBm like the rBergomi model in Example 2.5.1, the joint covariance between the RLfBm driving the volatility process and the Brownian motion driving asset prices (see Equations 2.5.3 and 2.5.4) allows us to jointly simulate asset prices and volatilities (see for example Bayer, Friz, and Gatheral (2016)).

5.5 THE FRACTIONAL BROWNIAN MOTION IN FINANCE

The introduction of the fBm in asset price modelling was proposed in Mandelbrot (1971) to better fit the long-memory properties of returns in real market data. Empirical studies documenting this long-run dependence include Greene and Fielitz (1977), Fama and French (1988), and Poterba and Summers (1988). Nevertheless, some papers call into question the statistical significance of these empirical results (see Lo (1991) and Jacobsen (1996)). Teverovsky, Taqqu, and Willinger (1999) and Willinger, Taqqu, and Teverovsky (1999) point out that the conclusions in Lo (1991) can simply come from the low power of the tests against moderate long-memory, and they agree it is difficult to establish from real market data the existence or absence of long memory properties. We remark that even there is not a consensus on the existence/absence of long memory in asset returns, this may be compatible with the existence of a long-memory behaviour in absolute returns or volatility (see for example Lovato and Velasco (2000) and Cont (2005)).

Long-memory asset price models based on a fBm with $H > \frac{1}{2}$ have been studied deeply in the literature (see Cheridito (2003), Guasoni (2006), Czichowsky, Peyre, Schachermayer, and Yang (2018), and Guasoni, Nika, and Rásony (2019), among others). One of the main criticism of these models is that (as the discounted prices are not martingales), they allow for arbitrage opportunities. Nevertheless, under some realistic hypotheses (as the existence of transaction costs or a minimum time lapse between transactions) these opportunities disappear (see for example Cheridito (2003)). A version of the fundamental theorem of asset pricing for the fBm under transaction costs has been obtained in Guasoni, Rásonyi, and Schachermayer (2010).

On the other side, fractional volatility models (that is, stochastic volatility models where the volatility is assumed to be driven by a fBm or a RLfBm) were introduced in Comte and Renault (1998) to describe the slow flattening of the implied volatility surface for long maturities.

In this seminal paper, it was proved that fractional volatility models with $H > \frac{1}{2}$ are able to describe the long-term properties of implied volatilities observed in real market data. More recently, Alòs, León, and Vives (2007) proved that fractional volatility models with $H < \frac{1}{2}$ fit some properties of the implied volatility for short maturities. This observation has led to the development of the so-called *rough* volatility models, as we will see in the next chapters. We would like to remark that both in Comte and Renault (1998) and in Alòs, León, and Vives (2007), fractional volatilities were not introduced to fit the behaviour of the volatility time series, but to reproduce the behavior of implied volatilities.

Even when some fractional volatility models presented in the literature are based on the fBm (see for example Fukasawa (2011, 2014)), most of them, as we pointed out in Section 5.2.1, are based on the RLfBm, because of the simplicity of its representation as an integral with respect to the Brownian motion. There are also other approaches where the volatility is not a function of a fBm or a RLfBm, but a fractional integral of a diffusion process. For example, in Comte, Coutin, and Renault (2012) the volatility σ is defined by

$$\sigma_t^2 = \theta + \frac{1}{\Gamma(a)} \int_{-\infty}^{t} (t-s)^{a-1} X_s ds,$$

where X is a CIR process and for some $a \in (0, \frac{1}{2})$. Notice that this defines a long-memory process σ via a Markovian process X, leading to a fractional Heston model that preserves the main properties of the model presented in Comte and Renault (1998), but is more tractable in applications. This family of models has also been considered in Alòs and Yan (2017) and in Guennoun, Jacquier, Roome, and Shi (2018). Finally, some other works consider volatilities that are the solution of a fractional SDE, as in El Euch, Fukasawa, and Rosenbaum (2018).

As we commented in Chapter 2, there has been in the last decade a big controversy over the long-memory or short-memory of volatility processes. Notice that both properties are not incompatible. A process can depend on both long-term and short-term components, and then each one of these components dominates at different scales and, in consequence, at different maturities in the implied volatility surface. We can see this effect in the following example.

Example 5.5.1 *Consider a process X of the form*

$$X_t = B_t^{0.3} + B_t^{0.7}, t \in [0, T]$$

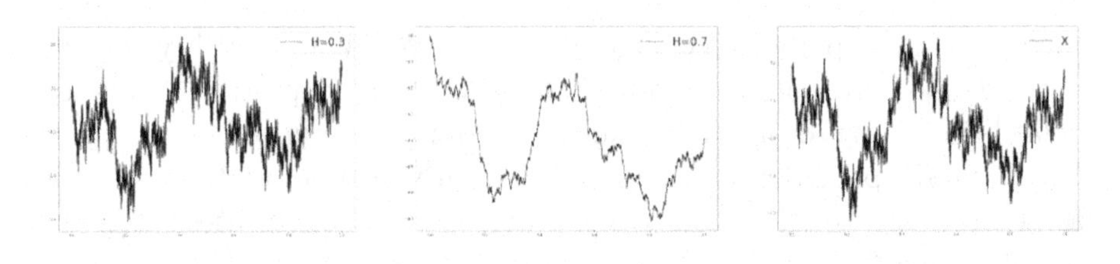

Figure 5.6 fBm simulations for $B^{0.3}$, $B^{0.7}$, and X in the time interval $[0, 1]$.

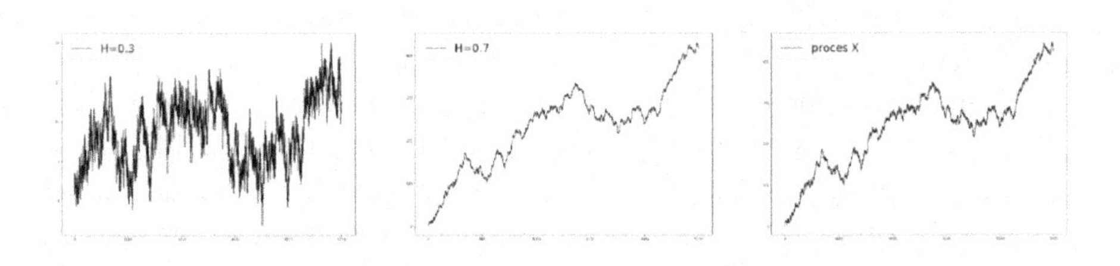

Figure 5.7 fBm simulations for $B^{0.3}, B^{0.7}$, and X in the time interval $[0, 500]$.

where $B^{0.3}$ is a fBm with Hurst parameter $H = 0.3$, $B^{0.7}$ is a fBm with $H = 0.7$ and $T > 0$. In Figure 5.6 we can see a simulation of the paths of $B^{0.3}, B^{0.7}$, and X in the interval $[0, 1]$. We can observe that the path of X seems to be 'rough' as the path of $B^{0.3}$. In Figure 5.7 we plot sample paths of the same three processes but in the time interval $[0, 5000]$. In this case, the path of X behaves more smooth and similar to the path of $B^{0.7}$.

5.6 THE MALLIAVIN DERIVATIVE OF FRACTIONAL VOLATILITIES

Our analysis of the properties of fractional volatility models is based on Malliavin calculus techniques. In this section, we compute the Malliavin derivative of volatility models based on the fBm/rLfBm. Towards this end, we make use of the representation (5.1.5), which gives us that a fBm B^H with Hurst parameter H admits a representation of the form

$$B_t^H = \int_0^t K_H(t, s) dB_s,$$

where B is a Brownian motion and K_H is the kernel defined in (5.1.6). Then, a direct computation gives us that the Malliavin derivative of B_t^H with respect to B is given by

$$D_s B_t^H = K_H(t,s),$$

for $s < t$. Similarly, the Malliavin derivative of a random variable of the form

$$\int_0^T (K_H^* \phi)(u) dB_u,$$

for some deterministic function ϕ, is given by

$$D_s \left(\int_0^T (K_H^* \phi)(u) dB_u \right) = (K_H^* \phi)(u),$$

for all $s < T$.

The same arguments give us that the Malliavin derivative of the (RLfBm) $\hat{B}$ defined in (5.2.1) with respect to B is given by

$$D_s \hat{B}_t^H = \frac{1}{\Gamma(H + \frac{1}{2})} (t-s)^{H-\frac{1}{2}},$$

for $s < 0$.

The above equalities allow us to compute the Malliavin derivative of most fractional volatility models in the literature, as we see in the following examples.

5.6.1 Fractional Ornstein-Uhlenbeck volatilities

Assume, as in Comte and Renault (1998) or in Alòs, León, and Vives (2007) that the volatility σ can be written as $\sigma_r = f(Y_r)$, where $f \in C_b^1(\mathbb{R})$ and Y_r is a *fractional Ornstein-Uhlenbeck* process of the form

$$Y_r = m + (Y_0 - m) e^{-\alpha r} + c\sqrt{2\alpha} \int_0^r e^{-\alpha(r-s)} d\hat{B}_s^H, \tag{5.6.1}$$

where $\hat{B}_s^H := \int_0^s (s-u)^{H-\frac{1}{2}} dB_u$ and Y_0, m, c, and α are positive constants.

Notice that the Malliavin derivative of the stochastic integral $\int_t^r e^{-\alpha(r-s)} d\hat{B}_s^H$ is given by

$$D_s \left(\int_0^r e^{-\alpha(r-s)} d\hat{B}_s^H \right) = (\hat{K}_H^* \mathbb{1}_{[0,r]}(\cdot) e^{-\alpha(r-\cdot)})(s), \tag{5.6.2}$$

for $s < r$. Then,

$$\begin{aligned} D_s\sigma_r &= f'(Y_r)D_sY_r \\ &= f'(Y_r)c\sqrt{2\alpha}(\hat{K}_H^* \mathbb{1}_{[0,r]}(\cdot)e^{-\alpha(r-\cdot)})(s). \end{aligned} \tag{5.6.3}$$

In the case $H > \frac{1}{2}$, the RHS in (5.6.3) takes the form

$$f'(Y_r)c\sqrt{2\alpha}\left(H-\frac{1}{2}\right)\left(\int_s^r e^{-\alpha(r-u)}(u-s)^{H-\frac{3}{2}}du\right),$$

while in the case $H < \frac{1}{2}$ it reads

$$\begin{aligned} f'(Y_r)c\sqrt{2\alpha}\Bigg[&\left(\frac{1}{2}-H\right)\left(\int_s^r \left[e^{-\alpha(r-u)} - e^{-\alpha(r-s)}\right](u-s)^{H-\frac{3}{2}}du\right) \\ &+e^{-\alpha(r-s)}(r-s)^{H-\frac{1}{2}}\Bigg]. \end{aligned}$$

Recall that, as $|r-s| \to 0$, the above Malliavin derivatives tend to zero if $H > \frac{1}{2}$, and to ∞ if $H < \frac{1}{2}$. This is a key property with relevant implications in the behaviour of the corresponding implied volatilities, as we see in the next chapters.

5.6.2 The rough Bergomi model

Consider the Gaussian process

$$Z_r := \int_0^r (r-s)^{H-\frac{1}{2}}dW_s,$$

where W denotes a Brownian motion. Then, under the rough Bergomi model (see Section 2.5.1), the volatility σ is defined as

$$\sigma_r^2 = \sigma_0^2 exp\left(\nu\sqrt{2H}Z_r - \frac{1}{2}\nu^2 r^{2H}\right), r \in [0,T], \tag{5.6.4}$$

for some positive real values σ_0^2, ν and for some $H < \frac{1}{2}$. Now, the chain rule and the fact that σ_r is a function of Z_r give us that

$$\begin{aligned} D_s\sigma_r^2 &= \nu\sqrt{2H}\sigma_r^2 D_s\left(\int_0^r (r-s)^{H-\frac{1}{2}}dW_s - \frac{1}{2}\nu^2 r^{2H}\right) \\ &= \nu\sqrt{2H}\sigma_r^2(r-s)^{H-\frac{1}{2}}. \end{aligned} \tag{5.6.5}$$

Notice that $|D_s\sigma_r^2| \to \infty$ as $|s-r| \to 0$.

5.6.3 A fractional Heston model

Consider a CIR process of the form

$$\tilde{\sigma}_t^2 = \theta + \left(\tilde{\sigma}_0^2 - \theta\right) e^{-\kappa t} + \nu \int_0^t \exp\left(-\kappa(t-u)\right) \sqrt{\tilde{\sigma}_u^2} dW_u, \tag{5.6.6}$$

where $\tilde{\sigma}_0^2$, θ, κ, and ν are positive constants satisfying the condition $\frac{2\kappa\theta}{\nu^2} \geq 1$. We denote

$$Y_t = \theta + \left(\tilde{\sigma}_0^2 - \theta\right) e^{-\kappa t} \tag{5.6.7}$$

and

$$Z_t = \int_0^t \exp\left(-\kappa(t-u)\right) \sqrt{\tilde{\sigma}_u^2} dW_u. \tag{5.6.8}$$

Now, we define the volatility process σ by

$$\sigma_t^2 = Y_t + c_1 \nu Z_t + c_2 \nu \frac{1}{\Gamma\left(H - \frac{1}{2}\right)} \int_0^t (t-r)^{H-\frac{1}{2}} Z_r dr,$$

for some $H \in (1/2, 1)$ and for some positive constants c_1, c_2 (see Alòs and Yang (2017)). Notice that if $c_1 = 1$ and $c_2 = 0$ this process is a Heston volatility. Then, a direct computation gives us that

$$D_s^W \sigma_t^2 = c_1 D_s^W \tilde{\sigma}_t^2 + \frac{c_2}{\Gamma(\alpha)} \int_s^t (t-r)^{\alpha-1} D_s^W \tilde{\sigma}_r^2 dr, \tag{5.6.9}$$

where $D_s^W \tilde{\sigma}_t^2$ is the Malliavin derivative of a CIR process

$$D_s^W \tilde{\sigma}_t^2 = \nu \sqrt{\tilde{\sigma}_t^2} \exp\left(\int_s^t \left(-\frac{\kappa}{2} - \left(\frac{\kappa\theta}{2} - \frac{\nu^2}{8} \right) \frac{1}{\tilde{\sigma}_u^2} \right) du \right)$$

(see Section 3.3.2). That is, in this case, the computation of the Malliavin derivative of this long-memory process is computed via the Malliavin derivative of a classical diffusion process.

5.7 CHAPTER'S DIGEST

A fractional Brownian motion (fBm) is a Gaussian self-similar process with stationary increments that allows the increments to be correlated. Its covariance function depends on a parameter $H \in (0,1)$ (the Hurst parameter). If $H = 1/2$, the increments are uncorrelated, and we recover the classical Brownian motion. If $H > \frac{1}{2}$ the increments are positively correlated, and the process exhibits long-memory properties, while if $H < \frac{1}{2}$ this correlation is negative and the fBm has short-memory properties.

The paths of a fBm are λ-Hölder continuous for all $\lambda < H$. The corresponding p-variation tends (as $n \to \infty$) to zero if $pH > 1$, and to infinity if $pH < 1$, from where we deduce that the fBm (with $H \neq \frac{1}{2}$) cannot be a semimartingale. The fBm can be represented in terms of a Brownian motion in the real line and in terms of a Brownian motion in the positive half-line. A truncated version of the fBm, the Riemann-Liouville fractional Brownian motion (RLfBm) has a simpler representation in terms of the Brownian motion and has been widely used in the construction of fractional volatilities.

There is a vast literature on numerical methods for the simulation of a fBm. Among the simplest ones, we quote the Cholesky method, which allows us to simulate Gaussian processes via the Cholesky decomposition of their covariance matrix. This method can be applied to both the fBm and the RLfBm.

The application of the fBm in finance was introduced by Mandelbrot in the 70s. Asset price models based on the fBm are not semimartingales, and they allow for arbitrage opportunities. Nevertheless, these opportunities disappear under some realistic hypotheses on the market, as if there is a minimum time interval between transactions, or if there are transaction costs. More recently, the fractional Brownian motion has been introduced to better explain the long-term and the short-term behaviour of the implied volatility surface, leading to the development of rough volatility models. Even when there has been a big controversy over the long-memory or short-memory of volatility processes, we remark that both properties can be compatible. A process can depend both on long-term and short-term components, and then each one of these components dominates at different scales and, in consequence, at different maturities in the implied volatility surface.

The Malliavin derivative of most fractional volatility models can be easily computed, as we see in the cases of fractional Ornstein-Uhlenbeck volatilities and the rough Bergomi model. We observe that, if $H < \frac{1}{2}$, the Malliavin derivatives $D_r\sigma_s$ tend to infinity as $|r - s| \to 0$. This is a key property with relevant implications in the behaviour of the corresponding implied volatilities, as we see in the next chapters.

III

Applications of Malliavin Calculus to the study of the implied volatility surface

CHAPTER 6

The ATM short-time level of the implied volatility

This chapter* is devoted to the study of the short-time limit of the at-the-money implied (ATMI) volatility. It is well known (see Durrleman (2008, 2010)) that, for continuous volatilities, the ATM implied volatility tends to the spot volatility as the time to maturity tends to zero. In this chapter, following and extending the work in Alòs and Shiraya (2019), we deep in this short-time behaviour, and we study the relationship between the ATM implied volatility, the volatility swap, the variance swap, and the spot volatility.

As the first step in our analysis, following Alòs (2006), we use the anticipating Itô formula presented in Section 4.3.1 to decompose option prices as the sum of the corresponding prices in the uncorrelated case (that is, the case where the volatility and the noise driving the stock prices are correlated), plus a term due to the correlation. This characterisation allows us to get a similar decomposition of the implied volatility as the sum of two terms: the implied volatility in the uncorrelated case, plus a term due to the correlation.

*Adapted with permission from Springer Nature: Springer. Finance and Stochastics. A generalization of the Hull and White formula with applications to option pricing approximation, Elisa Alòs © Springer-Verlag (2006), and Estimating the Hurst parameter from short term volatility swaps: a Malliavin calculus approach, Elisa Alòs and Kenichiro Shiraya © Springer-Verlag (2019).

We remark that the above formulas are not expansions, but exact decompositions. Starting from them, we make use of a Malliavin calculus approach to derive an expression for the difference between the volatility swap and the ATMI. In particular, we see that the order of convergence is different in the correlated and the uncorrelated cases and that it depends on the behaviour of the Malliavin derivative of the volatility process. In particular, we see that for volatilities driven by a fractional Brownian motion, this order depends on the corresponding Hurst parameter H. Using similar techniques, we can also study the difference between the volatility swap and the square root of a variance swap.

Based on the above decompositions, we can deduce simple approximation formulas for the ATMI that recover some well-known results in the literature, and that can be applied to the case of fractional volatilities.

6.1 BASIC DEFINITIONS AND NOTATION

Consider the following model for the log-price of a stock price under a risk-neutral probability P

$$dX_t = \left(r - \frac{1}{2}\sigma_t^2\right) dt + \sigma_t \left(\rho dW_t + \sqrt{1-\rho^2} B_t\right), \ t \in [0,T] \quad (6.1.1)$$

for some $T > 0$ and where r is the instantaneous interest rate (supposed to be constant), W and B are independent standard Brownian motions, $\rho \in [-1,1]$, and σ is a square integrable process adapted to the filtration generated by the Brownian motion W. In the sequel, we denote by $\mathcal{F}^W$ and $\mathcal{F}^B$ the filtrations generated by W and B, respectively. Moreover, we define $\mathcal{F} := \mathcal{F}^W \vee \mathcal{F}^B$.

The price of a European call option with payoff $(e^{X_T} - K)_+$, for some strike price K is given by

$$V_t = e^{-r(T-t)} E\left[\left(e^{X_T} - K\right)_+ \middle| \mathcal{F}_t\right], \quad (6.1.2)$$

for some strike K and where E denotes the expectation with respect to P. Our objective in this section is to decompose this price into two terms: the price in the uncorrelated case $\rho = 0$ plus a correction due to the correlation. Towards this end, we introduce the following notation:

- $v_t = \sqrt{\frac{Y_t}{T-t}}$, where $Y_t := \int_t^T \sigma_s^2 ds$ denotes the integrated volatility.

- $BS(t, x; \sigma)$ is the Black-Scholes of a European option with payoff function $h(X_T)$, volatility equal to σ, current log-stock price x, time to maturity $T - t$, and interest rate r. That is

$$BS(s, x, \sigma) = e^x N(d_+) + K e^{-r(T-t)} N(d_-),$$

where N denotes the normal cumulative probability function and

$$d_\pm := \frac{x - x_t^*}{\sigma\sqrt{T-t}} \pm \frac{\sigma}{2}\sqrt{T-t},$$

with $x_t^* := \ln K - r(T-t)$.

- $\mathcal{L}_{BS}(\sigma)$ denotes the Black-Scholes differential operator (in the log variable) with volatility σ

$$\mathcal{L}_{BS}(\sigma) = \frac{\partial}{\partial t} + \frac{1}{2}\sigma^2 \frac{\partial^2}{\partial x^2} + \left(r - \frac{1}{2}\sigma^2\right)\frac{\partial}{\partial x} - r\cdot$$

It is well known that $\mathcal{L}_{BS}(\sigma) BS(\cdot, \cdot; \sigma) = 0$.

- For any random variable A, we denote $E_t(A) := E[A|\mathcal{F}_t]$.

6.2 THE CLASSICAL HULL AND WHITE FORMULA

In this section, we prove the classical Hull and White formula in the uncorrelated case $\rho = 0$, as well as its extensions to the correlated case. The classical approach, based on conditional expectations, leads to an extension of this formula that is of practical use in simulation. We also show how it can be proved via Itô's formula, a technique that we will extend to the uncorrelated case via the anticipating Itô formula introduced in Section 4.3.

6.2.1 Two proofs of the Hull and White formula

6.2.1.1 Conditional expectations

Taking conditional expectations we can write

$$\begin{aligned} V_t &= e^{-r(T-t)} E_t(h(X_T)_+), \\ &= e^{-r(T-t)} E_t[E(h(X_T)_+ \big| \mathcal{F}_t \vee \mathcal{F}_T^B)]. \end{aligned} \tag{6.2.1}$$

In the uncorrelated case $\rho = 0$, the random variable X_T follows, conditioned to $\mathcal{F}_t^W \vee \mathcal{F}_T^B$, a normal distribution with mean $\left(r - \frac{v_t^2}{2}\right)(T-t)$ and variance $v_t^2(T-t)$. Then

$$e^{-r(T-t)} E(h(X_T)_+ | \mathcal{F}_t^W \vee \mathcal{F}_T^B)$$

can be computed as a Black-Scholes price with volatility v_t. That is

$$e^{-r(T-t)} E(h(X_T)_+ | \mathcal{F}_t^W \vee \mathcal{F}_T^B) = BS(t, X_t, v_t).$$

This, jointly with (6.2.1), gives us that the well-known Hull and White formula (see Hull and White (1987))

$$V_t = E_t(BS(t, X_t, v_t)). \tag{6.2.2}$$

Remark 6.2.1 *In the correlated case, we can see that X_T follows a normal distribution, but now with mean*

$$\left(r - \frac{v_t^2}{2}\right)(T-t) + \rho \int_t^T \sigma_s dW_s$$

and variance equal to $(1-\rho^2)v_t^2$. This leads to the following extension of the Hull and White formula (see Romano and Touzi (1997) and Willard (1997))

$$V_t = E_t[BS(t, X_t \xi_t, \sqrt{1-\rho^2} v_t)], \tag{6.2.3}$$

where $\xi_t := \rho \int_t^T \sigma_s dW_s - \frac{1}{2}\rho^2 v_t^2 (T-t)$. Equation (6.2.3) is of practical use in Monte Carlo simulations, as only one Brownian motion has to be simulated.

6.2.1.2 The Hull and White formula from classical Itô's formula

The Hull and White formula can also be proved via Itô's formula. Consider again the uncorrelated case $\rho = 0$. As $V_T = BS(T, X_T, v_T)$ we can write

$$\begin{aligned} V_t &= e^{-r(T-t)} E_t(V_T) \\ &= e^{-r(T-t)} E_t(BS(T, X_T, v_T)). \\ &= e^{rt} E_t(e^{-rT} BS(T, X_T, v_T)) \\ &= e^{rt} E_t[e^{-rT} BS(T, X_T, v_T) - e^{-rt} BS(t, X_t, v_t)] \\ &+ E_t(BS(t, X_t, v_t)) \\ &=: T_1 + T_2. \end{aligned} \tag{6.2.4}$$

Notice that T_2 is nothing more than the Hull and White term. Then, in order to prove (6.2.2), it suffices to prove that $T_1 = 0$. Towards this end, we have to see that

$$e^{-rT}BS(T, X_T, v_T) - e^{-rt}BS(t, X_t, v_t)$$

is a zero-expectation term. Notice that this is the difference of the same process, evaluated at two different time moments $T > t$. Is it natural to apply the classical Itô's formula to study this difference? As v_t is independent of the Brownian motion driving X we can write

$$\begin{aligned}
&e^{-rT}BS(T, X_T, v_T)\\
&= e^{-rt}BS(t, X_t, v_t)\\
&-r\int_t^T e^{-rs}BS(s, X_s, v_s)ds + \int_t^T e^{-rs}\frac{\partial BS}{\partial s}(s, X_s, v_s)ds\\
&+\int_t^T e^{-rs}\frac{\partial BS}{\partial \sigma}(s, X_s, v_s)dv_s + \int_t^T e^{-rs}\frac{\partial BS}{\partial x}(s, X_s, v_s)\left(r - \frac{1}{2}\sigma_s^2\right)ds\\
&+\int_t^T e^{-rs}\frac{\partial BS}{\partial x}(s, X_s, v_s)dB_s + \frac{1}{2}\int_t^T e^{-rs}\frac{\partial^2 BS}{\partial x^2}(s, X_s, v_s)ds\\
&= e^{-rt}BS(t, X_t, v_t)\\
&-r\int_t^T e^{-rs}BS(s, X_s, v_s)ds + \int_t^T e^{-rs}\frac{\partial BS}{\partial s}(s, X_s, v_s)ds\\
&+\int_t^T e^{-rs}\frac{\partial BS}{\partial \sigma}(s, X_s, v_s)\frac{1}{2v_t(T-t)}(v_s^2 - \sigma_s^2)ds\\
&+\int_t^T e^{-rs}\frac{\partial BS}{\partial x}(s, X_s, v_s)\left(r - \frac{1}{2}\sigma_s^2\right)ds + \int_t^T e^{-rs}\frac{\partial BS}{\partial x}(s, X_s, v_s)dB_s\\
&+\frac{1}{2}\int_t^T e^{-rs}\frac{\partial^2 BS}{\partial x^2}(s, X_s, v_s)\sigma_s^2 ds. \qquad (6.2.5)
\end{aligned}$$

Now, due to the *Delta-Gamma-Vega* relationship we have that

$$\frac{\partial BS}{\partial \sigma}(s, X_s, v_s)\frac{1}{v_t(T-t)} = \left(\frac{\partial^2}{\partial x^2} - \frac{\partial}{\partial x}\right)BS(s, X_s, v_s), \qquad (6.2.6)$$

and then we can write

$$
\begin{aligned}
&e^{-rT}BS(T, X_T, v_T)\\
&= e^{-rt}BS(t, X_t, v_t)\\
&-r\int_t^T e^{-rt}BS(s, X_s, v_s)ds + \int_t^T e^{-rs}\frac{\partial BS}{\partial s}(s, X_s, v_s)ds\\
&+r\int_t^T e^{-rs}\frac{\partial BS}{\partial x}(s, X_s, v_s)ds + \frac{1}{2}\int_t^T e^{-rs}\left(\frac{\partial^2}{\partial x^2} - \frac{\partial}{\partial x}\right)\\
&\quad \times BS(s, X_s, v_s)v_s^2 ds\\
&+\int_t^T e^{-rs}\frac{\partial BS}{\partial x}(s, X_s, v_s)dB_s\\
&= e^{-rt}BS(t, X_t, v_t)\\
&+\int_t^T e^{-rt}\mathcal{L}_{BS}(v_s)(s, X_s, v_s)ds + \int_t^T e^{-rs}\frac{\partial BS}{\partial x}(s, X_s, v_s)dB_s.
\end{aligned}
\tag{6.2.7}
$$

The second term in the RHS is zero since $\mathcal{L}_{BS}(v_s)(s, X_s, v_s) = 0$. The last one has zero expectation. Then the term T_1 in (6.2.4) is equal to zero, and this proves the Hull and White formula (6.2.2).

Now we would like to apply the same arguments to the correlated case, but the problem is that now, classical Itô's formula cannot be applied, due to the fact that v_t is not adapted, nor independent of the Brownian motion driving X. This is why we need to use the anticipating Itô's formula in Proposition 4.3.1, in a similar way as in Theorem 4.3.5, as we see in the next section.

6.3 AN EXTENSION OF THE HULL AND WHITE FORMULA FROM THE ANTICIPATING ITÔ'S FORMULA

In this section, we apply the anticipating Itô's formula to obtain an extension of the Hull and White formula different from (6.2.3), following the ideas in Alòs (2006) and Alòs, León, and Vives (2007). Towards this end, we need the following bound for the Greeks, inspired in Lemma 5.2 in Fouque, Papanicolau, Sircar, and Solna (2003).

Lemma 6.3.1 *Consider the model (6.1.1). Let* $0 \leq t \leq s \leq T$, $\rho \in (-1, 1)$ *and* $\mathcal{G}_t := \mathcal{F}_t \vee \mathcal{F}_T^W$. *Then for every* $n \geq 0$, *there exists* $C =$

$C(n,\rho)$ such that

$$\left| E\left(\frac{\partial^n G}{\partial x^n}(s, X_s, v_s) \middle| \mathcal{G}_t \right) \right| \leq C \left(\int_t^T \sigma_s^2 ds \right)^{-\frac{1}{2}(n+1)},$$

where $G(t,x,\sigma) := \left(\frac{\partial^2}{\partial x^2} - \frac{\partial}{\partial x}\right) BS(t,x,\sigma)$.

Proof. A direct computation gives us that

$$\frac{\partial BS}{\partial x}(s,x,\sigma) = e^x N(d_+) + e^x \frac{N'(d_+)}{\sigma\sqrt{T-t}} - Ke^{-r(T-t)} \frac{N'(d_-)}{\sigma\sqrt{T-t}}.$$

Notice that $e^x N'(d_+) = Ke^{-r(T-t)} N'(d_-)$. Then the above equality reads as

$$\frac{\partial BS}{\partial x}(s,x,\sigma) = e^x N(d_+).$$

Then

$$\frac{\partial^2 BS}{\partial x^2}(s,x,\sigma) = e^x N(d_+) + Ke^{-r(T-t)} \frac{N'(d_-)}{\sigma\sqrt{T-t}}.$$

As a consequence

$$G(s, X_s, v_s) = Ke^{-r(T-s)} p\left(X_s - \mu, v_s\sqrt{T-s}\right),$$

where $p(x,\tau)$ denotes the centered Gaussian kernel with variance τ^2 and $\mu = \ln K - (r - v_s^2/2)(T-s)$. This allows us to write

$$E\left(\frac{\partial^n G}{\partial x^n}(s, X_s, v_s)|\mathcal{G}_t\right) = (-1)^n Ke^{-r(T-s)} \frac{\partial^n}{\partial \mu^n} E(p(X_s - \mu, v_s\sqrt{T-s})|\mathcal{G}_t). \tag{6.3.1}$$

Now, notice that the conditional expectation of X_s given $\mathcal{G}_t$ is a normal random variable with mean equal to

$$\phi = X_t + \int_t^s \left(r - \sigma_\theta^2/2\right) d\theta + \rho \int_t^s \sigma_\theta dW_\theta$$

and variance equal to $(1-\rho^2)\int_t^s \sigma_\theta^2 d\theta$. Then we get

$$
\begin{aligned}
& E\left(p\left(X_s-\mu, v_s\sqrt{T-s}\right)\Big|\mathcal{G}_t\right) \\
= & \int_R p\left(y-\mu, v_s\sqrt{T-s}\right) p\left(y-\phi, \sqrt{(1-\rho^2)\int_t^s \sigma_\theta^2 d\theta}\right) dy \\
= & \; p\left(\phi-\mu, \sqrt{\int_s^T \sigma_s^2 ds + (1-\rho^2)\int_t^s \sigma_\theta^2 d\theta}\right) \\
= & \; p\left(\phi-\mu, \sqrt{(1-\rho^2)\int_t^T \sigma_\theta^2 d\theta + \rho^2\int_s^T \sigma_\theta^2 d\theta}\right).
\end{aligned}
$$

Putting this result in (6.3.1), we have

$$
\begin{aligned}
& E\left(\frac{\partial^n G}{\partial x^n}(s, X_s, v_s)|\mathcal{G}_t\right) \\
& = (-1)^n K e^{-r(T-s)} \frac{\partial^n p}{\partial \mu^n}\left(\phi-\mu, \sqrt{(1-\rho^2)\int_t^T \sigma_\theta^2 d\theta + \rho^2\int_s^T \sigma_\theta^2 d\theta}\right).
\end{aligned}
$$

Now, taking into account that, for every positive constants c, d, the function $x^c e^{-dx^2}$ is bounded, we can write

$$
\begin{aligned}
& \left|\frac{\partial^n p}{\partial \mu^n}\left(\phi-\mu, \sqrt{(1-\rho^2)\int_t^T \sigma_s^2 ds + \rho^2\int_s^T \sigma_\theta^2 d\theta}\right)\right| \\
& \leq C\left(\left(1-\rho^2\right)\int_t^T \sigma_s^2 ds + \rho^2\int_s^T \sigma_\theta^2 d\theta\right)^{-\frac{1}{2}(n+1)} \\
& \leq C\left(\int_t^T \sigma_s^2 ds\right)^{-\frac{1}{2}(n+1)},
\end{aligned}
$$

and now the proof is complete. ■

Now we are in a position to prove the main result of this section (see Alòs, León, and Vives (2007)), where we decompose option prices as the Hull and White term plus the contribution of the correlation.

Theorem 6.3.2 *Assume the model (6.1.1) holds with $\rho \in (-1,1)$ and $\sigma \in \mathbb{L}_W^{1,2}$. Then it follows that*

$$
V_t = E_t(BS(t, X_t, v_t)) + \frac{\rho}{2} E_t\left(\int_t^T e^{-r(s-t)} H(s, X_s, v_s)\Phi_s ds\right), \quad (6.3.2)
$$

where $H := \frac{\partial G}{\partial x}$ *and* $\Phi_s := \sigma_s \int_s^T D_s^W \sigma_u^2 du$.

Proof. Apart from technical points, that come from the singularities of BS, this proof follows the same scheme as the proof of Theorem 4.3.5, so we only sketch it. First notice that the second term in the right-hand side of (6.3.2) is well defined. In fact, Lemma 6.3.1 gives us that

$$E\left(\frac{\partial G}{\partial x}(s, X_s, v_s)|\mathcal{G}_t\right) \leq C\left(\int_t^T \sigma_s^2 ds\right)^{-1}, \tag{6.3.3}$$

for some positive constant C. Also, by Hölder's inequality we have

$$\begin{aligned}\left|\int_t^T \Phi_s ds\right| &\leq 2\int_t^T \sigma_s \int_s^T \sigma_\theta \left|D_s^W \sigma_\theta\right| d\theta ds \\ &\leq 2\left(\int_t^T \sigma_\theta^2 d\theta\right)\left(\int_t^T\int_t^T \left(D_r^W \sigma_\theta\right)^2 dr d\theta\right)^{\frac{1}{2}}. \quad (6.3.4)\end{aligned}$$

Then, (6.3.3) and (6.3.4) imply that the integral in (6.3.2) is well defined. Now, notice that

$$BS(T, X_T, v_T) = V_T.$$

Then, from (6.1.2) we have

$$e^{-rt}V_t = E_t(e^{-rT}BS(T, X_T, v_T)).$$

Now, applying Proposition 4.3.1 to the process $e^{-rt}BS(t, X_t, v_t)$ and grouping terms we get

$$\begin{aligned}&e^{-rT}BS(T, X_T, v_T) \\ &= e^{-rt}BS(t, X_t, v_t^{H-\frac{1}{2}}) + \int_t^T e^{-rs}\mathcal{L}_{BS}(\sigma_s)BS(s, X_s, v_s)ds \\ &-\frac{1}{2}\int_t^T e^{-rs}\frac{\partial BS}{\partial \sigma}(s, X_s, v_s)\frac{(\sigma_s^2 - (v_s)^2)}{v_s(T-s)}ds \\ &+\int_t^T e^{-rs}\frac{\partial BS}{\partial x}(s, X_s, v_s)\sigma_s(\rho dW_s + \sqrt{1-\rho^2}dB_s) \\ &+\frac{\rho}{2}\int_t^T e^{-rs}\frac{\partial^2 BS}{\partial\sigma\partial x}(s, X_s, v_s)\frac{1}{v_s(T-s)}\Phi_s ds.\end{aligned}$$

The stochastic integral in the above equation is a Skorohod integral, and then its expectation is zero. This allows us to write

$$\begin{aligned}
&e^{-rT}E_t(BS(T,X_T,v_T)) \\
&= E_t\left\{e^{-rt}BS(t,X_t,v_t) + \int_t^T e^{-rs}\mathcal{L}_{BS}(\sigma_s)BS(s,X_s,v_s)ds\right. \\
&-\frac{1}{2}\int_t^T e^{-rs}\frac{\partial BS}{\partial\sigma}(s,X_s,v_s)\frac{(\sigma_s^2-(v_s)^2)}{v_s(T-s)}ds \\
&\left.+\frac{\rho}{2}\int_t^T e^{-rs}\frac{\partial^2 BS}{\partial\sigma\partial x}(s,X_s,v_s)\frac{1}{v_s(T-s)}\Phi_s ds\right\}. \qquad (6.3.5)
\end{aligned}$$

Notice that

$$\begin{aligned}
&\mathcal{L}_{BS}(\sigma_s)BS(s,X_s,v_s) \\
&= \mathcal{L}_{BS}(v_s)BS(s,X_s,v_s) + \frac{1}{2}(\sigma_s^2-(v_s)^2)\left(\frac{\partial^2}{\partial x^2}-\frac{\partial}{\partial x}\right)BS(s,X_s,v_s) \\
&= \frac{1}{2}(\sigma_s^2-(v_s)^2)\left(\frac{\partial^2}{\partial x^2}-\frac{\partial}{\partial x}\right)BS(s,X_s,v_s). \qquad (6.3.6)
\end{aligned}$$

Moreover, because of the Delta-Gamma-Vega relationship (6.2.6),

$$\frac{\partial BS}{\partial\sigma}(s,X_s,v_s)\frac{1}{v_s(T-s)} = \left(\frac{\partial^2}{\partial x^2}-\frac{\partial}{\partial x}\right)(s,X_s,v_s).$$

Then, the second and the third terms in (6.3.5) cancel and we obtain that

$$\begin{aligned}
&e^{-rT}E_t(BS(T,X_T,v_T)) \\
&= E_t\left\{e^{-rt}BS(t,X_t,v_t)\right. \\
&\left.+\frac{\rho}{2}\int_t^T e^{-rs}\frac{\partial^2 BS}{\partial\sigma\partial x}(s,X_s,v_s)\frac{1}{v_s(T-s)}\Phi_s ds\right\}. \qquad (6.3.7)
\end{aligned}$$

Then, applying again the Delta-Gamma-Vega relationship (6.2.6), we have that

$$\frac{\partial^2 BS}{\partial\sigma\partial x}(s,X_s,v_s)\frac{1}{v_s(T-s)} = H(s,X_s,v_s).$$

This allows us to write

$$\begin{aligned}
&e^{-rT}E_t(BS(T,X_T,v_T)) \\
&= E_t\left\{e^{-rt}BS(t,X_t,v_t) + \frac{\rho}{2}\int_t^T e^{-rs}H(s,X_s,v_s)\Phi_s ds\right\},
\end{aligned}$$

and this allows us to complete the proof. ■

Remark 6.3.3 *Theorem 6.3.2 decomposes an option price into two terms. The first one is exactly the option price in the case $\rho = 0$. The effect of correlation is given by the second term*

$$\frac{\rho}{2} E_t \left(\int_t^T e^{-r(s-t)} H(s, X_s, v_s) \Phi_s ds \right).$$

Remark 6.3.4 *This decomposition formula does not need the volatility to be a semimartingale process nor to be Markovian. It can be applied to classical volatility models as Heston and SABR, but also to non-Markovian volatilities based on the fractional Brownian motion, as we see in the following section.*

Example 6.3.5 (The SABR model) *Assume the SABR volatility model as in 2.3.2. Then, the Malliavin derivative with respect to W (see Section 3.3.1) is given by*

$$D_r^W \sigma_t = \alpha \sigma_t \mathbf{1}_{[0,t]}(r).$$

Then, the chain rule gives us that

$$D_r^W \sigma_t^2 = 2\alpha \sigma_t^2 \mathbf{1}_{[0,t]}(r)$$

and 6.3.2 reads as

$$\begin{aligned} V_t &= E_t(BS(t, X_t, v_t)) \\ &\quad + \alpha \rho E_t \left(\int_t^T e^{-r(s-t)} H(s, X_s, v_s) \sigma_s \left(\int_s^T \sigma_u^2 du \right) ds \right). \end{aligned}$$

Example 6.3.6 (The rough Bergomi model) *Consider the rough Bergomi model introduced in Section 5.6.2. It was proved that*

$$D_s^W \sigma_r^2 = \nu \sqrt{2H} \sigma_r^2 (r-s)^{H-\frac{1}{2}}. \tag{6.3.8}$$

Then, 6.3.2 can be written as

$$\begin{aligned} V_t &= E_t(BS(t, X_t, v_t)) \\ &+ \frac{\nu \sqrt{2H} \rho}{2} E_t \left(\int_t^T e^{-r(s-t)} H(s, X_s, v_s) \sigma_s \right. \\ &\left. \left(\int_s^T \sigma_u^2 (u-s)^{H-\frac{1}{2}} du \right) ds \right). \end{aligned} \tag{6.3.9}$$

Notice that this expression is similar to the one obtained for the SABR model, except for the kernel $(u-s)^{H-\frac{1}{2}}$*.This kernel modifies the behaviour of the short-time implied volatility, as we will see in the following chapters.*

Remark 6.3.7 (Local volatilities) *The hypothesis* $\rho \in (-1,1)$ *in Theorem 6.3.2 has been taken for the sake of simplicity. This condition allows us to apply Lemma 6.3.1, and then the only required hypothesis for* σ *is to be in* $\mathbb{L}_W^{1,2}$*. The same result can be proved for* $\rho \in [-1,1]$ *provided the volatility process* σ_t *satisfies other integrability conditions. Then, we can apply this result for local volatility models under some adequate regularity hypotheses. Given a local volatility model of the form*

$$dS_t = rS_t + \sigma(t, S_t)dW_t,$$

and assuming that σ *is a continuous differentiable function, we can write*

$$D_s^W \sigma_t^2 = 2\sigma(t, S_t)\frac{\partial \sigma}{\partial x}(t, S_t)D_s^W S_t. \tag{6.3.10}$$

Now, $D_s^W S_t$ *can be computed by the method in Detemple, García, and Rindisbacher (2005) (see Section 3.2.2.2). Let* $F(t,x)$ *be a function such that* $\frac{\partial F}{\partial x} = \frac{1}{\sigma(t,x)}$ *and denote*

$$f(t,x) := \left(\frac{\partial F}{\partial t}(t, F^{-1}(t,x)) + \frac{r}{\sigma(t, F^{-1}(t,x))} - \frac{1}{2}\frac{\partial \sigma}{\partial x}(t, F^{-1}(t,x))\right).$$

Then

$$D_s^W S_t = \sigma(t, S_t) \exp\left(\int_s^t \frac{\partial f}{\partial z}(u, F_t(u, S_u))du\right). \tag{6.3.11}$$

Now, 6.3.2 allows us to write

$$\begin{aligned}
V_t &= E_t(BS(t, X_t, v_t)) \\
&\quad + E_t\left(\int_t^T e^{-r(s-t)} H(s, X_s, v_s)\sigma(s, S_s)\right. \\
&\quad \left.\times\left(\int_s^T \left(\sigma(u, S_u)\frac{\partial \sigma}{\partial x}(u, S_u) \exp\left(\int_s^u \frac{\partial f}{\partial z}(\theta, F_u(\theta, S_\theta))\right)\right)^2 du\right) ds\right).
\end{aligned}$$

Example 6.3.8 *Consider the CEV model for asset prices given in Example 2.2.6*

$$dS_t = rS_t dt + \sigma S_t^{\gamma} dW_t. \tag{6.3.12}$$

This is a local volatility model where $\sigma(t, S_t) = \sigma S_t^{\gamma-1}$. *Then, as computed in Example 3.2.4*

$$D_u^W S_t = \sigma S_t^{\gamma} \mathbf{1}_{[0,t]}(u) \exp\left((1-\gamma)\int_u^t \left(r + \frac{\gamma\sigma^2}{2} S_u^{2\gamma-2}\right) du\right),$$

and then, for $\gamma \neq 1$,

$$\begin{aligned} V_t &= E_t(BS(t, X_t, v_t)) \\ &\quad + \sigma^3 E_t \left(\int_t^T e^{-r(s-t)} H(s, X_s, v_s) S_s^{\gamma-1} \right. \\ &\quad \times \left. \left(\int_s^T \left(S_u^{3\gamma-3} \exp\left((1-\gamma) \int_s^u \left(r + \frac{\gamma\sigma^2}{2} S_\theta^{2\gamma-2} \right) d\theta \right) \right) du \right) ds \right). \end{aligned}$$

6.4 DECOMPOSITION FORMULAS FOR IMPLIED VOLATILITIES

Now let us consider the payoff of a call option $h(S_T) = (S_T - K)_+$ and denote by $BS(t, x, k, \sigma)$ the classical Black-Scholes price for a call option with time to maturity $T - t$, log-price x, log-strike k, and volatility σ. Then the implied volatility $I_t(k)$ is defined by

$$V_t(k) = BS(t, X_t, k, I_t(k)).$$

We also denote $BS^{-1}(t, x, k, a)$ the inverse of the Black-Scholes function in the volatility parameter. That is, $BS^{-1}(t, x, k, BS(t, x, k, \sigma)) = \sigma$. Notice that $I_t(k) = BS^{-1}(t, x, k, V_t(k))$.

Our objective in this section is to use the above results for option prices to get an adequate decomposition formula for the implied volatility. Our main result is the following theorem.

Theorem 6.4.1 *Consider the model (6.1.1) with* $\rho \in (-1, 1)$ *and* $\sigma \in \mathbb{L}_W^{1,2}$. *Then*

$$\begin{aligned} I_t(k) &= I_t^0(t, k) \qquad (6.4.1) \\ &\quad + \frac{\rho}{2} E_t \left[\int_t^T e^{-r(s-t)} \left(BS^{-1}\right)'(k, \Phi_s)\, H(s, X_s, k, v_s) \Gamma_s ds \right], \end{aligned}$$

provided the last term in (6.4.1) finite, and where $I_t^0(k)$ denotes the implied volatility in the uncorrelated case $\rho = 0$, and

$$\Gamma_s := E_t\left[BS(t, X_t, k, v_t)\right] + \frac{\rho}{2} E_t\left[\int_t^s e^{-r(u-t)} H(u, X_u, k, v_u)\Phi_u du\right].$$

Proof. The decomposition formula (6.3.2) allows us to write

$$V_t = E_t\left[BS(t, X_t, k, v_t)\right] + A_t^T,$$

where

$$A_t^T = \frac{\rho}{2} E_t\left[\int_t^T e^{-r(s-t)} H(s, X_s, k, v_s)\Phi_s ds\right].$$

Thus,

$$I_t(k_t) = BS^{-1}(k, V_t(k)) = E_t\left[BS^{-1}(k, V_t^0 + A_t^T)\right],$$

where $V_t^0 := E_t\left[BS(t, X_t, k, v_t)\right]$ denotes the option price in the uncorrelated case $\rho = 0$. Then, it follows that

$$\begin{aligned} & E_t\left[BS^{-1}(k, V_t^0 + A_t^T) - BS^{-1}(k, V_t^0)\right] \\ = & \frac{\rho}{2} E_t\left[\int_t^T e^{-r(s-t)} \left(BS^{-1}\right)'(k, \Gamma_s) H(s, X_s, k, v_s)\Phi_s ds\right], \end{aligned}$$

which proves (6.4.2). ■

Theorem 6.4.1 gives us a decomposition of the implied volatility as the sum of two terms: the implied volatility in the uncorrelated case, plus the effect of correlation. As this expression is exact, it gives us an interesting tool to understand the role of correlation on the implied volatility.

6.5 THE ATM SHORT-TIME LEVEL OF THE IMPLIED VOLATILITY

Now we apply the above decomposition result for the implied volatility to study the difference between the volatility swap and the ATMI. This representation of the ATMI allows to break down this analysis into two steps: the first one devoted to the study of the uncorrelated case, and the second one extending the results to the correlated case. In the sequel, we denote by $k_t^* = X_t + r(T-t)$ the ATM log-strike.

6.5.1 The uncorrelated case

Let us assume $\rho = 0$ and consider the following hypotheses:

(H1) There exist two positive constants a, b such that $a \leq \sigma_t \leq b$, for all $t \in [0, T]$.

(H2) $\sigma^2 \in \mathbb{L}_W^{1,2}$.

We notice that these hypotheses have been chosen for the sake of simplicity. The results in this section still hold for some other processes with adequate integrability conditions (see Alòs and Shiraya (2019)).

The key tool in our analysis is the following relationship between the ATMI and the volatility swap fair strike, which we deduce from a direct application of the Clark-Ocone-Haussman formula and the classical Itô formula.

Proposition 6.5.1 *Consider the model (6.1.1) with $\rho = 0$ and assume that hypotheses (H1) and (H2) hold. Then, the at-the-money implied volatility admits the representation*

$$I_t(k_t^*) = E_t[v_t] - \frac{1}{32(T-t)}$$

$$\times E_t\left[\int_t^T \frac{\Psi_r}{(N'(d_+(k_t^*, \Psi_r)))^2}\left(E_r\left[N'(d_+(k_t^*, v_t))\frac{\int_r^T D_r^W \sigma_s^2 ds}{v_t}\right]\right)^2 dr\right],$$

where

$$\Psi_r := BS^{-1}(k_t^*, \Lambda_r),$$

with

$$\Lambda_r := E_r\left[BS\left(t, X_t, k_t^*, v_t\right)\right].$$

Proof. In the uncorrelated case, the Hull and White formula allows us to write

$$V_t = E_t[BS(t, X_t, k_t^*, v_t)].$$

Then, the implied volatility satisfies that

$$\begin{aligned} I_t(k_t^*) &= BS^{-1}(k_t^*, V_t) \\ &= E_t[BS^{-1}(k_t^*, E_t[BS(t, X_t, k_t^*, v_t)])] \\ &= E_t[BS^{-1}(k_t^*, \Lambda_t)]. \end{aligned}$$

On the other hand, we can write

$$\begin{aligned} E_t[v_t] &= E_t[BS^{-1}(k_t^*, E_t[BS(t, X_t, k_t^*, v_t)]) \\ &= E_t[BS^{-1}(k_t^*, BS(t, X_t, k_t^*, v_t))] \\ &= E_t[BS^{-1}(k_t^*, \Lambda_T)]. \end{aligned}$$

That is, both the ATMI and the volatility swap are the expectation of a function of the same martingale Λ, evaluated at different moments $t < T$. Now, from the martingale representation theorem, we know that

$$\Lambda_T = \Lambda_t + \int_t^T U_s dW_s,$$

for some adapted process U. Then, a direct application of Itô's formula to the process Λ and the function $BS^{-1}(k, \cdot)$ gives us that

$$\begin{aligned} &E_t[BS^{-1}(k_t^*, \Lambda_t) - BS^{-1}(k_t^*, \Lambda_T)] \\ &= -E_t\left[\int_t^T (BS^{-1})'(k_t^*, \Lambda_r) U_r dW_r + \frac{1}{2}\int_t^T (BS^{-1})''(k_t^*, \Lambda_r) U_r^2 dr\right], \end{aligned} \tag{6.5.1}$$

where $(BS^{-1})'$ and $(BS^{-1})''$ denote, respectively, the first and second derivatives of BS^{-1} with respect to Λ. Then, taking expectations,

$$E_t[BS^{-1}(k_t^*, \Lambda_t) - BS^{-1}(k_t^*, \Lambda_T)] = -\frac{1}{2}E_t\left[\int_t^T (BS^{-1})''(k_t^*, \Lambda_r) U_r^2 dr\right].$$

Now, as

$$(BS^{-1})''(k_t^*, \Lambda_r) = \frac{\Psi_r}{4(\exp(X_t)N'(d_+(k_t^*, \Psi_r)))^2}, \tag{6.5.2}$$

and

$$\begin{aligned} U_r &= E_r[D_r^W(BS(t, X_t, k_t^*, v_t))] \\ &= E_r\left[\exp(X_t)N'(d_+(k_t^*, v_t))\frac{\int_r^T D_r^W \sigma_s^2 ds}{2\sqrt{T-t}v_t}\right], \end{aligned}$$

the proof is complete. ■

Remark 6.5.2 (Volatility and variance swaps) *As a consequence of the above result, it follows that, in the uncorrelated case, $E_t[v_t] \geq I_t(k_t^*)$. On the other hand, because of Jensen's inequality,*

$$E_t[v_t] \leq \sqrt{\int_t^T E_t(\sigma_s^2) ds}.$$

Then

$$\sqrt{\frac{1}{T-t} \int_t^T \sigma_s^2 ds} \geq E_t[v_t] \geq I_t(k_t^*),$$

and then the volatility swap is (in the uncorrelated case) a better approximation of the ATMI than the square root of a variance swap. Notice also that, from (4.1.22)

$$E_t \sqrt{\int_t^T \sigma_s^2 ds} = \sqrt{E_t \int_t^T \sigma_s^2 ds} - \frac{1}{8} E \int_t^T \frac{\left(E_r \int_r^T D_r^W \sigma_s^2 ds\right)^2 ds}{\left(E_r \int_t^T \sigma_s^2 ds\right)^{\frac{3}{2}}},$$

from where it follows that (see Alòs and Muguruza (2021))

$$\lim_{T \to t} \frac{E\left[\frac{1}{T-t}\sqrt{\int_t^T \sigma_s^2 ds}\right] - \sqrt{\frac{1}{T-t} \int_t^T E_t(\sigma_s^2) ds}}{(T-t)^{2H}}$$
$$= -\frac{1}{8\sigma_t^3 (T-t)^{2H+2}} \lim_{T \to t} E \int_t^T \left(E_r \int_r^T D_r^W \sigma_s^2 ds\right)^2 ds.$$

Then, the difference between the volatility swap and the variance swap fair prices is $O(T-t)^{2H}$. As an example, consider a rough Bergomi model as in Section 5.6.2

$$\sigma_t^2 = \sigma_0^2 \exp\left(\nu \sqrt{2H} W_t^H - \frac{1}{2} \nu^2 t^{2H}\right), r \in [0, T], \tag{6.5.3}$$

where $\sigma_0^2 = 0.05$, $\nu = 0.5$, and $H = 0.3$. In Figure 6.1 we can see the simulated difference between the volatility swap and the square root of the variance swap in this case.

Proposition 6.5.1 gives us an exact expression of the difference between the volatility swap and the ATMI. The next step in our analysis is to use this expression to study the rate of convergence of the difference between these two quantities. To prove our limit results, we need the following hypotheses.

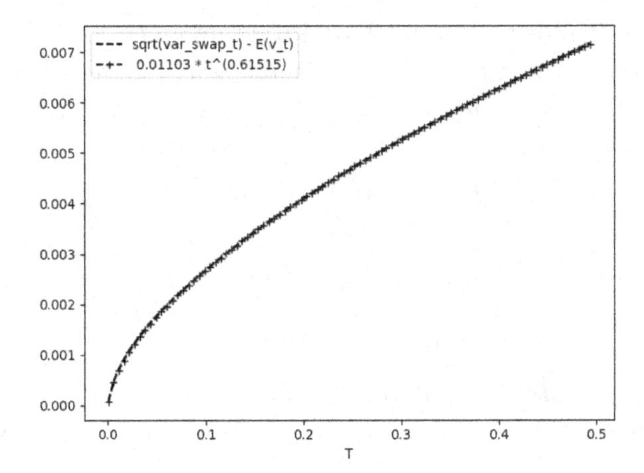

Figure 6.1 Difference between the square root of the variance swap and the volatility swap in the rBergomi model, with $\sigma_0^2 = 0.05$, $\nu = 0.5$, and $H = 0.3$.

(H3) Hypothesis (H2) holds, and there exist two constants $H \in (0,1)$ and $C > 0$ such that, for any $0 < r < s < T$,

$$E_r[D_r^W \sigma_s^2] \leq C\,(s-r)^{H-\frac{1}{2}}.$$

(H4) Hypotheses (H2) and (H3) hold and the term

$$\frac{1}{(T-t)^{2+2H}} E_t \left[\int_t^T \left(E_r \int_r^T D_r^W \sigma_s^2 ds \right)^2 dr \right]$$

has a finite limit as $T \to t$.

Hypotheses (H3) and (H4) are regularity conditions on the volatility process in the Malliavin calculus sense. We notice that these conditions can be satisfied for both processes based on the Brownian motion and processes based on fractional noises. Then, our analysis is not restricted to diffusion volatilities.

The following theorem gives the order of convergence of the difference between the ATMI and the volatility swap.

Theorem 6.5.3 *Consider the model* (4.3.7) *with* $\rho = 0$ *and assume that hypotheses (H1), (H2), (H3), and (H4) hold. Then*

$$\lim_{T\to t}\frac{I_t(k_t^*)-E_t[v_t]}{(T-t)^{1+2H}}$$
$$=-\frac{1}{32\sigma_t}\lim_{T\to t}\frac{1}{(T-t)^{2+2H}}E_t\left[\int_t^T\left(E_r\int_r^T D_r^W\sigma_s^2 ds\right)^2 dr\right].$$

Proof. Proposition 6.5.1 gives us that

$$\begin{aligned}
&I_t(k_t^*)\\
&\quad= E_t[v_t]-\frac{1}{32(T-t)}\\
&\quad\times E_t\left[\int_t^T\frac{\Psi_r}{(N'(d_+(k_t^*,\Psi_r)))^2}\left(E_r\left[N'(d_+(k_t^*,v_t))\frac{\int_r^T D_r^W\sigma_s^2 ds}{v_t}\right]\right)^2 dr\right]\\
&\quad= -\frac{1}{32}\lim_{T\to t}\frac{1}{(T-t)^{2+2H}}E_t\left[\int_t^T\Psi_r\right.\\
&\quad\left.\times\left(E_r\left[\exp\left(\frac{1}{8}(\Psi_r^2-v_t^2)(T-t)\right)\frac{\int_r^T D_r^W\sigma_s^2 ds}{v_t}\right]\right)^2 dr\right]. \qquad (6.5.4)
\end{aligned}$$

Now, similar arguments as in the proof of Proposition 6.5.1 give us that Ψ_r might be expanded as

$$\begin{aligned}
\Psi_r &= E_r[\Psi_r]\\
&= E_r\left[BS^{-1}(X_t,\Lambda_r)\right]\\
&= E_r\left[BS^{-1}(X_t,\Lambda_T-\int_r^T U_s dW_s)\right]\\
&= E_r[\Psi_T]-\frac{1}{2}E_r\left[\int_r^T(BS^{-1})''(k_t^*,\Lambda_\theta)U_\theta^2 d\theta\right]\\
&= E_r[v_t]-\frac{1}{2}E_r\left[\int_r^T(BS^{-1})''(k_t^*,\Lambda_\theta)U_\theta^2 d\theta\right].
\end{aligned}$$

This implies that $\exp\left(\frac{1}{8}(\Psi_r^2-v_t^2)(T-t)\right)$ tends to 1 as $r\to t$. Then, taking limits in (6.5.4) we get

$$\begin{aligned}
&\lim_{T\to t}\frac{I_t(k_t^*)-E_t[v_t]}{(T-t)^{1+2H}}\\
&\quad= -\frac{1}{32\sigma_t}\lim_{T\to t}\frac{1}{(T-t)^{2+2H}}E_t\left[\int_t^T\left(E_r\int_r^T D_r^W\sigma_s^2 ds\right)^2 dr\right],
\end{aligned}$$

and now the proof is complete. ■

Example 6.5.4 *Assume a fractional volatility model of the form* $\sigma_t = f(B_t^H)$*, where* $f \in C_b^1$ *and* B_t^H *is a RLfBm motion with Hurst parameter* H*. Then, (H1)–(H4) hold, and* $I_t(k_t^*) - E_t[v_t] = O((T-t)^{1+2H})$.

6.5.2 The correlated case

Now our objective is to study the difference between the ATMI and the volatility swap in the correlated case. Towards this end, we consider the following hypotheses:

(H2') $\sigma \in \mathbb{L}_W^{3,2}$.

(H3') Hypothesis (H2') holds and there exist two constants $H \in (0,1)$ and $C > 0$ such that, for any $0 < r < s < T$,

$$E_r[D_r^W \sigma_s^2] \leq C(s-r)^{H-\frac{1}{2}},$$

$$E_r[D_\theta^W D_r^W \sigma_s^2] \leq C(s-r)^{H-\frac{1}{2}}(s-\theta)^{H-\frac{1}{2}},$$

and

$$E_r[D_u^W D_\theta^W D_r^W \sigma_s^2] \leq C(s-r)^{H-\frac{1}{2}}(s-\theta)^{H-\frac{1}{2}}(s-u)^{H-\frac{1}{2}}.$$

(H5) Hypotheses (H1), (H2'), (H3'), and (H4) hold and the terms

$$\frac{1}{(T-t)^{\frac{3}{2}+H}} E_t\left[\int_t^T \int_s^T D_s^W \sigma_r^2 drds\right],$$

$$\frac{1}{(T-t)^{3+2H}} E_t\left[\left(\int_t^T \int_s^T D_s^W \sigma_r^2 drds\right)^2\right],$$

$$\frac{1}{(T-t)^{2+2H}} E_t\left[\int_t^T \int_s^T D_s^W \sigma_r \int_r^T D_r^W \sigma_u^2 dudrds\right],$$

and

$$\frac{1}{(T-t)^{2+2H}} E_t\left[\int_t^T \int_s^T \int_r^T D_s^W D_r^W \sigma_u^2 dudrds\right]$$

have a finite limit as $T \to t$.

Theorem 6.5.3 and Proposition 6.4.1 allow us to prove the following result.

Theorem 6.5.5 *Consider the model (4.3.7) and assume that hypotheses (H1), (H2'), (H3'), (H4), and (H5) hold for some $H \in (0,1)$. Then*

- *If $H - \frac{1}{2} < 0$*

$$
\begin{aligned}
\lim_{T\to t} \frac{I_t(k_t^*) - E_t[v_t]}{(T-t)^{2H}} \\
= \lim_{T\to t} \frac{3\rho^2}{8\sigma_t^3 (T-t)^{3+2H}} E_t\left[\left(\int_t^T \int_s^T D_s^W \sigma_r^2 dr ds\right)^2\right] \\
- \lim_{T\to t} \frac{\rho^2}{\sigma_t (T-t)^{2+2H}} E_t\left[\int_t^T \int_s^T D_s^W \sigma_r \int_r^T D_r^W \sigma_u du dr ds\right] \\
- \lim_{T\to t} \frac{\rho^2}{2\sigma_t (T-t)^{2+2H}} E_t\left[\int_t^T \int_s^T \int_r^T D_s^W D_r^W \sigma_u^2 du dr ds\right].
\end{aligned}
$$

- *If $H - \frac{1}{2} > 0$*

$$
\lim_{T\to t} \frac{I_t(k_t^*) - E_t[v_t]}{(T-t)^{\frac{1}{2}+H}} = \lim_{T\to t} \frac{\rho}{4(T-t)^{\frac{3}{2}+H}} E_t\left[\int_t^T \int_s^T D_s^W \sigma_r^2 dr ds\right].
$$

- *If $H - \frac{1}{2} = 0$*

$$
\begin{aligned}
\lim_{T\to t} \frac{I_t(k_t^*) - E_t[v_t]}{(T-t)} \\
= \lim_{T\to t} \frac{3\rho^2}{8\sigma_t^3 (T-t)^{4}} E_t\left[\left(\int_t^T \int_s^T D_s^W \sigma_r^2 dr ds\right)^2\right] \\
- \lim_{T\to t} \frac{\rho^2}{\sigma_t (T-t)^{3}} E_t\left[\int_t^T \int_s^T D_s^W \sigma_r \int_r^T D_r^W \sigma_u du dr ds\right] \\
- \lim_{T\to t} \frac{\rho^2}{2\sigma_t (T-t)^{3}} E_t\left[\int_t^T \int_s^T \int_r^T D_s^W D_r^W \sigma_u^2 du dr ds\right] \\
+ \lim_{T\to t} \frac{\rho}{4(T-t)^{2}} E_t\left[\int_t^T \int_s^T D_s^W \sigma_r^2 dr ds\right].
\end{aligned}
$$

Proof. The first step in our proof is to notice that Proposition 6.4.1 allows us to decompose this difference into two terms

$$
I_t(k_t^*) - E[v_t] = T_1 + T_2,
$$

where

$$\begin{aligned} T_1 &= I^0(t,T,X_t,k_t^*) - E[v_t], \\ T_2 &= \frac{\rho}{2} E_t \left[\int_t^T (BS^{-1})'(k_t^*, \Gamma_s) H(s, X_s, k_t^*, v_s) \Phi_s ds \right]. \end{aligned}$$

We know, from the previous section, that $T_1 = O((T-t)^{1+2H})$. Let us study the term T_2. Now, the idea is to expand T_2 around the term

$$\frac{\rho}{2} E_t[H(t, X_t, k_t^*, v_t) J_t],$$

where $J_t = \int_t^T (BS^{-1})'(k_t^*, \Gamma_u) \Phi_u du$. Towards this end, we follow an approach that we use several times in this book. It consists of applying again the anticipating Itô's formula (as in Remark 4.3.3) to the process that we consider 'the leading' term. In this case, to the process

$$A_s := H(s, X_s, k_t^*, v_s) J_s,$$

(notice that we have to consider two non-adapted processes, v and J). Then, taking into account that $A_T = 0$ we get

$$\begin{aligned} 0 = E_t \Bigg[& H(t, X_t, k_t^*, v_t) J_t \\ & + \int_t^T H(s, X_s, k_t^*, v_s) dJ_s & (6.5.5) \\ & + \int_t^T \frac{\partial^2 H}{\partial x \partial \sigma}(s, X_s, k_t^*, v_s) J_s \frac{\partial v}{\partial y}(D_s^W Y_s) \sigma_s ds \\ & + \int_t^T \frac{\partial H}{\partial x}(s, X_s, k_t^*, v_s)(D_s^W J_s) \sigma_s ds \\ & + \int_t^T \frac{\partial H}{\partial t}(s, X_s, k_t^*, v_s) J_s ds \\ & + \int_t^T \frac{\partial H}{\partial \sigma}(s, X_s, k_t^*, v_s) \frac{\partial v}{\partial t} J_s ds \\ & + \int_t^T \frac{\partial H}{\partial \sigma}(s, X_s, k_t^*, v_s) \frac{\partial v}{\partial y} J_s dY_s \\ & + \int_t^T \frac{\partial H}{\partial x}(s, X_s, k_t^*, v_s) J_s dX_s \\ & + \frac{1}{2} \int_t^T \frac{\partial^2 H}{\partial x^2}(s, X_s, k_t^*, v_s) J_s d\langle X \rangle_s \Bigg]. & (6.5.6) \end{aligned}$$

Even when this expression seems to be long and complex, it can be easily simplified using similar tricks as in the proof of the decomposition formula in Theorem 6.3.2. From one side, notice that $\frac{\partial H}{\partial t} - \frac{\sigma_s^2}{2}\frac{\partial H}{\partial x} + \frac{1}{2}\frac{\partial^2 H}{\partial x^2}\sigma_s^2 = \mathcal{L}_{BS}(\sigma_s)H$ and then the above equality can be written as

$$\begin{aligned} 0 &= E_t\Bigg[H(t, X_t, k_t^*, v_t)J_t \\ &\quad + \int_t^T H(s, X_s, k_t^*, v_s)dJ_s \\ &\quad + \int_t^T \frac{\partial^2 H}{\partial x \partial \sigma}(s, X_s, k_t^*, v_s)J_s\frac{\partial v}{\partial y}(D^- Y_s)\sigma_s ds \\ &\quad + \int_t^T \frac{\partial H}{\partial x}(s, X_s, k_t^*, v_s)(D^- J_s)\sigma_s ds \\ &\quad + \int_t^T \mathcal{L}_{BS}(\sigma_s)H(s, X_s, k_t^*, v_s)J_s ds \\ &\quad + \int_t^T \frac{\partial H}{\partial \sigma}(s, X_s, k_t^*, v_s)\frac{\partial v}{\partial t}J_s ds \\ &\quad + \int_t^T \frac{\partial H}{\partial \sigma}(s, X_s, k_t^*, v_s)\frac{\partial v}{\partial y}J_s dY_s\Bigg]. \end{aligned} \tag{6.5.7}$$

Now, notice that H also satisfies the Black-Scholes equation. Then we can write

$$\begin{aligned} &\mathcal{L}_H(\sigma_s)H(s, X_s, k_t^*, v_s) \\ &= \mathcal{L}_{BS}(v_s)H(s, X_s, k_t^*, v_s) + \frac{1}{2}(\sigma_s^2 - (v_s)^2)\left(\frac{\partial^2}{\partial x^2} - \frac{\partial}{\partial x}\right)H(s, X_s, k_t^*, v_s) \\ &= \frac{1}{2}(\sigma_s^2 - (v_s)^2)\left(\frac{\partial^2}{\partial x^2} - \frac{\partial}{\partial x}\right)H(s, X_s, k_t^*, v_s). \end{aligned} \tag{6.5.8}$$

Moreover,

$$\frac{\partial v}{\partial t} + \frac{\partial v}{\partial y}dY_s = -\frac{\sigma_s^2 - (v_s)^2}{2v_s(T-s)}. \tag{6.5.9}$$

Then, (6.5.8), (6.5.9), and the Delta-Gamma-Vega relationship (6.2.6) imply that the last three terms in (6.5.7) cancel. Then we get

$$\begin{aligned}
0 &= E_t\Bigg[H(t, X_t, k_t^*, v_t)J_t \\
&+ \int_t^T H(s, X_s, k_t^*, v_s)dJ_s \\
&+ \int_t^T \frac{\partial^2 H}{\partial x \partial \sigma}(s, X_s, k_t^*, v_s)J_s\frac{\partial v}{\partial y}(D^- Y_s)\sigma_s ds \\
&+ \int_t^T \frac{\partial H}{\partial x}(s, X_s, k_t^*, v_s)(D^- J_s)\sigma_s ds\Bigg].
\end{aligned} \tag{6.5.10}$$

Now, using again the Delta-Gamma-Vega relationship (6.2.6) and the equalities

$$D^- J_s = \rho \int_s^T (BS^{-1})'(k_t^*, \Gamma_r) D_s^W \Phi_r dr,$$

and

$$D^- Y_s = \rho \int_s^T D_s^W \sigma_r^2 dr,$$

Equation (6.5.10) reduces to

$$\begin{aligned}
0 &= E_t\Bigg[H(t, X_t, k_t^*, v_t)J_t \\
&- \int_t^T H(s, X_s, k_t^*, v_s)(BS^{-1})'(X_t, \Gamma_s)\Phi_s ds \\
&+ \frac{\rho}{2}\int_t^T \left(\frac{\partial^3}{\partial x^3} - \frac{\partial^2}{\partial x^2}\right) H(s, X_s, k_t^*, v_s)J_s\Phi_s ds \\
&+ \rho \int_t^T \frac{\partial H}{\partial x}(s, X_s, k_t^*, v_s)\left(\int_s^T (BS^{-1})'(k_t^*, \Gamma_r)(D_s^W \Phi_r)dr\right)\sigma_s ds\Bigg],
\end{aligned}$$

which implies that

$$\begin{aligned}
T_2 &= E_t\Bigg[\frac{\rho}{2}H(t, X_t, k_t^*, v_t)J_t \\
&+ \frac{\rho^2}{4}\int_t^T \left(\frac{\partial^3}{\partial x^3} - \frac{\partial^2}{\partial x^2}\right) H(s, X_s, k_t^*, v_s)J_s\Phi_s ds \\
&+ \frac{\rho^2}{2}\int_t^T \frac{\partial H}{\partial x}(s, X_s, k_t^*, v_s)\left(\int_s^T (BS^{-1})'(k_t^*, \Gamma_r)(D_s^W \Phi_r)dr\right)\sigma_s ds\Bigg] \\
&= T_2^1 + T_2^2 + T_2^3.
\end{aligned}$$

Notice that now we have decomposed T_2 as the sum of

$$T_2^1 = \frac{\rho}{2} E_t[H(t, X_t, k_t^*, v_t) J_t]$$

plus two other terms. Notice that T_2^1 is (as a function of the correlation) of order ρ, while the other two terms are of order ρ^2. Nevertheless, being the leading term as a function of ρ does not imply being the leading term as a function of time to maturity $T - t$, so we need to study the short-end limit of these three terms. Now, the proof will be decomposed into four steps.

Step 1. Let us see that T_2^1 is of the order $O((T-t)^{\frac{1}{2}+H})$. As

$$(BS^{-1})'(k_t^*, \Gamma_s) = \frac{1}{e^{X_t} N'(d_+(k_t^*, BS^{-1}(k_t^*, \Gamma_s)))\sqrt{T-t}},$$

and

$$H(t, X_t, k_t^*, v_t) = \frac{e^{X_t} N'(d_+(k_t^*, v_t))}{v_t \sqrt{T-t}} \left(1 - \frac{d_+(k_t^*, v_t)}{v_t \sqrt{T-t}}\right),$$

we can write

$$\begin{aligned}
&\lim_{T \to t} \frac{T_2^1}{(T-t)^{\frac{1}{2}+H}} \\
&= \lim_{T \to t} \frac{1}{(T-t)^{\frac{1}{2}+H}} E_t \left[\frac{\rho}{2} H(t, X_t, k_t^*, v_t) J_t \right] \\
&= \lim_{T \to t} \frac{\rho}{2(T-t)^{\frac{1}{2}+H}} E_t \Bigg[\frac{e^{X_t} N'(d_+(k_t^*, v_t))}{v_t \sqrt{T-t}} \left(1 - \frac{d_+(k_t^*, v_t)}{v_t \sqrt{T-t}}\right) \\
&\qquad \times \int_t^T \frac{1}{e^{X_t} N'(d_+(k_t^*, BS^{-1}(k_t^*, \Gamma_s)))\sqrt{T-t}} \Phi_s ds \Bigg] \\
&= \lim_{T \to t} \frac{\rho}{4(T-t)^{\frac{3}{2}+H}} E_t \left[\int_t^T \frac{\Phi_s}{v_s} ds \right] \\
&= \lim_{T \to t} \frac{\rho}{4(T-t)^{\frac{3}{2}+H}} E_t \left[\int_t^T \frac{\sigma_s}{v_s} \int_s^T D_s^W \sigma_r^2 dr ds \right] \\
&= \lim_{T \to t} \frac{\rho}{4(T-t)^{\frac{3}{2}+H}} E_t \left[\int_t^T \int_s^T D_s^W \sigma_r^2 dr ds \right]. \qquad (6.5.11)
\end{aligned}$$

Step 2. Here we see that T_2^2 is the sum of terms of the order $O((T-t)^{2H})$ plus terms of the order $O((T-t)^{3H})$. To this end, we expand T_2^2 around the term

$$\left(\frac{\partial^3}{\partial x^3}-\frac{\partial^2}{\partial x^2}\right)H(t,X_t,k_t^*,v_t)Z_t,$$

where $Z_t:=\int_t^T \Phi_u J_u du$. Once again, the technique is to apply the anticipating Itô's formula to the process

$$\left(\frac{\partial^3}{\partial x^3}-\frac{\partial^2}{\partial x^2}\right)H(s,X_s,k_t^*,v_s)Z_s$$

(notice that Z is anticipating). Then, using again the Delta-Gamma-Vega relationship and the Black-Scholes equation, we get

$$\begin{aligned}T_2^2 &= \frac{\rho^2}{4}E_t\Bigg[\left(\frac{\partial^3}{\partial x^3}-\frac{\partial^2}{\partial x^2}\right)H(t,X_t,k_t^*,v_t)Z_t \\ &+\frac{\rho}{2}\int_t^T\left(\frac{\partial^3}{\partial x^3}-\frac{\partial^2}{\partial x^2}\right)^2 H(s,X_s,k_t^*,v_s)Z_s\Phi_s ds \\ &+\rho\int_t^T\frac{\partial}{\partial x}\left(\frac{\partial^3}{\partial x^3}-\frac{\partial^2}{\partial x^2}\right)H(s,X_s,k_t^*,v_s)(D_s^W Z_s)\sigma_s ds\Bigg].\end{aligned} \tag{6.5.12}$$

Now, the bound in Lemma 6.3.1 allows us to see that the last two terms in the above equation are $O((T-t)^{\frac{3}{2}+3H-\frac{1}{2}})$. This implies that only the first term contributes to the limit. That is,

$$\begin{aligned}&\lim_{T\to t}\frac{T_2^2}{(T-t)^{2H}} \\ &= \lim_{T\to t}\frac{\rho^2}{4(T-t)^{2H}}E_t\left[\left(\frac{\partial^3}{\partial x^3}-\frac{\partial^2}{\partial x^2}\right)H(t,X_t,k_t^*,v_t)Z_t\right]\end{aligned} \tag{6.5.13}$$

Now, as

$$\left(\frac{\partial^3}{\partial x^3}-\frac{\partial^2}{\partial x^2}\right)H(t,X_t,k_t^*,v_t)=-\frac{1}{16}\frac{e^{X_t}N'(d_+(k_t^*,v_t))}{(v_t\sqrt{T-t})^5}(v_t^4(T-t)^2-48),$$

we deduce that

$$\begin{aligned}
&\lim_{T\to t}\frac{T_2^2}{(T-t)^{2H}}\\
&=\lim_{T\to t}\frac{\rho^2}{4(T-t)^{2H}}E_t\left[\left(\frac{\partial^3}{\partial x^3}-\frac{\partial^2}{\partial x^2}\right)H(t,X_t,k_t^*,v_t)Z_t\right]\\
&=\lim_{T\to t}\frac{\rho^2}{4(T-t)^{3+2H}}E_t\Bigg[-\frac{1}{16}\frac{e^{X_t}N'(d_+(k_t^*,v_t))}{v_t^5}(v_t^4(T-t)^2-48)\\
&\quad\times\int_t^T\sigma_s\left(\int_t^T D_s^W\sigma_r^2dr\right)\left(\int_s^T\frac{\Phi_r}{e^{X_t}N'(d_+(k_t^*,BS^{-1}(k_t^*,\Gamma_r)))}dr\right)ds\Bigg].
\end{aligned}\tag{6.5.14}$$

Then, straightforward computations lead to

$$\begin{aligned}
&\lim_{T\to t}\frac{T_2^2}{(T-t)^{2H}}\\
&=\lim_{T\to t}\frac{3\rho^2}{4\sigma_t^5(T-t)^{3+2H}}E_t\left[\int_t^T\left(\int_s^T D_s^W\sigma_r^2dr\right)\left(\int_s^T\Phi_rdr\right)\sigma_sds\right]\\
&=\lim_{T\to t}\frac{3\rho^2}{4\sigma_t^5(T-t)^{3+2H}}\\
&\quad\times E_t\left[\int_t^T\left(\int_s^T D_s^W\sigma_r^2dr\right)\left(\int_s^T\sigma_r\int_r^T D_r^W\sigma_\theta^2d\theta dr\right)\sigma_sds\right]\\
&=\lim_{T\to t}\frac{3\rho^2}{4\sigma_t^3(T-t)^{3+2H}}E_t\left[\int_t^T\left(\int_s^T D_s^W\sigma_r^2dr\right)\left(\int_s^T\int_r^T D_r^W\sigma_\theta^2d\theta dr\right)ds\right]\\
&=\lim_{T\to t}\frac{3\rho^2}{8\sigma_t^3(T-t)^{3+2H}}E_t\left[\left(\int_t^T\int_s^T D_s^W\sigma_r^2drds\right)^2\right].
\end{aligned}\tag{6.5.15}$$

Step 3. Here we see that T_2^3 is the sum of terms of the order $O((T-t)^{2H})$ plus terms of the order $O((T-t)^{3H})$. The analysis is very similar to the one in the last step. Applying again the anticipating Itô's formula to the process

$$\frac{\partial H}{\partial x}(s,X_s,k_t^*,v_s)R_s,$$

where

$$R_s := \int_s^T\left(\int_u^T(BS^{-1})'(k_t^*,\Gamma_r)(D_u^W\Phi_r)dr\right)\sigma_udu$$

(notice that R is an anticipating process), we get (using again the Delta-Gamma-Vega relationship and the Black-Scholes equation)

$$
\begin{aligned}
T_2^3 &= \frac{\rho^2}{2} E_t \Bigg[\frac{\partial H}{\partial x}(t, T, X_t, k_t^*, v_t) R_t \\
&\quad + \frac{\rho}{2} \int_t^T \left(\frac{\partial^3}{\partial x^3} - \frac{\partial^2}{\partial x^2} \right) \frac{\partial H}{\partial x}(s, X_s, k_t^*, v_s) R_s \Phi_s ds \\
&\quad + \rho \int_t^T \frac{\partial^2 H}{\partial x^2}(s, X_s, k_t^*, v_s) \\
&\quad \times \left(\int_s^T \int_r^T (BS^{-1})'(k_t^*, \Gamma_u)(D_s^W D_r^W \Phi_u) du dr \right) \sigma_s ds \Bigg].
\end{aligned}
\tag{6.5.16}
$$

Again, Lemma 6.3.1 gives us that the last two terms in (6.5.12) and (6.5.16) are $O((T-t)^{3H})$. This implies that

$$
\begin{aligned}
&\lim_{T \to t} \frac{T_2^3}{(T-t)^{2H}} \\
&= \lim_{T \to t} \frac{\rho^2}{2(T-t)^{2H}} E_t \left[\frac{\partial H}{\partial x}(t, T, X_t, k_t^*, v_t) R_t \right].
\end{aligned}
\tag{6.5.17}
$$

Now,

$$
\frac{\partial H}{\partial x}(t, T, X_t, k_t^*, v_t) = \frac{1}{4} \frac{e^{X_t} N'(d_+(k_t^*, v_t))}{(v_t \sqrt{T-t})^3} (v_t^2 (T-t) - 4),
$$

which allows us to write

$$
\begin{aligned}
&\lim_{T \to t} \frac{T_2^3}{(T-t)^{2H}} \\
&= \lim_{T \to t} \frac{\rho^2}{2(T-t)^{2H}} E_t \left[\frac{\partial H}{\partial x}(t, T, X_t, k_t^*, v_t) R_t \right] \\
&= \lim_{T \to t} \frac{\rho^2}{2(T-t)^{2H}} E_t \Bigg[\frac{1}{4} \frac{e^{X_t} N'(d_+(k_t^*, v_t))}{(v_t \sqrt{T-t})^3} (v_t^2 (T-t) - 4) \\
&\quad \times \int_t^T \int_s^T \frac{1}{e^{X_r} N'(d_+(k_t^*, BS^{-1}(k_t^*, \Gamma_r))) \sqrt{T-t}} \\
&\quad \times \left(D_s^W \left(\sigma_r \int_r^T D_s^W \sigma_u^2 du \right) \right) dr \sigma_s ds \Bigg].
\end{aligned}
\tag{6.5.18}
$$

Then, a direct computation gives us that

$$
\begin{aligned}
&\lim_{T\to t}\frac{T_2^3}{(T-t)^{2H}}\\
&= -\lim_{T\to t}\frac{\rho^2}{2\sigma_t^2(T-t)^{2+2H}}E_t\Bigg[\int_t^T\int_s^T D_s^W\sigma_r\int_r^T D_r^W\sigma_u^2 dudrds\\
&\quad+\int_t^T\int_s^T\sigma_r\int_r^T D_s^W D_r^W\sigma_u^2 dudrds\Bigg]\\
&= -\lim_{T\to t}\frac{\rho^2}{\sigma_t(T-t)^{2+2H}}E_t\left[\int_t^T\int_s^T D_s^W\sigma_r\int_r^T D_r^W\sigma_u dudrds\right]\\
&\quad-\lim_{T\to t}\frac{\rho^2}{2\sigma_t(T-t)^{2+2H}}E_t\left[\int_t^T\int_s^T\int_r^T D_s^W D_r^W\sigma_u^2 dudrds\right].
\end{aligned}
\tag{6.5.19}
$$

Step 4. From the results in the last steps, we deduce that $I_t(k_t^*)-E[v_t]$ is the sum of terms of the orders $O((T-t)^{\frac{1}{2}+H})$, $O((T-t)^{2H})$, and higher-order terms. Then,

- For $H-\frac{1}{2}<0$, $I_t(k_t^*)-E[v_t]$ is of the order $O((T-t)^{2H})$, and the terms that contribute to the limit are (6.5.15) and (6.5.19).
- If $H-\frac{1}{2}>0$, this difference is of the order $O((T-t)^{\frac{1}{2}+H})$ and the term that contributes to the limit is (6.5.11).
- If $H=\frac{1}{2}$, we have that $2H=\frac{1}{2}+H=1$. Then, this difference is of the order $O(T-t)$ and the terms that contribute to the limit are (6.5.11), (6.5.15), and (6.5.19).

Now the proof is complete. ■

Remark 6.5.6 (Malliavin expansion technique) *The approach in the proof of the above theorem can be seen as a general expansion technique. Let us review it. Consider an expectation of the form*

$$
E_t\left(\int_t^T \Sigma(s,X_s,k,v_s)A_s ds\right),
$$

where Σ denotes some Greek and A is some process adapted to the filtration generated by the Brownian motion W that satisfies some conditions *in the Malliavin calculus sense. Then, we can apply the anticipating Itô formula to the process*

$$\Sigma(t, X_t, k, v_t) \int_t^T A_s ds.$$

This allows us to prove an equality of the form

$$E_t\left(\int_t^T \Sigma(s, X_s, k, v_s) A_s ds\right) = E_t\left(\Sigma(t, X_t, k, v_t) \int_t^T A_s ds\right) + \text{Other terms}.$$

Then, since Σ *satisfies the Black-Scholes equation and because of the Delta-Gamma-Vega relationship, all the other terms in the expansion cancel except those containing the Malliavin derivative of the non-adapted processes* Y *and* $\int_t^T A_s ds$*, leading to an expression of the type*

$$\begin{aligned}
&E_t\left(\int_t^T \Sigma(s, X_s, k, v_s) A_s ds\right) \\
&= E_t\left(\Sigma(t, X_t, k, v_t) \int_t^T A_s ds\right) \\
&+E_t\left(\int_t^T D_s^W\left(\frac{\partial \Sigma}{\partial x}(s, X_s, k, v_s) \int_s^T A_u du\right) \sigma_s ds\right) \\
&= E_t\left(\Sigma(t, X_t, k, v_t) \int_t^T A_s ds\right) \\
&+\frac{\rho}{2} E_t\left(\int_t^T \left(\frac{\partial^3}{\partial x^3} - \frac{\partial^2}{\partial x^2}\right) \Sigma(s, X_s, k, v_s) \left(\int_s^T A_u du\right) \sigma_s D^- Y_s ds\right) \\
&+\rho E_t\left(\int_t^T \frac{\partial \Sigma}{\partial x}(s, X_s, k, v_s) \left(\int_s^T D_s^W A_u du\right) \sigma_s ds\right). \qquad (6.5.20)
\end{aligned}$$

Notice that Equality (6.5.20) is not just an approximation but an exact formula. This procedure can be iterated, leading to an expansion for expectations of the form

$$E_t\left(\int_t^T \Sigma(s, X_s, k, v_s) A_s ds\right),$$

in terms of higher-order Greeks and the Malliavin derivatives of the volatility process. This technique will be used to analyse the ATM skew and curvature in Chapters 7 and 8, respectively. It can also be applied to other problems as the study of expansions for the discretisation errors in stochastic integrals (see Alòs and Fukasawa (2021)).

Remark 6.5.7 *Hypotheses (H1)–(H5) have been chosen for the sake of simplicity. The same results can be extended to other stochastic volatility models (see Alòs and Shiraya (2019)).*

Corollary 6.5.8 *Assume that $\sigma_t = f(B_t^H)$, where $f \in C_b^3$ and B_t^H is a RLfBm with Hurst parameter H. Then, the above result proves that, in the correlated case*

- *If $H \leq 1/2$, then $I_t(k_t^*) - E_t[v_t] = O((T-t)^{2H})$.*
- *If $H \geq 1/2$, then $I_t(k_t^*) - E_t[v_t] = O((T-t)^{H+1/2})$.*

Notice that, the order of convergence is different in the uncorrelated and the correlated case. This is in line with Fukasawa (2014), where it was established that the correlation plays a crucial role in the ATM short-time behaviour of the implied volatility.

Remark 6.5.9 (The zero-vanna implied volatility) *As the ATM implied volatility depends on the correlation parameter, the estimation of the volatility swap strike by the ATMI is sensible to ρ. Immune-to-correlation approximations of the volatility swap have been studied in Carr and Lee (2008), where an approximated replicating portfolio has been constructed. Another approach has been proposed recently in Rolloos and Arslan (2017), based on the so-called zero-vanna implied volatility. Denote by $\hat{k}_t$ the strike such that*

$$\frac{X_t - \hat{k}_t + r(T-t)}{I_t(\hat{k}_t)\sqrt{T-t}} - \frac{I_t(\hat{k}_t)}{2}\sqrt{T-t} = 0.$$

Then, the zero-vanna implied volatility is defined as $I_t(\hat{k}_t)$. The approximation of the volatility swap by the zero-vanna implied volatility is model-free and, due to its lower sensitivity to the correlation parameter, it is a better estimator of the volatility swap fair price than the at-the-money implied volatility. The rigorous relationship between the zero vanna-implied volatility and the volatility swap has been studied recently in Alòs, Rolloos, and Shiraya (2020). The results in this paper confirm that the zero-vanna approximation outperforms the AMTI as an estimator of the volatility swap strike.

Remark 6.5.10 *We proved in Remark 6.5.2 the following relationship between the volatility swap and the variance swap*

$$\lim_{T\to t}\frac{E[v_t]-\sqrt{\frac{1}{T-t}\int_t^T E_t(\sigma_s^2)ds}}{(T-t)^{2H}}$$
$$=-\frac{1}{8(T-t)^{2H+2}}\lim_{T\to t}E_t\int_t^T\left(E_r\int_r^T D_r^W\sigma_s^2 ds\right)^2 ds.$$

This, jointly with Theorem 6.5.5, allows us to write the following expressions for the difference between the ATMI and the square root of a variance swap:

- *If* $H-\frac{1}{2}<0$

$$\lim_{T\to t}\frac{I_t(k_t^*)-\sqrt{\frac{1}{T-t}\int_t^T E_t(\sigma_s^2)ds}}{(T-t)^{2H}}$$
$$= -\lim_{T\to t}\frac{1}{8\sigma_t^3(T-t)^{2H+2}}E_t\left[\int_t^T\left(E_r\int_r^T D_r^W\sigma_s^2 ds\right)^2 ds\right]$$
$$+\lim_{T\to t}\frac{3\rho^2}{8\sigma_t^3(T-t)^{3+2H}}E_t\left[\left(\int_t^T\int_s^T D_s^W\sigma_r^2 drds\right)^2\right]$$
$$-\lim_{T\to t}\frac{\rho^2}{\sigma_t(T-t)^{2+2H}}E_t\left[\int_t^T\int_s^T D_s^W\sigma_r\int_r^T D_r^W\sigma_u dudrds\right]$$
$$-\lim_{T\to t}\frac{\rho^2}{2\sigma_t(T-t)^{2+2H}}E_t\left[\int_t^T\int_s^T\int_r^T D_s^W D_r^W\sigma_u^2 dudrds\right].$$

- *If* $H-\frac{1}{2}>0$

$$\lim_{T\to t}\frac{I_t(k_t^*)-\sqrt{\frac{1}{T-t}\int_t^T E_t(\sigma_s^2)ds}}{(T-t)^{\frac{1}{2}+H}}$$
$$=\lim_{T\to t}\frac{\rho}{4(T-t)^{\frac{3}{2}+H}}E_t\left[\int_t^T\int_s^T D_s^W\sigma_r^2 drds\right].$$

- *If* $H - \frac{1}{2} = 0$

$$
\begin{aligned}
&\lim_{T\to t} \frac{I_t(k_t^*) - \sqrt{\frac{1}{T-t}\int_t^T E_t(\sigma_s^2)ds}}{(T-t)} \\
&= -\lim_{T\to t} \frac{1}{8\sigma_t^3(T-t)^3} E_t\left[\int_t^T \left(E_r \int_r^T D_r^W \sigma_s^2 ds\right)^2 ds\right] \\
&\quad + \lim_{T\to t} \frac{3\rho^2}{8\sigma_t^3(T-t)^4} E_t\left[\left(\int_t^T \int_s^T D_s^W \sigma_r^2 drds\right)^2\right] \\
&\quad - \lim_{T\to t} \frac{\rho^2}{\sigma_t(T-t)^{2+2H}} E_t\left[\int_t^T \int_s^T D_s^W \sigma_r \int_r^T D_r^W \sigma_u dudrds\right] \\
&\quad - \lim_{T\to t} \frac{\rho^2}{2\sigma_t(T-t)^3} E_t\left[\int_t^T \int_s^T \int_r^T D_s^W D_r^W \sigma_u^2 dudrds\right] \\
&\quad + \lim_{T\to t} \frac{\rho}{4(T-t)^2} E_t\left[\int_t^T \int_s^T D_s^W \sigma_r^2 drds\right].
\end{aligned}
$$

Notice that these limits prove that the difference between the ATMI and the square root of a variance swap is of the order $O((T-t)^{2H})$ *if* $H \leq \frac{1}{2}$, *and of the order* $H + \frac{1}{2}$ *if* $H \geq \frac{1}{2}$, *according to the results in El Euch, Fukasawa, Gatheral, and Rosenbaum (2018).*

6.5.3 Approximation formulas for the ATMI

Theorem 6.5.5 suggests the following approximations for the ATM implied volatility:

- If $H - \frac{1}{2} < 0$

$$
\begin{aligned}
&I_t(k_t^*) \\
&\approx E_t[v_t] + \frac{3\rho^2}{8\sigma_t^3(T-t)^3} E_t\left[\left(\int_t^T \int_s^T D_s^W \sigma_r^2 drds\right)^2\right] \\
&\quad - \lim_{T\to t} \frac{\rho^2}{\sigma_t(T-t)^{2+2H}} E_t\left[\int_t^T \int_s^T D_s^W \sigma_r \int_r^T D_r^W \sigma_u dudrds\right] \\
&\quad - \lim_{T\to t} \frac{\rho^2}{2\sigma_t(T-t)^2} E_t\left[\int_t^T \int_s^T \int_r^T D_s^W D_r^W \sigma_u^2 dudrds\right].
\end{aligned}
\tag{6.5.21}
$$

- If $H - \frac{1}{2} > 0$

$$I_t(k_t^*) \approx E_t[v_t] + \frac{\rho}{4(T-t)} E_t \left[\int_t^T \int_s^T D_s^W \sigma_r^2 dr ds \right]. \tag{6.5.22}$$

- If $H - \frac{1}{2} = 0$

$$\begin{aligned} & I_t(k_t^*) \\ & \approx E_t[v_t] + \frac{3\rho^2}{8\sigma_t^3 (T-t)^3} E_t \left[\left(\int_t^T \int_s^T D_s^W \sigma_r^2 dr ds \right)^2 \right] \\ & - \frac{\rho^2}{\sigma_t (T-t)^2} E_t \left[\int_t^T \int_s^T D_s^W \sigma_r \int_r^T D_r^W \sigma_u du dr ds \right] \\ & - \frac{\rho^2}{2\sigma_t (T-t)^2} E_t \left[\int_t^T \int_s^T \int_r^T D_s^W D_r^W \sigma_u^2 du dr ds \right] \\ & + \frac{\rho}{4(T-t)} E_t \left[\int_t^T \int_s^T D_s^W \sigma_r^2 dr ds \right]. \end{aligned} \tag{6.5.23}$$

In Chapter 7 we will see that the term of order ρ in the above approximations in the cases $H = \frac{1}{2}$ and $H > \frac{1}{2}$ can be expressed in terms of the ATM skew slope, according to Carr and Lee (2008). This will allow us to prove a model-free approximation of the volatility swap in terms of the ATM implied volatility level and skew.

Similarly, from Remark 6.5.10 we get the following approximations in terms of the variance swap:

- If $H - \frac{1}{2} < 0$

$$\begin{aligned} I_t(k_t^*) & \approx \sqrt{\frac{1}{T-t} \int_t^T E_t(\sigma_s^2) ds} \\ & - \frac{1}{8\sigma_t^3 (T-t)^2} \lim_{T \to t} E_t \int_t^T \left(E_r \int_r^T D_r^W \sigma_s^2 ds \right)^2 ds \\ & + \frac{3\rho^2}{8\sigma_t^3 (T-t)^3} E_t \left[\left(\int_t^T \int_s^T D_s^W \sigma_r^2 dr ds \right)^2 \right] \end{aligned}$$

$$
\begin{aligned}
&- \lim_{T\to t} \frac{\rho^2}{\sigma_t(T-t)^2} E_t \left[\int_t^T \int_s^T D_s^W \sigma_r \int_r^T D_r^W \sigma_u du dr ds \right] \\
&- \lim_{T\to t} \frac{\rho^2}{2\sigma_t(T-t)^2} E_t \left[\int_t^T \int_s^T \int_r^T D_s^W D_r^W \sigma_u^2 du dr ds \right].
\end{aligned}
\tag{6.5.24}
$$

- If $H - \frac{1}{2} > 0$

$$
I_t(k_t^*) \approx \sqrt{\frac{1}{T-t} \int_t^T E_t(\sigma_s^2) ds + \frac{\rho}{4(T-t)} E_t \left[\int_t^T \int_s^T D_s^W \sigma_r^2 dr ds \right]}.
\tag{6.5.25}
$$

- If $H - \frac{1}{2} = 0$

$$
\begin{aligned}
I_t(k_t^*) &\approx \sqrt{\frac{1}{T-t} \int_t^T E_t(\sigma_s^2) ds} \\
&- \frac{1}{8\sigma_t^3(T-t)^3} E_t \left[\int_t^T \left(E_r \int_r^T D_r^W \sigma_s^2 ds \right)^2 ds \right] \\
&+ \frac{3\rho^2}{8\sigma_t^3(T-t)^3} E_t \left[\left(\int_t^T \int_s^T D_s^W \sigma_r^2 dr ds \right)^2 \right] \\
&- \lim_{T\to t} \frac{\rho^2}{\sigma_t(T-t)^2} E_t \left[\int_t^T \int_s^T D_s^W \sigma_r \int_r^T D_r^W \sigma_u du dr ds \right] \\
&- \lim_{T\to t} \frac{\rho^2}{2\sigma_t(T-t)^2} E_t \left[\int_t^T \int_s^T \int_r^T D_s^W D_r^W \sigma_u^2 du dr ds \right] \\
&+ \frac{\rho}{4(T-t)} E_t \left[\int_t^T \int_s^T D_s^W \sigma_r^2 dr ds \right].
\end{aligned}
\tag{6.5.26}
$$

Consider the case $H = \frac{1}{2}$. Approximating $D_s^W \sigma_r$, $D_s^W \sigma_r^2$ and, $D_s^W D_r^W \sigma_u^2$ by their limits when $r, s, u \to t$ (that we denote by $D_t^+ \sigma_t$, $D_t^+ \sigma_t^2$ and, $(D_t^+)^2 \sigma_t^2$, respectively), (6.5.23) leads to the following approximation:

$$
\begin{aligned}
&I_t(k_t^*) \\
&\approx E_t[v_t] \\
&+(T-t)\left[\frac{3\rho^2}{32\sigma_t^3}(D_t^+\sigma_t^2)^2 - \frac{\rho^2}{6\sigma_t}(D_t^+\sigma_t)^2 - \frac{\rho^2}{12\sigma_t}(D_t^+)^2\sigma_t^2 + \frac{\rho}{8}D_t^+\sigma_t^2\right].
\end{aligned}
\tag{6.5.27}
$$

In a similar way, (6.5.26) gives us that

$$
\begin{aligned}
I_t(k_t^*) \approx& \sqrt{\frac{1}{T-t}\int_t^T E_t(\sigma_s^2)ds + (T-t)} \\
\times& \left[\frac{1}{\sigma_t^3}\left(\frac{3\rho^2}{32} - \frac{1}{24}\right)(D_t^+\sigma_t^2)^2 - \frac{\rho^2}{6\sigma_t}(D_t^+\sigma_t)^2 - \frac{\rho^2}{12\sigma_t}(D_t^+)^2\sigma_t^2 + \frac{\rho}{8}D_t^+\sigma_t^2\right].
\end{aligned}
\tag{6.5.28}
$$

6.5.4 Examples

6.5.4.1 *Diffusion models*

Assume that the volatility σ_r is of the form $\sigma_r = f(Y_r)$, where $f \in C_b^1(\mathbb{R})$ and Y_r is the solution of a stochastic differential equation

$$
dY_r = a\,(r, Y_r)\,dr + b\,(r, Y_r)\,dW_r, \tag{6.5.29}
$$

for some real functions $a, b \in C_b^1(\mathbb{R})$. Then, the results in Section 3.2.2 give us that $Y \in \mathbb{L}_W^{1,2}$ and that

$$
D_s^W Y_r = \int_s^r \frac{\partial a}{\partial x}(u, Y_u) D_s^W Y_u du + b(s, Y_s) + \int_s^r \frac{\partial b}{\partial x}(u, Y_u) D_s^W Y_u dW_u. \tag{6.5.30}
$$

Then, it follows that

$$
D_t^+\sigma_t = f'(Y_t)b(t, Y_t), \;\; D_t^+\sigma_t^2 = 2\sigma_t f'(Y_t)b(t, Y_t),
$$

and similar arguments allow us to see that

$$
\left(D_t^+\right)^2\sigma_t^2 := 2\left[\left((f'(Y_t))^2 + \sigma_t f''(Y_t)\right)b^2(t, Y_t) + \sigma_t f'(Y_t)\frac{\partial b}{\partial x}(t, Y_t)b(t, Y_t)\right].
$$

Then, (6.5.27) reads as

$$
\begin{aligned}
& I_t(k_t^*) \\
\approx\ & E_t[v_t] + \frac{3\rho^2(T-t)}{32\sigma_t^3}(2\sigma_t f'(Y_t)b(t,Y_t))^2 \\
& -\frac{\rho^2(T-t)}{6\sigma_t}(f'(Y_t)b(t,Y_t))^2 \\
& -\frac{\rho^2(T-t)}{12\sigma_t}2\left[\left((f'(Y_t))^2+\sigma_t f''(Y_t)\right)b^2(t,Y_t)+\sigma_t f'(Y_t)\frac{\partial b}{\partial x}(t,Y_t)b(t,Y_t)\right] \\
& +\frac{\rho(T-t)}{8}(2\sigma_t f'(Y_t)b(t,Y_t) \\
=\ & E_t[v_t] + \frac{5\rho^2(T-t)}{24\sigma_t}(f'(Y_t)b(t,Y_t))^2 \\
& -\frac{\rho^2(T-t)}{6\sigma_t}\left[\left((f'(Y_t))^2+\sigma_t f''(Y_t)\right)b^2(t,Y_t)+\sigma_t f'(Y_t)\frac{\partial b}{\partial x}(t,Y_t)b(t,Y_t)\right] \\
& +\frac{\rho(T-t)}{4}(\sigma_t f'(Y_t)b(t,Y_t)).
\end{aligned}
\tag{6.5.31}
$$

Now, in the particular case $f(x) = x$, $\sigma_t = Y_t$ and the above approximation reads

$$
\begin{aligned}
I_t(k_t^*) \approx\ & E_t[v_t] + \frac{\rho^2(T-t)}{24\sigma_t}b^2(t,Y_t) \\
& -\frac{\rho^2(T-t)}{6\sigma_t}\frac{\partial b}{\partial x}(t,Y_t)b(t,Y_t) + \frac{\rho(T-t)}{4}(\sigma_t b(t,Y_t)).
\end{aligned}
\tag{6.5.32}
$$

Now, as

$$
\sigma_s^2 = \sigma_t^2 + 2\int_t^s b(u,Y_u)Y_u dW_u + \int_t^s [2a(u,Y_u)Y_u + b^2(u,Y_u)]du
$$

it follows that

$$
\begin{aligned}
& E_t[v_t] \\
&= \sqrt{\frac{1}{T-t}\int_t^T \left(\sigma_t^2 + 2\int_t^s b(u,Y_u)Y_u dW_u + \int_t^s [2a(u,Y_u)Y_u + b^2(u,Y_u)]du\right) ds} \\
&= \sqrt{\sigma_t^2 + \frac{1}{T-t}\int_t^T \left(2\int_t^s b(u,Y_u)Y_u dW_u + \int_t^s [2a(u,Y_u)Y_u + b^2(u,Y_u)]du\right) ds}.
\end{aligned}
$$

Then, a second-order Taylor expansion gives us that

$$
\begin{aligned}
E_t[v_t] &\approx \sigma_t + \frac{1}{2\sigma_t(T-t)} E_t \int_t^T \left(\int_t^s [2a(u,Y_u)Y_u + b^2(u,Y_u)]du \right) ds \\
&- \frac{4}{8\sigma_t^3(T-t)^2} E_t \left(\int_t^T \left(\int_t^s b(u,Y_u)Y_u dW_u \right) ds \right)^2 \\
&\approx \sigma_t + \frac{1}{2\sigma_t(T-t)} E_t \int_t^T \left(\int_t^s [2a(u,Y_u)Y_u + b^2(u,Y_u)] \right) ds \\
&- \frac{1}{2\sigma_t^3(T-t)^2} E_t \left(\int_t^T b(u,Y_u)Y_u(T-u)dW_u \right)^2 \\
&\approx \sigma_t + \frac{1}{2\sigma_t(T-t)} E_t \int_t^T \left(\int_t^s [2a(u,Y_u)Y_u + b^2(u,Y_u)]du \right) ds \\
&- \frac{1}{2\sigma_t^3(T-t)^2} E_t \left(\int_t^T b^2(u,Y_u)Y_u^2(T-u)^2 du \right).
\end{aligned}
$$

Now, approximating Y_u by Y_t in the right-hand side of the above equality and using the fact that $\sigma = Y$ we get

$$
\begin{aligned}
E_t[v_t] &\approx \sigma_t + \frac{(T-t)}{4\sigma_t}[2a(t,Y_t)\sigma_t + b^2(t,Y_t)] \\
&- \frac{(T-t)}{6\sigma_t} b^2(t,Y_t) \\
&= \sigma_t + \frac{(T-t)}{2} a(t,Y_t) + \frac{(T-t)}{12\sigma_t} b^2(t,Y_t).
\end{aligned}
$$

This, jointly with (6.5.28), gives the following approximation for the ATM implied volatility

$$
\begin{aligned}
I_t(k_t^*) &\approx \sigma_t + \frac{(T-t)}{2} a(t,Y_t) + \frac{(T-t)}{12\sigma_t} b^2(t,Y_t) \\
&+ \frac{\rho^2(T-t)}{24\sigma_t} b^2(t,Y_t) \\
&- \frac{\rho^2(T-t)}{6\sigma_t} \frac{\partial b}{\partial x}(t,Y_t)b(t,Y_t) + \frac{\rho(T-t)}{4}(\sigma_t b(t,Y_t)),
\end{aligned}
\tag{6.5.33}
$$

according to the results in Mendevev and Scaillet (2003).

Remark 6.5.11 *Notice that (6.5.33) can also be obtained from the approximation (6.5.28). Assume that $f(x) = x$. Then, the same arguments* *as in (6.5.32) and the fact that* $D_t^+ \sigma_t^2 = 2\sigma_t f'(Y_t) b(t, Y_t) = 2\sigma_t b(t, Y_t)$ *lead to*

$$
\begin{aligned}
I_t(k_t^*) &\approx \sqrt{\frac{1}{T-t}\int_t^T E_t(\sigma_s^2)ds} - \frac{(T-t)}{6\sigma_t} b^2(t, Y_t) + \frac{\rho^2 (T-t)}{24\sigma_t} b^2(t, Y_t) \\
&\quad - \frac{\rho^2 (T-t)}{6\sigma_t} \frac{\partial b}{\partial x}(t, Y_t) b(t, Y_t) + \frac{\rho(T-t)}{4}(\sigma_t b(t, Y_t)).
\end{aligned} \tag{6.5.34}
$$

Then, as

$$
\sqrt{\frac{1}{T-t}\int_t^T E_t(\sigma_s^2)ds} = \sqrt{\sigma_t^2 + E_t \int_t^s [2a(u, Y_u)Y_u + b^2(u, Y_u)]du},
$$

a second-order Taylor expansion gives that

$$
\begin{aligned}
&\sqrt{\frac{1}{T-t}\int_t^T E_t(\sigma_s^2)ds} \\
&\approx \sigma_t + \frac{1}{2\sigma_t (T-t)} E_t \int_t^T \left(\int_t^s [2a(u, Y_u)Y_u + b^2(u, Y_u)]du \right) ds.
\end{aligned}
$$

Then, approximating Y_u *by* $Y_t = \sigma_t$ *in the right-hand side of the above equality we obtain*

$$
\sqrt{\frac{1}{T-t}\int_t^T E_t(\sigma_s^2)ds} \approx \sigma_t + \frac{(T-t)}{4\sigma_t}[2a(t, Y_t)\sigma_t + b^2(t, Y_t)].
$$

This, jointly with (6.5.34), recovers (6.5.33).

Example 6.5.12 (The SABR model) *Consider a SABR model as in Section 3.3.1. This is a particular case of a diffusion model of the form (6.5.29), with* $f(x) = x$*, and where* $a(u, Y_u) = 0$ *and* $b(u, Y_u) = \alpha Y_u$*. Then, (6.5.32) reads as*

$$
I_t(k_t^*) \approx E_t[v_t] + \left(\frac{\rho \alpha \sigma_t^2}{4} - \frac{\rho^2 \alpha^2 \sigma_t}{8} \right)(T-t). \tag{6.5.35}
$$

Now, as

$$
E_t[v_t] \approx \sigma_t + \frac{\alpha^2 \sigma_t (T-t)}{4} - \frac{\alpha^2 \sigma_t (T-t)}{6} = \sigma_t + \frac{\alpha^2 \sigma_t (T-t)}{12}, \tag{6.5.36}
$$

we obtain the following approximation for the ATM implied volatility

$$I_t(k_t^*) \approx \sigma_t + \left(\frac{\alpha^2 \sigma_t}{12} + \frac{\rho \alpha \sigma_t^2}{4} - \frac{\rho^2 \alpha^2 \sigma_t}{8} \right) (T - t), \tag{6.5.37}$$

that coincides with the results in Hagan, Kumar, Lesniewski, and Woodward (2002).

Example 6.5.13 (The Heston model) *Consider a Heston model as in Section 3.3.2. This is a particular case of a diffusion model of the form (6.5.29), with $f(x) = x$, and where $a(u, Y_u) = \left(\frac{k\theta}{2} - \frac{\nu^2}{8}\right) \frac{1}{Y_u} - \frac{k}{2} Y_u$ and $b(u, Y_u) = \frac{\nu}{2}$. Then, (6.5.32) can be written as*

$$I_t(k_t^*) \approx E_t[v_t] + \frac{\rho^2 \nu^2 (T-t)}{96 \sigma_t} + \frac{\rho \nu \sigma_t (T-t)}{8}. \tag{6.5.38}$$

Now, as

$$E_t[v_t] \approx \sigma_t + \frac{(T-t)}{2} \left[\left(\frac{k\theta}{2} - \frac{\nu^2}{8} \right) \frac{1}{\sigma_t} - \frac{k}{2} \sigma_t \right] + \frac{\nu^2 (T-t)}{48 \sigma_t}$$

we obtain the following approximation for the ATM implied volatility

$$I_t(k_t^*) \approx \sigma_t + (T-t) \left[\left(\frac{k\theta}{2} - \frac{2\nu^2}{24} \right) \frac{1}{2\sigma_t} - \frac{k}{4} \sigma_t + \frac{\rho^2 \nu^2}{96 \sigma_t} + \frac{\rho \nu \sigma_t}{8} \right].$$

6.5.4.2 Local volatility models

Consider a local volatility model where $\sigma_t = \sigma(t, S_t)$, for some function σ. We have seen in Section 7.6.2 that

$$D_t^+ \sigma_t = \frac{\partial \sigma}{\partial x}(t, S_t) \sigma(t, S_t) S_t,$$

which implies that

$$D_t^+ \sigma_t^2 = 2\sigma_t \frac{\partial \sigma}{\partial x}(t, S_t) \sigma(t, S_t) S_t = 2 \frac{\partial \sigma}{\partial x}(t, S_t) \sigma^2(t, S_t) S_t.$$

In a similar way,

$$\begin{aligned}(D_t^+)^2 \sigma_t^2 =& 2 \frac{\partial^2 \sigma}{\partial x^2}(t, S_t) \sigma^3(t, S_t) S_t^2 \\ &+ 4 \left(\frac{\partial \sigma}{\partial x}(t, S_t) \right)^2 \sigma^2(t, S_t) S_t^2 + 2 \frac{\partial \sigma}{\partial x}(t, S_t) \sigma^3(t, S_t) S_t.\end{aligned} \tag{6.5.39}$$

Then, the expression (6.5.27) reads

$$
\begin{aligned}
&I_t(k_t^*) \\
&\approx E_t[v_t] + \frac{3}{8}\left(\frac{\partial \sigma}{\partial x}(t,S_t)\right)^2 \sigma(t,S_t)S_t^2(T-t) \\
&-\frac{1}{6}\left(\frac{\partial \sigma}{\partial x}(t,S_t)\right)^2 \sigma(t,S_t)S_t^2(T-t) - \frac{1}{6}\frac{\partial^2 \sigma}{\partial x^2}(t,S_t)\sigma^2(t,S_t)S_t^2(T-t) \\
&-\frac{1}{3}\left(\frac{\partial \sigma}{\partial x}(t,S_t)\right)^2 \sigma(t,S_t)S_t^2(T-t) - \frac{1}{6}\frac{\partial \sigma}{\partial x}(t,S_t)\sigma^2(t,S_t)S_t(T-t) \\
&+\frac{1}{4}\frac{\partial \sigma}{\partial x}(t,S_t)\sigma^2(t,S_t)S_t(T-t) \\
&= E_t[v_t] - \frac{1}{8}\left(\frac{\partial \sigma}{\partial x}(t,S_t)\right)^2 \sigma(t,S_t)S_t^2(T-t) \\
&-\frac{1}{6}\frac{\partial^2 \sigma}{\partial x^2}(t,S_t)\sigma^2(t,S_t)S_t^2(T-t) + \frac{1}{12}\frac{\partial \sigma}{\partial x}(t,S_t)\sigma^2(t,S_t)S_t(T-t).
\end{aligned} \tag{6.5.40}
$$

Now, as

$$
\begin{aligned}
\sigma(u,S_u) &= \sigma(t,S_t) + \int_t^u \frac{\partial \sigma}{\partial r}(r,S_r)dr + \int_t^u \frac{\partial \sigma}{\partial x}(r,S_r)\sigma(r,S_r)S_r dW_r \\
&+\frac{1}{2}\int_t^u \frac{\partial^2 \sigma}{\partial x^2}(r,S_r)\sigma^2(r,S_r)S_r^2 dr
\end{aligned} \tag{6.5.41}
$$

it follows that

$$
\begin{aligned}
\sigma^2(u,S_u) = \sigma^2(t,S_t) &+ 2\int_t^u \frac{\partial \sigma}{\partial r}(r,S_r)dr + 2\int_t^u \frac{\partial \sigma}{\partial x}(r,S_r)\sigma^2(r,S_r)S_r dW_r \\
&+ \int_t^u \frac{\partial^2 \sigma}{\partial x^2}(r,S_r)\sigma^3(r,S_r)S_r^2 dr \\
&+ \int_t^u \left(\frac{\partial \sigma}{\partial x}(r,S_r)\sigma^2(r,S_r)S_r\right)^2 dr.
\end{aligned} \tag{6.5.42}
$$

Then

$$
E_t[v_t] = \sqrt{\sigma_t^2 + \frac{1}{T-t}\int_t^T \left(\int_t^s \eta(u)dW_u + \lambda(u)du\right) ds},
$$

where

$$
\eta(u) := 2\frac{\partial \sigma}{\partial x}(u,S_u)\sigma^2(u,S_u)S_u
$$

and

$$\lambda(u) := 2\frac{\partial \sigma}{\partial u}(u, S_u) + \frac{\partial^2 \sigma}{\partial x^2}(r, S_r)\sigma^3(u, S_u)S_u^2 + \left(\frac{\partial \sigma}{\partial x}(u, S_u)\sigma^2(u, S_u)S_u\right)^2.$$

Now, a second-order Taylor expansion gives us that

$$\begin{aligned}
E_t[v_t] &\approx \sigma_t + \frac{1}{2\sigma_t(T-t)}E_t\int_t^T\left(\int_t^s \lambda(u)du\right)ds \\
&- \frac{4}{8\sigma_t^3(T-t)^2}E_t\left(\int_t^T\left(\int_t^s \eta^2(u)dW_u\right)ds\right)^2 \\
&\approx \sigma_t + \frac{1}{2\sigma_t(T-t)}E_t\int_t^T\left(\int_t^s \lambda(u)du\right)ds \\
&- \frac{\alpha^2}{2\sigma_t^3(T-t)^2}E_t\left(\int_t^T \eta(u)(T-u)dW_u\right)^2 \\
&\approx \sigma_t + \frac{1}{2\sigma_t(T-t)}E_t\int_t^T\left(\int_t^s \lambda(u)du\right)ds \\
&- \frac{1}{2\sigma_t^3(T-t)^2}E_t\left(\int_t^T \gamma^2(u)(T-u)^2du\right).
\end{aligned}$$

Then,

$$\begin{aligned}
E_t[v_t] &\approx \sigma(t, S_t) + \frac{(T-t)}{4\sigma_t}\lambda(t) - \frac{(T-t)}{6\sigma_t^3}\gamma^2(t) \\
&= \sigma(t, S_t) + \frac{(T-t)}{4\sigma_t} \\
&\times \left[2\frac{\partial \sigma}{\partial t}(t, S_t) + \frac{\partial^2 \sigma}{\partial x^2}(r, S_r)\sigma^3(t, S_t)S_t^2 dr + \left(\frac{\partial \sigma}{\partial x}(t, S_t)\sigma^2(t, S_t)S_t\right)^2\right] \\
&- \frac{2(T-t)}{3\sigma_t^3}\left(\frac{\partial \sigma}{\partial x}(t, S_t)\sigma^2(t, S_t)S_t\right)^2 \\
&= \sigma(t, S_t) + \frac{(T-t)}{4} \\
&\times \left[2\frac{\partial \sigma}{\partial t}(t, S_t) + \frac{\partial^2 \sigma}{\partial x^2}(t, S_t)\sigma^2(t, S_t)S_t^2 + \left(\frac{\partial \sigma}{\partial x}(t, S_t)\right)^2 \sigma^3(t, S_t)S_t^2\right] \\
&- \frac{2(T-t)}{3}\left(\frac{\partial \sigma}{\partial x}(t, S_t)\right)^2 \sigma(t, S_t)S_t^2
\end{aligned}$$

$$\begin{aligned}
&= \sigma(t, S_t) \\
&\times \left[1 + (T-t)S_t^2\left(\frac{1}{4}\frac{\partial^2 \sigma}{\partial x^2}(t, S_t)\sigma(t, S_t) + \left(\frac{\partial \sigma}{\partial x}(t, S_t)\right)^2\left(\frac{1}{4}\sigma^2(t, S_t) - \frac{2}{3}\right)\right)\right] \\
&+ \frac{(T-t)}{2}\frac{\partial \sigma}{\partial t}(t, S_t).
\end{aligned} \tag{6.5.43}$$

Example 6.5.14 *Consider the particular case of the CEV model $\sigma(t,S_t)=\sigma S_t^{\gamma-1}$, for some $\sigma>0$ and $\gamma>1$. Then*

$$\frac{\partial\sigma}{\partial t}(t,S_t)=0,$$

$$\frac{\partial\sigma}{\partial x}(t,S_t)=\sigma(\gamma-1)S_t^{\gamma-2},$$

and

$$\frac{\partial^2\sigma}{\partial x^2}(t,S_t)=\sigma(\gamma-1)(\gamma-2)S_t^{\gamma-3}.$$

Then (6.5.40) reads

$$\begin{aligned} I_t(k_t^*) &\approx E_t[v_t]-\frac{\sigma^3(\gamma-1)^2}{8}S_t^{3\gamma-3}(T-t)\\ &- \frac{\sigma^3(\gamma-1)(\gamma-2)}{6}S_t^{3\gamma-3}(T-t)\\ &+ \frac{\sigma^3(\gamma-1)}{12}S_t^{3\gamma-3}(T-t)\\ &= E_t[v_t]\\ &+ \sigma^3(\gamma-1)\left[-\frac{(\gamma-1)}{8}-\frac{(\gamma-2)}{6}+\frac{1}{12}\right]S_t^{3\gamma-3}(T-t). \end{aligned} \tag{6.5.44}$$

Moreover, from (6.5.45) it follows that

$$\begin{aligned} &E_t[v_t]\approx\sigma S_t^{\gamma-1}\\ &\times\left[1+(T-t)S_t^2\left(\frac{1}{4}\sigma^2(\gamma-1)(\gamma-2)S^{2\gamma-3}\right.\right.\\ &\left.\left.+\sigma^2(\gamma-1)^2S^{2\gamma-2}\left(\frac{1}{4}\sigma^2S_t^{2\gamma-2}-\frac{2}{3}\right)\right)\right]. \end{aligned} \tag{6.5.45}$$

Then, (6.5.44) and (6.5.45) give us the following approximation for the ATMI

$$\begin{aligned} &I_t(k_t^*)\approx\sigma S_t^{\gamma-1}\\ &+(T-t)\sigma^2(\gamma-1)\left[\frac{1}{4}(\gamma-2)S_t^{2\gamma-1}+\frac{1}{4}\sigma^2(\gamma-1)S_t^{4\gamma-2}-\frac{2}{3}(\gamma-1)S_t^{2\gamma-2}\right]\\ &+(T-t)\sigma^3(\gamma-1)\left[-\frac{(\gamma-1)}{8}-\frac{(\gamma-2)}{6}+\frac{1}{12}\right]S_t^{3\gamma-3}. \end{aligned}$$

6.5.4.3 Fractional volatilities

We have seen in Corollary 6.5.8 that the order of convergence of the difference between the volatility swap and the ATMI depends strongly on the Hurst parameter H. This section is devoted to the study some examples of models with fractional volatilities, both with $H > \frac{1}{2}$ and $H < \frac{1}{2}$.

Example 6.5.15 (The case $H > \frac{1}{2}$) *Assume as in Section 5.6.1 that the squared volatility σ^2 is of the form $\sigma_r^2 = f(Y_r)$, where $f \in \mathcal{C}_b^1(\mathbb{R})$ and Y_r is a fractional Ornstein-Uhlenbeck process of the form*

$$Y_r = m + (Y_0 - m)\, e^{-\alpha r} + c\sqrt{2\alpha} \int_0^r e^{-\alpha(r-s)} d\hat{W}_s^H, \tag{6.5.46}$$

where $\hat{W}_s^H = \int_0^t (t-s)^{H-\frac{1}{2}} dW_s$, with $H > \frac{1}{2}$. We proved that, in this case

$$D_r^W \sigma_t^2 = f'(Y_t) h(t,r),$$

where

$$h(t,r) = c\sqrt{2\alpha}\left(H - \frac{1}{2}\right)\left(\int_t^r e^{-\alpha(r-u)}(u-t)^{H-\frac{3}{2}} du\right).$$

Then, Equality (6.5.22) gives us that

$$\begin{aligned} &I_t(k_t^*) \\ &\approx E_t[v_t] + \frac{\rho}{4(T-t)} E_t\left[\int_t^T \int_s^T D_s^W \sigma_r^2 drds\right] \\ &\approx E_t[v_t] \\ &+ \frac{\rho c\sqrt{2\alpha}\left(H - \frac{1}{2}\right) f'(Y_t)}{4(T-t)} E_t\left[\int_t^T \int_s^T \left(\int_s^r e^{-\alpha(r-u)}(u-s)^{H-\frac{3}{2}} du\right) drds\right], \end{aligned} \tag{6.5.47}$$

which is $O(T-t)^{H+\frac{1}{2}}$.

Example 6.5.16 (The rough volatility case) *Consider the following model, similar to the rough Bergomi model*

$$\sigma_t^2 = \sigma_0^2 \exp\left(2\nu\sqrt{2H} W_t^H - \nu^2 t^{2H}\right), r \in [0,T], \tag{6.5.48}$$

for some constant $\nu > 0$ *and where* W^H *is a RLfBm with Hurst parameter* $H < \frac{1}{2}$. *Then*

$$D_r^W \sigma_t^2 = 2\nu\sqrt{2H}\sigma_t^2 (t-r)^{H-\frac{1}{2}},$$

$$D_r^W \sigma_t = \nu\sqrt{2H}\sigma_t (t-r)^{H-\frac{1}{2}},$$

and

$$D_u^W D_r^W \sigma_t^2 = 4\nu^2 (2H)\sigma_t^2 (t-r)^{H-\frac{1}{2}} (t-u)^{H-\frac{1}{2}}.$$

Then (6.5.21) gives us that

$$\begin{aligned}
&I_t(k_t^*) \\
&\approx E_t[v_t] + \frac{3H\nu^2\rho^2\sigma_t}{(H+\frac{1}{2})^2(H+\frac{3}{2})^2}(T-t)^{2H} \\
&- \frac{2H\nu^2\rho^2\sigma_t B(H+3/2, H+3/2)}{(H+\frac{1}{2})^2}(T-t)^{2H} \\
&- \frac{\rho^2\nu^2 H\sigma_t}{(H+\frac{1}{2})^2(H+1)}(t-T)^{2H} \\
&= \nu^2 H\rho^2\sigma_t (T-t)^{2H} \\
&\times \left(\frac{3}{(H+\frac{1}{2})^2(H+\frac{3}{2})^2} - \frac{2B(H+3/2, H+3/2)}{(H+\frac{1}{2})^2} - \frac{1}{(H+\frac{1}{2})^2(H+1)} \right),
\end{aligned} \tag{6.5.49}$$

and (6.5.24) reads as

$$\begin{aligned}
&I_t(k_t^*) \\
&\approx \sqrt{\frac{1}{T-t}\int_t^T E_t(\sigma_s^2)ds} \\
&- \frac{\nu^2 H\sigma_t}{2(H+\frac{1}{2})^2(H+1)}(T-t)^{2H} \\
&+ \nu^2 H\rho^2\sigma_t (T-t)^{2H} \\
&\times \left(\frac{3}{(H+\frac{1}{2})^2(H+\frac{3}{2})^2} - \frac{2B(H+3/2, H+3/2)}{(H+\frac{1}{2})^2} - \frac{1}{(H+\frac{1}{2})^2(H+1)} \right).
\end{aligned} \tag{6.5.50}$$

6.5.5 Numerical experiments

In this section we check the above approximations for the ATM implied volatility. In all the examples, the ATM implied volatility has been computed from a Monte Carlo simulation with $n = 100000$ and a time step $\Delta_t = \frac{1}{100}$. We have considered $t = 0$ and ATM options with maturity $T = \frac{7}{365}, \cdots, \frac{30}{365}$.

Example 6.5.17 (The SABR model) *In this example, we check the asymptotic behaviour described in 6.5.10 for the SABR model. The parameters are*

F_0	α	ρ	σ_0
100.0	0.6	-0.6	0.3

With this set of values, the approximation (6.5.35) reads as

$$\begin{aligned} I(0,T,X_0,k_0^*) - E[v_0] &\approx \left(\frac{\rho\alpha\sigma_t^2}{4} - \frac{\rho^2\alpha^2\sigma_t}{8}\right)T \\ &= -0.01296T. \end{aligned} \tag{6.5.51}$$

The simulated values of the ATM implied volatility, as well as their linear fit as a function of T, are shown in Figure 6.2. We can see that, in the short end, the difference between the ATM implied volatility and the volatility swap fits the approximation (6.5.51).

Example 6.5.18 (The Heston model) *Now we check the asymptotic behaviour described in (6.5.38) for the Heston model. Consider the parameters*

S_0	v_0	k	θ	ν	ρ
100	0.05	1.0	0.06	0.3	-0.9

Then, equality (6.5.38) reduces to

$$I(0,T,X_0,k^*0) - E[v_0] \approx \left(\frac{\rho^2\nu^2}{96\sigma_0} + \frac{\rho\nu\sigma_0}{8}\right)T = -0.00415T. \tag{6.5.52}$$

The simulated ATM values and their linear fit can be observed in Figure 6.3. We can see again that the simulated values fit the corresponding approximation formula.

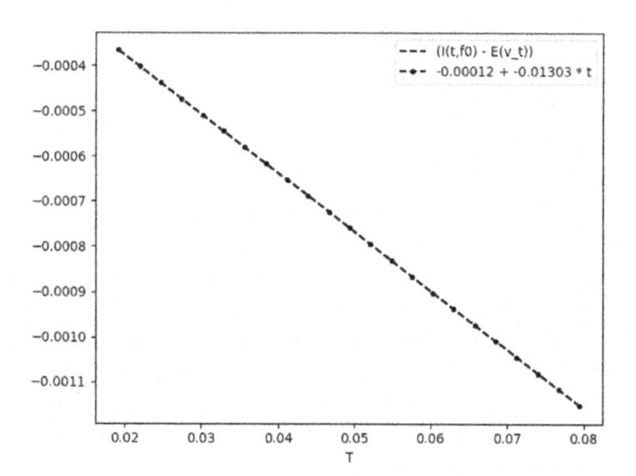

Figure 6.2 Asympthotic behaviour of the level ATM implied volatility SABR model with $F_0 = 100, \alpha = 0.6, \rho = -0.6$, and $\sigma_0 = 0.3$.

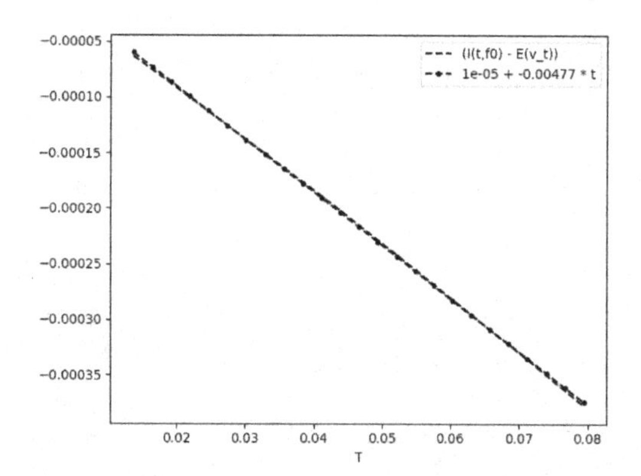

Figure 6.3 Asympthotic behaviour of the ATMI for the Heston model with $v_0 = 0.05, k = 1, \theta = 0.06, \nu = 0.3$, and $\rho = -0.9$.

Example 6.5.19 *Now we study the case of the CEV model, with the following parameters:*

F_0	σ	γ
100	0.3	0.4

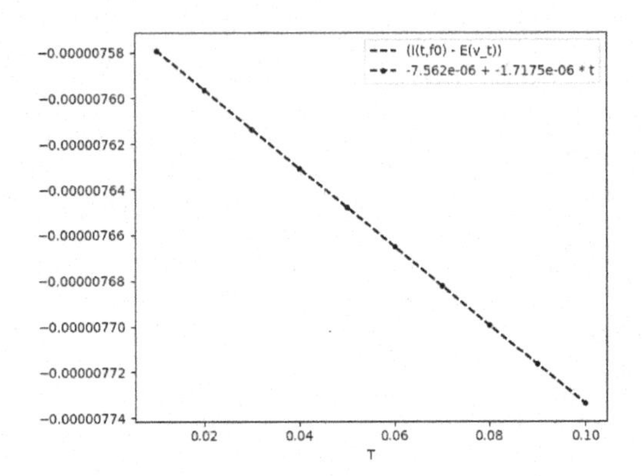

Figure 6.4 Asympthotic behaviour of the ATMI for the CEV model with $F_0 = 100, \sigma = 0.3$, and $\gamma = 0.4$.

Then, the approximation formula (6.5.44) takes the form

$$I(0,T,X_0,k_0^*) - E[v_0] \approx \sigma^3(\gamma-1)\left[-\frac{(\gamma-1)}{8} - \frac{(\gamma-2)}{6} + \frac{1}{12}\right] S_t^{3\gamma-3} T$$
$$= -1.7294 \times 10^{-6} T. \quad (6.5.53)$$

The simulation results are shown in Figure 6.4. Again, we can see a perfect fit for the short-end implied volatility and the approximation.

Example 6.5.20 *In this example, we check the asymptotic behaviour described in 6.5.10 for the rBergomi model (6.5.48). The considered parameters are*

v_0	ν	ρ	H
0.05	0.5	-0.6,0	0.3

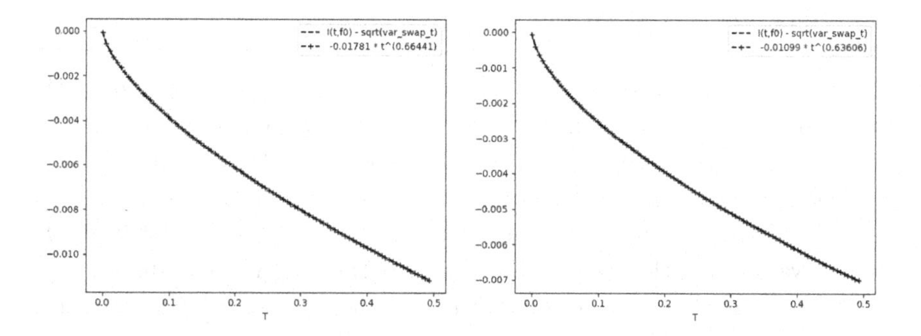

Figure 6.5 Asymphthotic behaviour of the level ATM implied volatility rBergomi model for $v_0 = 0.05, \nu = 0.5, H = 0.3$, and $\rho = -0.6$ (left) and $\rho = 0$ (right).

If $\rho = -0.6$, Equation (6.5.50) reads as

$$\begin{aligned}
&I(0,T,X_0,k_0^*) - \sqrt{\frac{1}{T-t}\int_t^T E_t(\sigma_s^2)ds} \\
&\approx -0.3\sqrt{0.05}0.5^2 \frac{T^{0.6}}{2(H+\frac{1}{2})^2(H+1)} \\
&+0.5^2 0.6^2 \sqrt{0.05} \\
&\times \left(\frac{3}{0.8^2 1.8^2} - \frac{2B(1.8,1.8)}{0.8^2} - \frac{1}{0.8^2 1.3}\right) T^{0.6} \\
&= -0.013003 T^{0.6},
\end{aligned} \tag{6.5.54}$$

while in the case $\rho = 0$ we get

$$\begin{aligned}
&I(0,T,X_0,k_0^*) - \sqrt{\frac{1}{T-t}\int_t^T E_t(\sigma_s^2)ds} \\
&\approx -0.3\sqrt{0.05}0.5^2 \frac{T^{0.6}}{2(H+\frac{1}{2})^2(H+1)} \\
&= -0.010078 T^{0.6}.
\end{aligned} \tag{6.5.55}$$

The simulated ATM implied volatilities and the corresponding power law fit can be seen in Figure 6.5.

6.6 CHAPTER'S DIGEST

The anticipating Itô's formula allows us to prove an extension of the Hull and White formula where option prices are decomposed as the sum of the corresponding prices if asset prices and volatilities are uncorrelated, and a term due to the correlation. Based on this representation, we easily deduce a similar decomposition of the implied volatility. The main advantage of these expressions is that they are not expansions, but exact formulas that can be applied to a wide class of volatility models, including the case of non-Markovian volatilities. In this chapter, we use them to study the ATMI of vanilla options. In the subsequent chapters, we extend this analysis to the study of the ATMI skew and curvature, both for vanilla and for exotic options.

A key observation in our analysis is the fact that we can write the fair strike of a volatility swap as $E_t(BS^{-1}(k_t^*, \Lambda_T))$, where $\Lambda_r := E_r\left[BS\left(t, X_t, k_t^*, v_t\right)\right]$. As, in the uncorrelated case, the ATMI is given by $BS^{-1}(k_t^*, \Lambda_t)$, the classical Itô formula allows us to evaluate the difference between both quantities as the expectation of a quadratic variation. Moreover, this quadratic variation can be represented in terms of the Malliavin derivative of the volatility process due to the Clark-Ocone-Haussman formula. From this representation, we deduce that the difference between the volatility swap and the ATMI is $O((T-t)^{2H+1})$, where H denotes the Hurst parameter of the volatility.

The previous result for the uncorrelated case, and the decomposition formula for the implied volatility, allows us to address the problem when the correlation is not zero. In this case, the Malliavin expansion technique allows us to identify the leading terms. Then we see that in the correlated case, this difference is $O((T-t)^{2H})$ if $H < \frac{1}{2}$, and $O((T-t)^{H+\frac{1}{2}})$ for $H > \frac{1}{2}$. That is, the convergence in the rough volatility case ($H < \frac{1}{2}$) is slower than for classical volatility models like Heston of SABR (where $H = \frac{1}{2}$).

Notice also that the square root of a variance swap can be written as the volatility swap plus a positive quantity that depends on the Malliavin derivative of the volatility process, from where we deduce that the difference between the volatility swap and the square root of the variance swap is $O((T-t)^{2H})$.

Our approach also allows us to deduce short-time approximation formulas for the ATMI, written as the volatility swap (or the square root of a variance swap) plus some terms depending on the Malliavin derivative of the volatility process. These short-time approximations are different in the cases $H > \frac{1}{2}$, $H = \frac{1}{2}$, and $H < \frac{1}{2}$, and they cover the cases of local, stochastic, and rough volatilities. The explicit computation of the Malliavin derivatives for each model leads us to simple expressions for most models in the literature. In particular, in the case $H = \frac{1}{2}$, our results fit some well-known formulas, as the Medeved-Scailled expansion for the case of diffusion volatilities.

CHAPTER 7

The ATM short-time skew

In Chapter 6, we have studied the short-time behaviour of the ATMI and its relationship with the volatility swap and the variance swap. Now, following Alòs, León, and Vives (2007), our objective is to study the short-time behaviour of the at-the-money implied volatility skew. In particular, we establish under which conditions the ATM skew slope tends to infinity (as observed in real market data, see Section 1.4.1), or to a constant value. Our approach can be applied to local, stochastic, and rough models, as it allows us to recover some well-known results in the literature, which appear as a consequence of general results*.

7.1 THE TERM STRUCTURE OF THE EMPIRICAL IMPLIED VOLATILITY SURFACE

In empirical implied volatility surfaces, smiles and skews are much more pronounced for short maturities. We can observe this phenomenon in Figure 7.1, corresponding to the EURO STOXX50 implied volatility surface, as of August 27, 2020. A popular rule-of-thumb, based on empirical observations, states that the skew slope is approximately $O((T-t)^{-\frac{1}{2}})$. We can see this blow-up in Figure 7.2, where we plot the EURO STOXX50 ATM skew slope as a function of time to maturity, as

*Adapted with permission from Springer Nature: Springer. Finance and Stochastics. On the short-time behavior of the implied volatility for jump-diffusion models with stochastic volatility, Elisa Alòs, Jorge A. León, and Josep Vives © Springer-Verlag (2007).

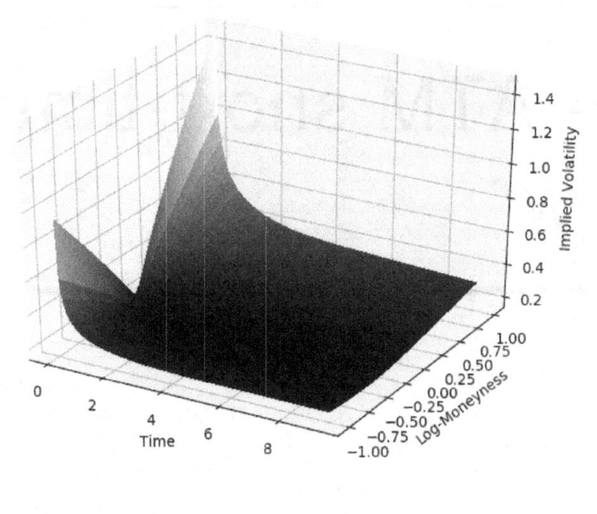

Figure 7.1 EURO STOXX50 implied volatility surface, August 17, 2020.

of November 19, 2020. Notice that these empirical skews follow a power law with an exponent equal to -0.42538.

The term structure of the implied volatility surface is not easily explained by classical stochastic volatility models, whose corresponding rate is $O(1)$ (see Lewis (2000), Lee (2005), or Medveved and Scaillet (2003)). Then, classical volatility models fail in adequately reproducing the short-end of the implied volatility surface.

There have been several attempts to construct a stochastic volatility model describing this phenomenon. Some works in the literature focus on the introduction of more volatility factors (as in the two-factor Bergomi model). These models allow for two scales of mean-reversion and lead to a power-like behaviour over a sufficiently wide range of maturities (see Bergomi (2009b)). Another approach has been introduced in Fouque, Papanicolau, Sircar, and Solna (2004), where the authors propose a model with time-varying coefficients, that depend on the option maturity dates. Other approaches are based on the introduction of jump processes. Even when the rate of the skew slope for most models with jumps is still $O(1)$ (as it is shown by Medveved and Scaillet (2003)), they allow flexible modelling, and generate skews and smiles similar to those observed in market data (see Bates (1996), Barndorff-Nielsen and

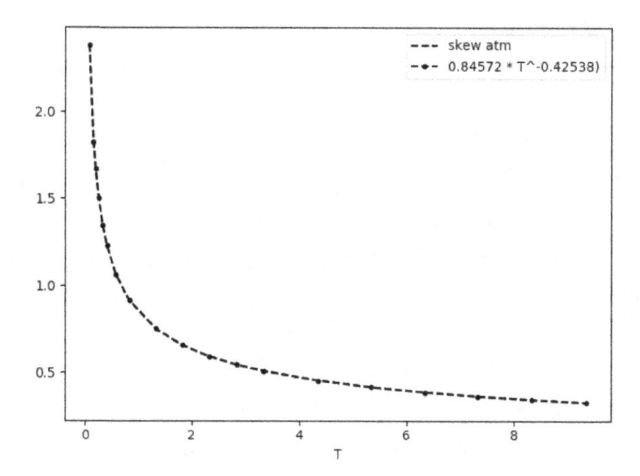

Figure 7.2 EURO STOXX50 ATM skew slope as a function of time to maturity, as of November 19, 2020.

Shephard (2001, 2002), Kijima (2002), Carr and Wu (2003), and Cont and Tankov (2004)).

Our analysis in this chapter relies on the explicit computation of the limit of the at-the-money skew slope for stochastic volatility models, in terms of the Malliavin derivative of the volatility process. This approach allows us to analyse the short-end ATMI skew of local, stochastic, and rough volatilities. Moreover, we can compare these limit skews with the ones corresponding to jump-diffusion models.

7.2 THE MAIN PROBLEM AND NOTATIONS

Through this chapter, we assume the same model (6.1.1) for log-prices and the same notations as in Chapter 6.

Let $I_t(k)$ denote the implied volatility process defined by

$$V_t(k) = BS(t, X_t, k, I_t(k)), \tag{7.2.1}$$

where $BS(t, x, k, \sigma)$ denotes the classical Black-Scholes price for a call option with time to maturity $T - t$, log-price X_t, log-strike k, and volatility σ. The first step in our approach is to take implicit derivatives in (7.2.1)

$$\frac{\partial V_t}{\partial k}(k) = \frac{\partial BS}{\partial k}(t, X_t, k, I_t(k)) + \partial_\sigma BS(t, X_t, k, I_t(k))\frac{\partial I_t}{\partial k}(k), \quad (7.2.2)$$

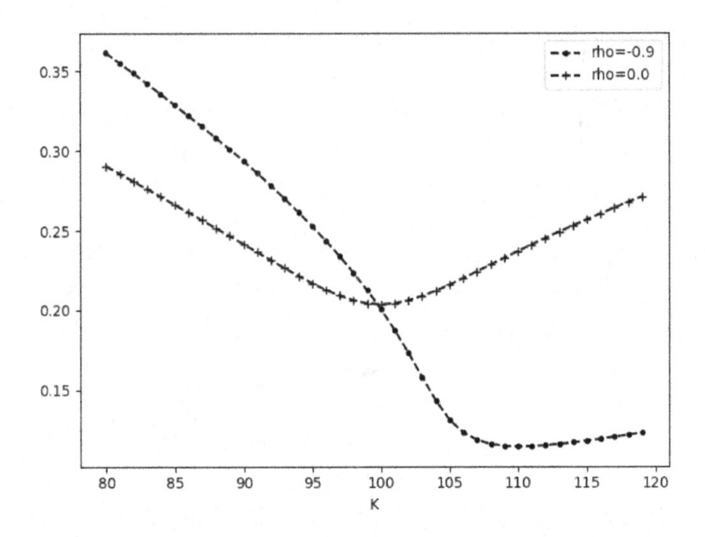

Figure 7.3 Heston smiles with $S_0 = 100, r = 0, v_0 = 0.05, \theta = 0.05, k = 0.5$, $\nu = 1.1$, and correlation values $\rho = -0.9$ and $\rho = 0$.

from where we develop adequate expressions for $\frac{\partial I_t}{\partial k}(k)$ that allow us to estimate the ATM short-end limit of this skew slope

$$\lim_{T \downarrow t} \frac{\partial I_t}{\partial k}(k_t^*),$$

where $k_t^* := X_t + r(T-t)$ denotes the at-the-money strike. More precisely, we will see that in the uncorrelated case $\rho = 0$ this slope is zero, and that all the contribution to the skew comes from the correlation.

Before starting our analysis, we can observe this dependence on the correlation parameter in Figure 7.3, corresponding to the simulated implied volatilities of a 1-year European call option under a Heston model with $S_0 = 100, r = 0, v_0 = 0.05, \theta = 0.05, k = 0.5$, $\nu = 1.1$, and correlation values $\rho = -0.9$ and $\rho = 0$. Notice that, ATM, the implied volatility is decreasing when $\rho = -0.9$, while it is locally convex if $\rho = 0$.

As we did in the study of the ATMI short-end level, we start with the analysis of the uncorrelated case.

7.3 THE UNCORRELATED CASE

Let us consider first the case $\rho = 0$. The following result (see Proposition 3.2 in Alòs and León (2017)) gives au that the at-the-money implied volatility skew slope is zero in this case. The proof is a direct consequence of Equation (7.2.2) and the Hull and White formula (6.2.2).

Proposition 7.3.1 (The ATMI volatility skew in the uncorrelated case) Consider the model (6.1.1) with $\rho = 0$ and assume that $\sigma_t^1 \leq \sigma_t$, where $\sigma^1 : [0, T] \to (0, \infty)$ is an integrable function. Then, for all $t \in [0, T]$, $\frac{\partial I_t}{\partial k}(k_t^*) = 0$.

Proof. From the definition of the implied volatility and the Hull and White formula (6.2.2) we can write

$$E_t\left(BS(t, X_t, k, v_t)\right) = BS(t, X_t, k, I(k)).$$

Then

$$\begin{aligned} E_t\left(\frac{\partial BS}{\partial k}(t, X_t, k, v_t)\right) &= \frac{\partial BS}{\partial k}(t, X_t, k, I(t,k)) \\ &+ \frac{\partial BS}{\partial \sigma}(t, X_t, k, I(t,k))\frac{\partial I}{\partial k}(k). \end{aligned} \tag{7.3.1}$$

Now, notice that, for any positive constant a

$$\frac{\partial BS}{\partial k}(t, X_t, k_t^*, a) = -\frac{1}{2}\left(\exp(X_t) - BS(t, X_t, k_t^*, a)\right).$$

Then,

$$\begin{aligned} E_t\left(\frac{\partial BS}{\partial k}(t, X_t, k_t^*, v_t)\right) &= \frac{1}{2}\left(E_t\left(BS(t, X_t, k_t^*, v_t)\right) - \exp(X_t)\right) \\ &= \frac{1}{2}\left(BS(t, X_t, k_t^*, I(k^*)) - \exp(X_t)\right) \\ &= \frac{\partial BS}{\partial k}(t, X_t, k_t^*, I(k^*)), \end{aligned}$$

which, jointly with (7.3.1) and the fact that the vega $\frac{\partial BS}{\partial \sigma} \neq 0$, implies that $\frac{\partial I}{\partial k}(k^*) = 0$. Now the proof is complete. ■

Remark 7.3.2 *The above result proves that, fixed $t \in [0, T]$, the implied volatility $I_t(k)$ has, in the uncorrelated case, a stationary point at $k = k_t^*$. Notice that this result is independent of the stochastic volatility* *model and agrees with Theorem 4.2 in Renault and Touzi (1996), where it is established that the implied volatility, as a function of the strike, is continuously differentiable, decreasing for in-the-money options, and increasing for out-of-the-money options (see also Proposition 5 in Renault (1997)).*

Remark 7.3.3 *From the same arguments as in the proof of Proposition 7.3.1 we deduce that, under the same hypotheses, the derivative with respect to the log-asset price* $\frac{\partial I_t}{\partial X_t}(x_t^*) = 0$, *where* $x_t^* = k - r(T-t)$.

7.4 THE CORRELATED CASE

Let us now study the correlated case. Now our analysis is based on Equation (7.2.2) and on the decomposition formula in Theorem (6.3.2). Our first step is the proof of the following expression of the at-the-money implied volatility skew slope. Even when this expression is not straightforward, it is an exact expression, that will be a useful tool in the computation of the short-time limit $\lim_{T\downarrow t} \frac{\partial I_t}{\partial k}(k_t^*)$.

Proposition 7.4.1 *Assume the model (1) holds with* $\rho \in (-1,1)$ *and* $\sigma \in \mathbb{L}_W^{1,2}$, *and that for every fixed* $t \in [0,T]$, $E_t\left(\int_t^T \sigma_s^2 ds\right)^{-1} < \infty$ *a.s. Then, it follows that*

$$\frac{\partial I_t}{\partial k}(k_t^*) = \frac{E_t(\int_t^T (\frac{\partial F}{\partial k}(s, X_s, k^*, v_s) - \frac{1}{2}F(s, X_s, k^*, v_s))ds)}{\frac{\partial BS}{\partial \sigma}(t, X_t, k^*, I_t(k^*))}, \quad a.s.$$

where

$$F(s, X_s, k, v_s) \quad := \quad \frac{\rho}{2} e^{-r(s-t)} H(s, X_s, k, v_s)\Phi_s,$$

with $H := \left(\frac{\partial^3}{\partial x^3} - \frac{\partial^2}{\partial x^2}\right) BS$ *and* $\Phi_s := \sigma_s \int_s^T D_s^W \sigma_u^2 du$.

Proof. Taking partial derivatives with respect to k on the expression $V_t = BS(t, X_t, k, I_t(X_t, k))$ we get

$$\frac{\partial V_t}{\partial k} = \frac{\partial BS}{\partial k}(t, X_t, k, I(t,k)) + \frac{\partial BS}{\partial \sigma}(t, X_t, k, I(t,k))\frac{\partial I}{\partial k}(t,k). \quad (7.4.1)$$

Now, Theorem 6.3.2 gives us that

$$V_t = E_t(BS(t, X_t, v_t)) + E_t\left(\int_t^T F(s, X_s, k, v_s)ds\right).$$

This, jointly with (7.4.1), implies that

$$\frac{\partial V_t}{\partial k} = E_t\left(\frac{\partial BS}{\partial k}(t, X_t, v_t)\right) + E_t\left(\int_t^T \frac{\partial F}{\partial k}(s, X_s, k, v_s)ds\right). \quad (7.4.2)$$

Notice that the conditional expectation $E_t\left(\int_t^T \frac{\partial F}{\partial k}(s, X_s, v_s)ds\right)$ is well defined and finite a.s. We can prove it in the following way. From one hand, the same arguments as in Proposition 6.3.1 allow us to write

$$E\left(\frac{\partial F}{\partial k}(s, X_s, k, v_s | \mathcal{G}_t\right) \leq C\left(\int_t^T \sigma_s^2 ds\right)^{-1},$$

where $\mathcal{G}_t = \mathcal{F}_t^B \vee \mathcal{F}^W$. Now, as $E\left(\int_t^T \sigma_s^2 ds \middle| \mathcal{F}_t\right)^{-1} < \infty$, the expectation $E_t\left(\int_t^T \frac{\partial F}{\partial k}(s, X_s, v_s)ds\right)$ is finite.

Thus, (7.4.1) and (8.3.4) imply

$$\frac{\partial I_t}{\partial k}(k_t^*) \quad (7.4.3)$$
$$= \frac{E_t(\frac{\partial BS}{\partial k}(t, X_t, k_t^*, v_t)) - \frac{\partial BS}{\partial k}(t, X_t, k_t^*, I_t(X_t, k_t^*)) + E_t(\int_t^T \frac{\partial F}{\partial k}(s, X_s, k_t^*, v_s)ds)}{\frac{\partial BS}{\partial \sigma}(t, X_t, k_t^*, I_t(X_t, k_t^*))}.$$

Now, notice that

$$\begin{aligned} E_t\left(\frac{\partial BS}{\partial k}(t, X_t, k_t^*, v_t)\right) &= \frac{\partial}{\partial k} E_t(BS(t, X_t, k, v_t)\Big|_{x=x_t^*} \\ &= \frac{\partial BS}{\partial k}(t, X_t, k, I_t^0(X_t, k))|_{k=k_t^*}, \end{aligned} \quad (7.4.4)$$

where $I_t^0(X_t, k)$ is the implied volatility in the case $\rho = 0$. Moreover, the classical Hull and White formula gives us

$$\frac{\partial BS}{\partial k}(t, X_t, k, I_t^0(X_t, k))\Big|_{k=k_t^*}$$
$$= \frac{\partial BS}{\partial k}(t, X_t, k_t^*, I_t^0(X_t, k^*)) + \frac{\partial BS}{\partial \sigma} BS(t, X_t, k_t^*, I_t^0(X_t, k^*)) \frac{\partial I_t^0}{\partial k}(X_t, k_t^*). \quad (7.4.5)$$

From Proposition 7.3.1 we know that $\frac{\partial I_t^0}{\partial k}(k_t^*) = 0$. Then, (7.4.4) and

(7.4.5) imply that

$$
\begin{aligned}
&E_t\left(\frac{\partial BS}{\partial k}(t, X_t, k_t^*, v_t\right) - \frac{\partial BS}{\partial k}(t, X_t, k_t^*, I_t(k_t^*)) \\
&= \frac{\partial BS}{\partial k}(t, x_t^*, I_t^0(x_t^*)) - \frac{\partial BS}{\partial k}(t, X_t, k_t^*, I_t(k_t^*)). \qquad (7.4.6)
\end{aligned}
$$

Now, direct computations lead us to, for every $\sigma > 0$

$$
\frac{\partial BS}{\partial k}(t, X_t, k_t^*, \sigma) = -\frac{1}{2}\left(e^{X_t} - BS(t, X_t, k_t^*, \sigma)\right),
$$

which implies

$$
\begin{aligned}
&\frac{\partial BS}{\partial k}(t, X_t, k_t^*, I_t^0(k_t^*)) - \frac{\partial BS}{\partial k}(t, X_t, k_t^*, I_t(k_t^*)) \\
&= \frac{1}{2}(BS(t, X_t, k_t^*, I_t^0(k_t^*)) - BS(t, X_t, k_t^*, I_t(k_t^*))) \\
&= \frac{1}{2}(E_t(BS(t, x_t^*, v_t) - V_t(x_t^*)) \\
&= -\frac{1}{2}E_t\left(\int_t^T F(s, X_s, k_t^*, v_s)ds\right).
\end{aligned}
$$

This, jointly with (7.4.3) and (7.4.6), allows us to complete the proof. ■

Remark 7.4.2 *The result in Proposition 7.4.1 is not an asymptotic expansion but an exact formula.*

Remark 7.4.3 *As in Theorem 6.3.2, the hypothesis $\rho \in (-1, 1)$ has been taken for the sake of simplicity, and the same result applies for $\rho \in [-1, 1]$ provided the volatility process σ satisfies some adequate integrability conditions.*

7.5 THE SHORT-TIME LIMIT OF IMPLIED VOLATILITY SKEW

Here, our purpose is to use the result in Proposition 7.4.1 to study the limit of $\frac{\partial I_t}{\partial k}(k_t^*)$ when $T \downarrow t$.

Henceforth, we will consider the following hypotheses:
(H1) $\sigma \in \mathbb{L}_W^{2,2}$
(H2) There exists a constant $a > 0$ such that $\sigma > a > 0$.
(H3) There exists a constant $H \in (0, 1)$ such that, for all $0 < t < s < r < T$,

$$
E_t\left(\left(D_s^W \sigma_r\right)^2\right) \leq C\,(r - s)^{2H-1} \qquad (7.5.1)
$$

$$E_t\left(\left(D_\theta^W D_s^W \sigma_r\right)^2\right) \leq C\,(r-s)^{2H-1}\,(r-\theta)^{2H-1}. \tag{7.5.2}$$

As in Chapter 6, these hypotheses have been chosen for the sake of simplicity of the exposition, but they can be replaced by other adequate integrability conditions.

In the next proposition, we identify the leading term in the computation of the short-time limit of the skew slope. In order to identify this term, we apply the same expansion technique introduced in the proof of Theorem 6.5.5.

Proposition 7.5.1 *Assume that the model (1) with $\rho \in [-1,1]$ and hypotheses (H1)–(H3) hold. Then*

$$\begin{aligned}&\frac{\partial BS}{\partial \sigma}(t, X_t, k_t^*, I_t(k_t^*))\frac{\partial I_t}{\partial k}(k_t^*)\\ =\;& \frac{\rho}{2}E_t\left(L(t, X_t, k_t^*, v_t)\int_t^T \Phi_s ds|\right) + O(T-t)^{(2H)\wedge 1},\end{aligned}$$

as $T \to t$ and where $L(t, x_t^*, v_t) = (\partial_k - \frac{1}{2})H(t, x_t^*, v_t)$.

Proof. Straightforward computations show that the results in Theorem 6.3.2 and Proposition 7.4.1 also hold for $\rho \in [-1,1]$ provided the stronger hypotheses (H1)–(H3) are satisfied. Then we have that

$$\begin{aligned}&\frac{\partial BS}{\partial \sigma}(t, X_t, k_t^*, I_t(k_t^*))\frac{\partial I_t}{\partial k}(k_t^*)\\ =\;& \frac{\rho}{2}E_t\left(\int_t^T e^{-r(s-t)}\left(\frac{\partial}{\partial k}-\frac{1}{2}\right)\partial_x G(s, X_s, k_t^*, v_s)\Phi_s ds\right)\end{aligned} \tag{7.5.3}$$

Now we see that the term in the right-hand side of the above equation can be written as

$$\frac{\rho}{2}E_t\left(L(t, X_t, k_t^*, v_t)\int_t^T \Phi_s ds\right) + O\,(T-t)^{2H}, \tag{7.5.4}$$

where $L(s, X_s, k_t^*, v_s) = (\partial_k - \frac{1}{2})H(t, x_t^*, v_t)$. Towards this end, we do a Malliavin expansion as described in Remark 6.5.6. That is, we apply the anticipating Itô formula (see Section 4.3) to the process

$$A_s := \frac{\rho}{2}e^{-r(s-t)}L(s, X_s, k_t^*, v_s)\left(\int_s^T \Phi_r dr\right).$$

Then, after taking conditional expectations with respect to $\mathcal{F}_t$, we get

$$
\begin{aligned}
&\frac{\rho}{2}E_t\left(\int_t^T e^{-r(s-t)}L(s,X_s,k_t^*,v_s)\Phi_s ds\right) = \frac{\rho}{2}E_t\left(L(t,X_t,k_t^*,v_t)\int_t^T \Phi_s ds\right)\\
&+\frac{\rho^2}{4}E_t\left(\int_t^T e^{-r(s-t)}\left(\frac{\partial^3}{\partial x^3}-\frac{\partial^2}{\partial x^2}\right)L(s,X_s,k_t^*,v_s)(\int_s^T \Phi_r dr)\Phi_s ds\right)\\
&+\frac{\rho^2}{2}E_t\left(\int_t^T e^{-r(s-t)}\frac{\partial L}{\partial x}(s,X_s,k_t^*,v_s)(\int_s^T D_s^W\Phi_r dr)\sigma_s ds\right)\\
&=\frac{\rho}{2}E_t\left(L(t,X_t,k_t^*,v_t)\int_t^T \Phi_s ds\right)+S_1+S_2.
\end{aligned}
$$

Lemma 6.3.1 gives us that

$$
\begin{aligned}
S_1 &= \frac{\rho^2}{4}E_t\left(\int_t^T e^{-r(s-t)}E\left[\left(\frac{\partial^3}{\partial x^3}-\frac{\partial^2}{\partial x^2}\right)L(s,X_s,k_t^*,v_s)\bigg|\mathcal{G}_t\right](\int_s^T \Phi_r dr)\Phi_s ds\right)\\
&\leq C\sum_{k=4}^{6}E_t\left[\left(\int_t^T \sigma_s^2 ds\right)^{-\frac{k}{2}}\int_t^T |(\int_s^T \Phi_r dr)\Phi_s| ds\right].
\end{aligned}
$$

Now, hypotheses (H2) and (H3) allow us to write

$$
\begin{aligned}
S_1 &\leq C\sum_{k=0}^{2}E_t\left[\left(\int_t^T \sigma_\theta^2 d\theta\right)^{-\frac{k}{2}}\left(\int_t^T\int_\theta^T \left(D_r^W\sigma_\theta\right)^2 dr d\theta\right)\right]\\
&\leq C(T-t)^{-1}\left(\int_t^T\int_t^\theta (\theta-r)^{2H-1}dr d\theta\right)\leq C(T-t)^{2H}.
\end{aligned}
$$

Using similar arguments it follows that $S_2 = O(T-t)^{2H}$, which proves (7.5.4). Therefore, the proof is complete. ■

Now we can state the main result of this chapter, which is an adaptation of Theorem 6.2 in Alòs, León, and Vives (2007). We will consider the following hypotheses:

(H4) σ has a.s. right-continuous trajectories.
(H5) There exists $H \in (0,1)$ such that

$$
\frac{1}{(T-t)^{\frac{3}{2}+H}}\int_t^T\left(\int_s^T D_s^W\sigma_r^2 dr\right)ds \to C
$$

a.s., for some constant C, as $T \to t$.

Theorem 7.5.2 *Consider the model (1) holds with* $\rho \in [-1,1]$ *and suppose that hypotheses (H1)–(H5) hold. Then*

$$\lim_{T\to t}(T-t)^{\frac{1}{2}-H}\frac{\partial I_t}{\partial k}(k_t^*) = \frac{\rho}{2\sigma_t^2}\lim_{T\to t}\frac{1}{(T-t)^{\frac{3}{2}+H}}E_t\int_t^T\left(D_s^W\int_s^T\sigma_r^2 dr\right)ds. \tag{7.5.5}$$

Proof. Using Proposition 7.5.1 and the facts that

$$\frac{\partial BS}{\partial\sigma}(t,X_t,k_t^*,I_t(k_t^*)) = \frac{e^k e^{-r(T-t)}e^{\frac{-I_t(x_t^*)^2(T-t)}{8}}\sqrt{T-t}}{\sqrt{2\pi}},$$

and

$$L(t,x_t^*,v_t) = e^k e^{-r(T-t)}\frac{1}{\sqrt{2\pi}}e^{-\frac{v_t^2(T-t)}{8}}v_t^{-3}(T-t)^{-\frac{3}{2}}$$

we can write

$$\begin{aligned}
&\frac{\partial I_t}{\partial k}(k_t^*)\\
=\;& \frac{\rho}{2}e^{\frac{I_t(x_t^*)^2(T-t)}{8}}(T-t)^{-2}E_t\left(e^{-\frac{v_t^2(T-t)}{8}}v_t^{-3}\int_t^T\Phi_s ds\right)\\
&+O(T-t)^{\left(2H-\frac{1}{2}\right)\wedge\frac{1}{2}}\\
=:\;& S_1+O(T-t)^{\left(2H-\frac{1}{2}\right)\wedge 1}.
\end{aligned}$$

From the results in Chapter 6, we know that $I_t(x_t^*)^2(T-t)\to 0$ as $T\to t$. Then,

$$\lim_{T\to t}(T-t)^{\frac{1}{2}-H}S_1 = \frac{\rho}{2}\lim_{T\to t}(T-t)^{\frac{1}{2}-H}\left[(T-t)^{-2}E_t\left(e^{-\frac{v_t^2(T-t)}{8}}v_t^{-3}\int_t^T\Phi_s ds\right)\right].$$

Then, the results follow from (H4) and (H5). Now the proof is complete. ■

Remark 7.5.3 *In a similar way, we can prove the following result on the derivative with respect to the log-asset price.*

$$\lim_{T\to t}(T-t)^{\frac{1}{2}-H}\frac{\partial I_t}{\partial X_t}(x_t^*) = -\frac{\rho}{2\sigma_t^2}\lim_{t\to T}\frac{1}{(T-t)^{\frac{3}{2}+H}}E_t\int_t^T\left(D_s^W\int_s^T\sigma_r^2 dr\right)ds, \tag{7.5.6}$$

where $x_t^* = X_t - r(T-t)$ *(see Alòs, León, and Vives (2007)).*

Remark 7.5.4 (Multifactor volatilities) *In multifactor volatility models, option prices satisfy an equation of the form*

$$dX_t = \left(r - \frac{\sigma_t^2}{2}\right) dt + \sigma_t \left(\sum_{i=1}^{d} \rho_i dW_t^i + \sqrt{1 - \sum_{i=1}^{d} \rho_i^2 dB_t}\right),$$

where $W = (W^1, ..., W^d)$ is a d-dimensional Brownian motion and σ is a square integrable process adapted to the filtration generated by W. Then straightforward computations give us that Theorem 7.5.2 also holds, replacing $\rho \int_s^T D_s^W \sigma_s^2 ds$ by $\sum_{i=1}^d \rho_i \int_s^T D_s^{W^i} \sigma_s^2 ds$. That is,

$$\lim_{T \to t} (T-t)^{\frac{1}{2}-H} \frac{\partial I_t}{\partial k}(x_t^*) = \frac{1}{2\sigma_t^2} \lim_{t \to T} \frac{1}{(T-t)^{\frac{3}{2}+H}} \sum_{i=1}^{d} \rho_i E_t \int_t^T \left(D_s^{W^i} \int_s^T \sigma_r^2 dr\right) ds. \tag{7.5.7}$$

Remark 7.5.5 *Assume that there exists a $\mathcal{F}_t$-measurable random variable $D_t^+ \sigma_t$ such that, for every $t > 0$,*

$$sup_{s,r \in [t,T]} \left| E_t \left(D_s^W \sigma_r - D_t^+ \sigma_t\right)\right| \to 0, \tag{7.5.8}$$

a.s. as $T \to t$. Then, Theorem 7.5.2 implies that

$$\lim_{T \to t} \frac{\partial I_t}{\partial k}(k_t^*) = \frac{\rho D_t^+ \sigma_t}{2\sigma_t}. \tag{7.5.9}$$

This is the case of classical diffusion volatility models, as we discuss in detail in Section 7.6.1.

Remark 7.5.6 (The short-end blow-up of the skew slope) The main contribution of Theorem 7.5.2 is that it proves that if we want the at-the-money skew slope $\frac{\partial I_t}{\partial k}(k_t^*)$ to tend to infinity as observed in real market practice, the stochastic volatility process has to satisfy (H5) with $H < \frac{1}{2}$. This property is not satisfied by classical stochastic volatility models but it is satisfied by models based on the fractional Brownian motion with $H < \frac{1}{2}$, as we see in the following examples.

Remark 7.5.7 (A model-free approximation for the volatility swap) In Theorem 6.5.5 we proved that, for $H \geq \frac{1}{2}$,

$$\lim_{T \to t} \frac{I_t(k_t^*) - E_t[v_t]}{(T-t)^{\frac{1}{2}+H}} = \frac{\rho}{4} \lim_{T \to t} \frac{1}{(T-t)^{\frac{3}{2}+H}} \int_t^T \int_s^T D_s^W \sigma_r^2 dr ds + O(\rho^2).$$

Now, Theorem 7.5.2 gives us that the limit in the right-hand side of this equality can be written in terms of the skew slope. That is, we can write

$$\lim_{T\to t}\frac{I_t(k_t^*)-E_t[v_t]}{(T-t)^{\frac{1}{2}+H}}=\frac{\sigma_t^2}{2}\lim_{T\to t}\frac{1}{(T-t)^{H-\frac{1}{2}}}\frac{\partial I_t}{\partial k}(k_t^*)+O(\rho^2). \quad (7.5.10)$$

This equality is in line with the previous results in Section 6.5 of Carr and Lee (2008) on the impact of correlation on volatility swap prices. Moreover, (7.5.10) gives us, in the case $H\geq 1/2$ (replacing σ_t by the ATMI $I_t(k_t^*)$), the following model-free approximation formula for the volatility swap

$$E_t[v_t]\approx I_t(k_t^*)-\frac{I_t(k_t^*)^2}{2}\frac{\partial I_t}{\partial k}(k_t^*)(T-t), \quad (7.5.11)$$

that is similar to the first-order vol-of-vol expansion around the variance swap in Bergomi and Guyon (2012). We can analyse the goodness of the approximation 7.5.11 in Figure 7.4, where we plot the log-ratio

$$\ln\left(\frac{|E_t[v_t]-I_t(k_t^*)|}{|E_t[v_t]-I_t(k_t^*)-\frac{I(t,T,X_t,k_t^*)^2}{2}\frac{\partial I}{\partial k}(t,T,X_t,k_t^*)(T-t)|}\right)$$

for a SABR model with $S_0=100, \rho=-0.6$, $\alpha=0.5$, and $\sigma_0=0.5$, where we take $t=0$ and varying values of T. We can observe that this log-ratio is always positive, meaning that the error of the approximation 7.5.11 is less than the error we do in approximating the volatility swap by the ATMI.

The goodness of the approximation (7.5.11) relies on the fact that it is immune to the correlation up to the order $O(\rho^2)$. As pointed out in Remark 6.5.9, other immune-to-correlation approximations of the volatility swap have been studied in Carr and Lee (2008) and Rolloos and Arslan (2017). Analytical results prove that the zero-vanna approach in Rolloos and Arslan (2017) outperforms the approximation (7.5.11) (see Alòs, Rolloos, and Shiraya (2020)).

Remark 7.5.8 *Theorem 7.5.2 suggests the following approximation for the ATM skew slope*

$$\frac{\partial I_t}{\partial k}(k_t^*)\approx\frac{\rho}{2\sigma_t^2}\frac{1}{(T-t)^2}\int_t^T\left(D_s^W\int_s^T\sigma_r^2dr\right)ds. \quad (7.5.12)$$

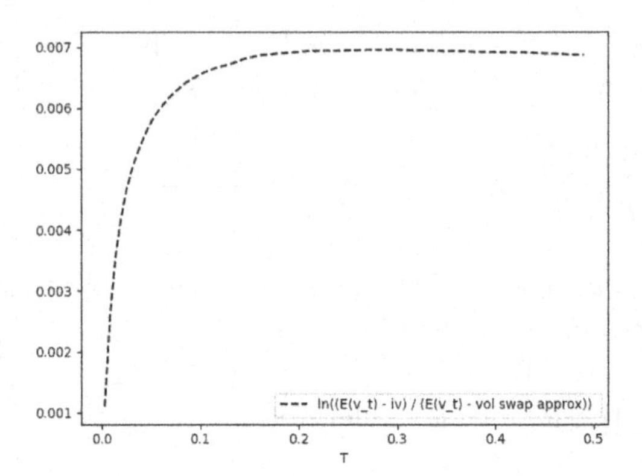

Figure 7.4 Log-ratio $\ln\left(\frac{|E[v_0]-I(0,T,X_0,k_0^*)|}{|E[v_0]-I(0,T,X_0,k_0^*)-\frac{I(0,T,X_0,k_0^*)^2}{2}\frac{\partial I}{\partial k}(0,T,X_0,k_0^*)T|}\right)$ for a SABR model with $S_0 = 100, \rho = -0.6$, $\alpha = 0.5$, and $\sigma_0 = 0.5$.

7.6 APPLICATIONS

In this section, we apply the above results to analyse the short-end skew slope for different types of models as local, stochastic, or rough volatilities, among others.

7.6.1 Diffusion stochastic volatilities: finite limit of the ATM skew slope

Assume that the volatility σ_r is of the form $\sigma_r = f(Y_r)$, where $f \in C_b^1(\mathbb{R})$ and Y_r is the solution of a stochastic differential equation

$$dY_r = a(r, Y_r)\, dr + b(r, Y_r)\, dW_r, \tag{7.6.1}$$

for some real functions $a, b \in C_b^1(\mathbb{R})$. Then, the results in Section 3.2.2 give us that $Y \in \mathbb{L}_W^{1,2}$ and that

$$D_s^W Y_r = \int_s^r \frac{\partial a}{\partial x}(u, Y_u) D_s^W Y_u du + b(s, Y_s) + \int_s^r \frac{\partial b}{\partial x}(u, Y_u) D_s^W Y_u dW_u. \tag{7.6.2}$$

Taking now into account that $D_s^W \sigma_r = f'(Y_r) D_s^W Y_r$ it can be easily deduced from (7.6.2) that

$$sup_{s,r\in[t,T]} \left| E\left(\left(D_s^W \sigma_r - f'(Y_t)b(t, Y_t)\right) \middle| \mathcal{F}_t\right)\right| \to 0$$

as $T \to t$. Then, Theorem 7.5.2 gives us that

$$\lim_{T \to t} \frac{\partial I_t}{\partial k}(k_t^*) = \frac{\rho}{2\sigma_t} f'(Y_t) b(t, Y_t), \tag{7.6.3}$$

that agrees with the results in Medveved and Scaillet (2004). Notice that this limit is finite, except if b has a singularity at (t, S_t). Models with this kind of singularities are studied in Section 7.6.4.3.

Let us see some particular examples that include some of the best-known classical volatility models in the literature.

7.6.1.1 Models based on the Ornstein-Uhlenbeck process

Assume that Y_r is a stationary Ornstein-Uhlenbeck process satisfying the equation

$$dY_r = \alpha(m - Y_r)dt + c\sqrt{2\alpha}dW_r,$$

for some positive constants m, c, and α. Then it follows that

$$Y_r = m + (Y_t - m)\, e^{-\alpha(r-t)} + c \int_t^r \sqrt{2\alpha} \exp\left(-\alpha\left(r - s\right)\right) dW_s. \tag{7.6.4}$$

Then, $D_s^W Y_r = c\sqrt{2\alpha} \exp\left(-\alpha\left(r - s\right)\right)$ for all $t < r < s$, which allows us to write

$$\lim_{T \to t} \frac{\partial I_t}{\partial k}(k_t^*) = c\sqrt{2\alpha} \frac{\rho}{2\sigma_t} f'(Y_t).$$

Notice that this limit is an increasing function of the correlation ρ, the vol-of-vol c, and the rate of mean reversion α.

7.6.1.2 The SABR model

We have seen in Section 3.3.1 that in the SABR volatility model

$$d\sigma_t = \alpha\sigma_t dW_t, \qquad t \in [0, T].$$

The Malliavin derivative is given by

$$D_r^W \sigma_t = \alpha\sigma_t \mathbf{1}_{[0,t]}(r),$$

which implies that

$$\lim_{T \to t} \frac{\partial I_t}{\partial k}(k_t^*) = \frac{\rho\alpha}{2},$$

according to the results in Section 2.3.2. Notice that, in this case, the short-time limit of the skew slope does not depend on the spot volatility σ_t.

7.6.1.3 The Heston model

The Heston model for $v_t = \sigma_t^2$ is given by

$$dv_t = k(\theta - v_t)dt + \nu\sqrt{v_t}dW_t$$

(see Section 3.3.2). We have seen that its Malliavin derivative is given by

$$D_r^W \sigma_t = \frac{\nu}{2}\mathbf{1}_{[0,t]}(r)\exp\left(\int_r^t \left(-\left(\frac{k\theta}{2} - \frac{\nu^2}{8}\right)\frac{1}{\sigma_s^2} - \frac{k}{2}\right) ds\right),$$

from where we get

$$D_t^+ \sigma_t = \frac{\nu}{2}.$$

Then,

$$\lim_{T\to t} \frac{\partial I_t}{\partial k}(k_t^*) = \frac{\rho\nu}{4\sigma_t}.$$

Notice that, as $I_t(k_t^*) \to \sigma_t$ (see Durrleman (2008)), the above equation implies that

$$\lim_{T\to t} \frac{\partial I_t^2}{\partial k}(k_t^*) = \lim_{T\to t} 2I_t(k_t^*)\frac{\partial I_t}{\partial k}(k_t^*) = \frac{\rho\nu}{2},$$

according to Gatheral (2006).

7.6.1.4 The two-factor Bergomi model

Assume, as in Bergomi (2005)) that the volatility is a process that depends on a two-dimensional Brownian motion $W = (W^1, W^2)$. More precisely, we assume that the squared volatility $v_t = \sigma_t^2$ can be written as

$$\begin{aligned} v_t &= v_0 \mathcal{E}(\nu x_t) \\ &= v_0 \exp\left(\nu x_t - \frac{\nu^2}{2}(\delta^2 + (1-\delta)^2 + 2\rho_{12}\delta(1-\delta))t\right) \end{aligned} \tag{7.6.5}$$

with $\nu > 0$ and $x_t := \delta Y_t^1 + (1-\delta)Y_t^2$ for some $\delta \in (0,1)$, where Y^1 and Y^2 are two processes of the form

$$dY_r^1 = -\alpha_1 Y_r^1 dt + dW_r^1,$$

$$dY_r^2 = -\alpha_2 Y_r^2 dt + \rho_{12} dW_r^1 + \sqrt{1-\rho_{12}^2} dW_r^2.$$

Notice that $Z := \rho_{12} W^1 + \sqrt{1-\rho_{12}^2} W^2$ is a Brownian motion and that $d\langle W^1, Z\rangle_t = \rho_{12} dt$. Then, we can write

$$D_r^{W^1} v_t = v_t \left(\nu D_r^{W^1} x_t\right) = v_t \nu \left(\delta e^{-\alpha_1(t-r)} + (1-\delta)\rho_{12} e^{-\alpha_2(t-r)}\right) \tag{7.6.6}$$

and

$$D_r^{W^2} v_t = v_t \left(\nu D_r^{W^2} x_t\right) = v_t \nu \delta \sqrt{1-\rho_{12}^2} e^{-\alpha_2(t-r)}. \tag{7.6.7}$$

Now, with the notations in Remark 7.5.4, we get

$$\begin{aligned} \lim_{T\to t} \frac{\partial I_t}{\partial k}(k_t^*) &= \frac{1}{4}\left[\rho_1 \nu(\delta + (1-\delta)\rho_{12}) + \rho_2 \nu \delta \sqrt{1-\rho_{12}^2}\right] \\ &= \frac{1}{4}\nu[\delta\rho_1 + (1-\delta)\rho_1\rho_{12} + \rho_2\sqrt{1-\rho_{12}^2}], \end{aligned} \tag{7.6.8}$$

and then the skew slope is $O(1)$.

Remark 7.6.1 *The above examples show that, in diffusion volatility models, the at-the-money volatility skew slope is not $O(T-t)^{-\frac{1}{2}}$ but $O(1)$. To reproduce a power-like phenomena, one should take a big vol-of-vol and over parametrise the model to reproduce market data (see for example Bergomi (2009b)).*

7.6.2 Local volatility models: the one-half rule and dynamic inconsistency

In local volatility models, the volatility is assumed to be a function of the underlying option price. That is,

$$\sigma_t = \sigma(t, S_t),$$

for some real function σ. This corresponds to model 6.1.1 with $\rho = 1$. If $\sigma(t,\cdot) \in C_b^1$, we have that

$$D_s^W \sigma_t = \frac{\partial \sigma}{\partial S}(t, S_t) D_s^W S_t,$$

and then (7.5.6) gives us that

$$\lim_{T\to t} \frac{\partial I_t}{\partial k}(k_t^*) = \frac{1}{2\sigma_t}\frac{\partial \sigma}{\partial S}(t, S_t)\sigma(t, S_t) S_t = \frac{1}{2}\frac{\partial \sigma}{\partial S}(t, S_t) S_t. \tag{7.6.9}$$

Notice that $\frac{\partial \sigma}{\partial S}(t, S_t)S_t = \frac{\partial \hat{\sigma}}{\partial X}(t, X_t)$, where $\hat{\sigma}$ denotes the local volatility function written in terms of the log-price X. This allows us to write

$$\lim_{T \to t} \frac{\partial I_t}{\partial k}(k_t^*) = \frac{1}{2} \frac{\partial \hat{\sigma}}{\partial X}(t, X_t). \tag{7.6.10}$$

That is, for short maturities, the ATM implied volatility skew is approximately half the skew of the local volatility. This agrees with the well-known **one-half rule**, a heuristical relationship introduced by Derman, Kani, and Zou (1996) that stablishes that, for short and intermediate maturities and for skewed option markets, the local volatility varies with the asset price twice as fast as the implied volatility varies with the strike (see for example Figure 2.5). Classical proofs of this property are based on the expression of the implied volatility as averaged local volatility (see Derman, Kani, and Zou (1996) or Gatheral (2006), among others). Another approach can be found in Lee (2001), where this rule is proved by a direct comparison of implied and local volatility expansions.

Notice that, if $\sigma(t, \cdot) \in C_b^1$, the skew slope of the implied volatility is $O(1)$, according to the results in Fukasawa (2017). To be consistent with the short-end blow-up of the implied volatility skew slope, $\sigma(t, \cdot)$ has to have a singularity at S_t. Discontinuous local volatility models have been studied recently in Pigato (2019). Notice that this discontinuity property cannot be satisfied for all (t, S_t). That is, even when local volatilities can reproduce the implied volatility surface at some fixed moment t, the same model cannot reproduce the surface for every t, and then these models are dynamically inconsistent and have to be re-calibrated from day to day. This implies that local volatilities are not satisfactory to hedging purposes, as we pointed out in Remark 2.2.7 (see also Fukasawa (2017)).

Example 7.6.2 *Consider a CEV model as in Example 2.2.6, where the volatility takes the form $\sigma_t = \sigma S_t^{\gamma-1}$, for some $\sigma, \gamma > 0$. Then (7.9.1) reads*

$$\lim_{T \to t} \frac{\partial I_t}{\partial k}(k_t^*) = \frac{\sigma(\gamma - 1)}{2} S_t^{\gamma-1}, \tag{7.6.11}$$

according to Section 3 in Hagan, Kumar, Lesniewski, and Woodward (2014). Notice that this limit is positive for $\gamma > 1$ and negative for $\gamma < 1$. We remark that the CEV model has not a singularity at (t, S_t). Then, it cannot replicate the short-end blow-up of the skew slope.

7.6.3 Stochastic-local volatility models

Consider a stochastic-local volatility model of the form

$$\sigma_t = \sigma(t, S_t) f(Y_t),$$

where σ and f are real functions and Y is a diffusion process of the form (9.5.6). Then

$$D_s^W \sigma_t = \frac{\partial \sigma}{\partial x}(t, S_t)(D_s^W S_t) + \sigma(t, S_t) D_s^W Y_t,$$

and the same arguments as before give us that

$$\lim_{T \to t} \frac{\partial I_t}{\partial k}(k_t^*) = \frac{1}{2} \frac{\partial \sigma}{\partial S}(t, S_t) S_t + \frac{\rho}{2\sigma_t} f'(Y_t) b(t, Y_t). \qquad (7.6.12)$$

That is, a stochastic-local volatility model can reproduce the short-end blow-up of the skew slope if σ has a singularity at (t, S_t) or b has a singularity at (t, Y_t). As these properties cannot hold for every (t, S_t, Y_t), stochastic-local volatility models have the same dynamical problems as local volatilities, and they have to be recalibrated from day to day.

7.6.4 Fractional stochastic volatility models

Remark 7.5.5 gives us that, if $D_s^W \sigma_r$ has a finite limit as $s, r \to t$, then the skew slope is $O(1)$ and the model cannot reproduce the empirical short-end blow-up of the implied volatility surface. But this phenomenon could be reproduced (see Theorem 7.5.2) if $D_s^W \sigma_r$ tends to infinity. As the Malliavin derivative D_s^W of a Gaussian process of the form $\int_0^r K(r, s) dW_s$ (for some deterministic kernel K) is given by $K(r, s)$, a natural candidate for our purposes should be a volatility depending on a Gaussian process with $K(r, s) \to \infty$ as $s, r \to 0$, like the fBm/RLfBm with $H < \frac{1}{2}$.

In this section, we study the short-time behaviour of the implied volatility for models based on the RLfBm. A similar analysis could be run for models based on the fractional Brownian motion, or for models based on general Gaussian processes with similar kernel singularities.

7.6.4.1 Fractional Ornstein-Uhlenbeck volatilities

Assume as in Section 5.6.1 that the volatility σ is of the form $\sigma_r = f(Y_r)$, where $f \in C_b^1(\mathbb{R})$ and Y_r is a *fractional Ornstein-Uhlenbeck* process of the form

$$Y_r = m + (Y_0 - m) e^{-\alpha r} + c\sqrt{2\alpha} \int_0^r e^{-\alpha(r-s)} d\hat{W}_s^H, \qquad (7.6.13)$$

where $\hat{W}_s^H := \int_0^s (s-u)^{H-\frac{1}{2}} dW_u$ and Y_0, m, c, and α are positive constants.

- Case $H > \frac{1}{2}$.

  Assume, as in Comte and Renault (1998), the model (7.6.4) with $H > \frac{1}{2}$. Then, it follows that

$$D_s^W \sigma_r = f'(Y_r) c\sqrt{2\alpha} \left(H - \frac{1}{2}\right) \left(\int_s^r e^{-\alpha(r-u)} (u-s)^{H-\frac{3}{2}} du\right),$$

  which implies that $sup_{s,r\in[t,T]} |E(D_s\sigma_r)| \to 0$ as $T \to t$. Then,

$$\lim_{T\to t} \frac{\partial I_t}{\partial k}(k_t^*) = 0.$$

  That is, the at-the-money short-dated skew slope of the implied volatility is not affected by the correlation in this case.

- Case $H < \frac{1}{2}$. Assume again the model (7.6.4), taking $0 < H < 1/2$. Then, as

$$\begin{aligned} D_s^W \sigma_r &= f'(Y_r) c\sqrt{2\alpha} \left[\left(\frac{1}{2} - H\right)\right. \\ &\times \left(\int_s^r \left[e^{-\alpha(r-u)} - e^{-\alpha(r-s)}\right] (u-s)^{H-\frac{3}{2}} du\right) \\ &\left.+ e^{-\alpha(r-s)} (r-s)^{H-\frac{1}{2}}\right], \end{aligned}$$

  we get

$$\lim_{T\to t} (T-t)^{\frac{1}{2}-H} \frac{\partial I_t}{\partial k}(k_t^*) = c\sqrt{2\alpha} \frac{\rho}{\sigma_t} f'(Y_t).$$

  That is, the introduction of fractional components with Hurst parameter $H < 1/2$ in the definition of the volatility process allows us to reproduce a skew slope of order $O(T-t)^\delta$, for every $\delta > -1/2$ (see Alòs, León, and Vives (2007)).

7.6.4.2 *The rough Bergomi model*

We have seen in Section 5.6.2 that the rough Bergomi model, where

$$\sigma_r^2 = E(\sigma_r^2) \exp\left(\nu\sqrt{2H} W_t^H - \frac{1}{2}\nu^2 r^{2H}\right), r \in [0, T], \tag{7.6.14}$$

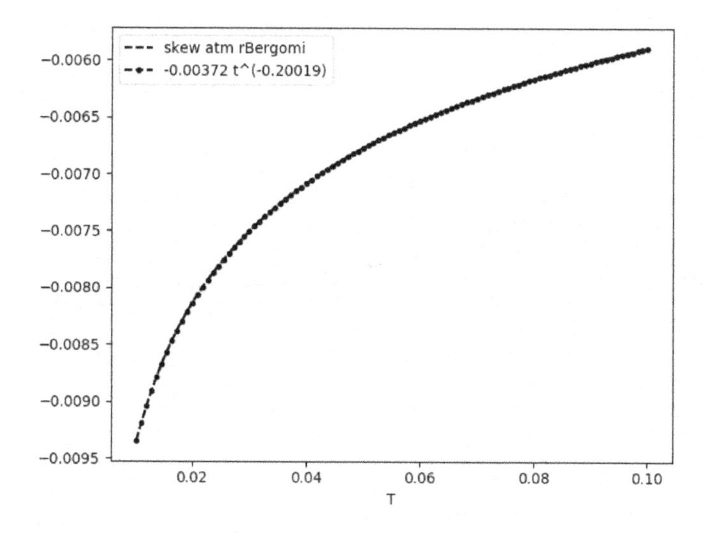

Figure 7.5 ATM skew for the rBergomi model as a function of time to maturity, with $S_0 = 100, \nu = 0.5$, $\rho = -0.5, \sigma_0^2 = 0.05$, and $H = 0.3$.

for some constant $\nu > 0$ and where W^H is a RLfBm with Hurst parameter H, satisfies that

$$D_r^W v_t = \nu\sqrt{2H} v_t (r - s)^{H-\frac{1}{2}}.$$

Then, the short-time skew slope is again $O((T - t)^{H-\frac{1}{2}})$, as we can see in Figures 7.5 and 7.6, where we plot the simulated ATM skew slopes corresponding to a rBergomi model with $S_0 = 100, \nu = 0.5$, $\rho = -0.5, \sigma_0^2 = 0.05$, $H = 0.3$, and $H = 0.7$, respectively. We can observe this skew is $O((T - t)^{H-\frac{1}{2}}) = O((T - t)^{-0.2})$ in the case $H = 0.3$, and $O(T^{H-\frac{1}{2}}) = O(T^{0.2})$ in the case $H = 0.7$.

Remark 7.6.3 *We can also define the rough Bergomi model replacing the RLfBm W^H by a fractional Brownian motion $\hat{W}^H$. Then, $D_r^W v_t = \nu\sqrt{2H} v_t K_H(r, s)$. Then, taking into account the properties of the kernel $K_H(t, s)$ (see Equation (5.1.6)), we can prove that the short-time skew slope is again $O(T - t)^{H-\frac{1}{2}}$.*

7.6.4.3 The approximation of fractional volatilities by Markov processes

Models based on the fBm/RLfBm are non-Markovian, and then several technical difficulties arise when it comes to derivative pricing. One way

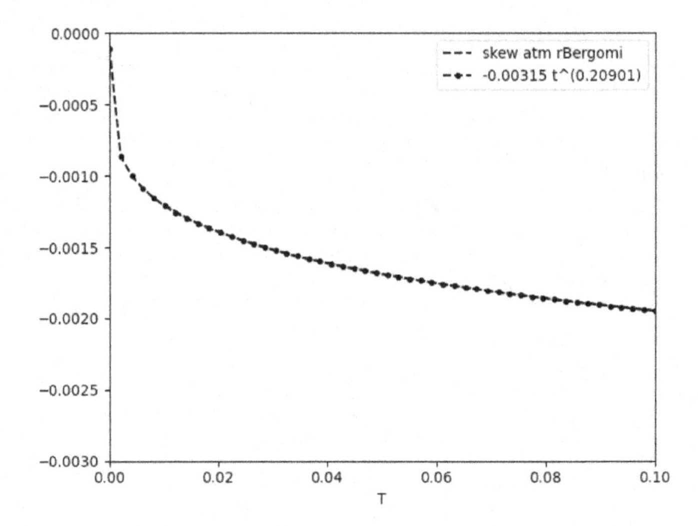

Figure 7.6 ATM skew for the rBergomi model as a function of time to maturity, with $S_0 = 100, \nu = 0.5$, $\rho = -0.5, \sigma_0^2 = 0.05$, and $H = 0.7$.

to overcome this problem that has been presented in the literature (see Carr and Itkin (2019) is to approximate the fBm by a semimartingale. More precisely, in Carr and Itkin (2019)) the volatility is driven by a semimartingale whose diffusion term is of the form $M_t := c \int_0^t s^{H-\frac{1}{2}} dW_s$. Then $D_s^W M_t = s^{H-\frac{1}{2}}$, which implies that $D_s^W M_t \to t^{H-\frac{1}{2}}$ as $s \to t$. As a consequence, $D_s^W M_t$ tends to a finite limit if $t \neq 0$, while it tends to infinity if $t = 0$. This implies that stochastic volatilities driven by this model satisfy that

$$\left| \lim_{T \to 0} \frac{\partial I_0}{\partial k}(k_0^*) \right| = \infty,$$

but not that $\lim_{T \to t} \frac{\partial I_t}{\partial k}(k_t^*) = \infty$ **for all** $t \in [0, T]$. If $\gamma = H - \frac{1}{2}$, then $\frac{\partial I_0}{\partial k}(k_0^*) = O(T^{H-\frac{1}{2}})$.

7.6.5 Time-varying coefficients

Fouque, Papanicolau, Sircar, and Solna (2004) introduced a different approach to capture the maturity-dependent behaviour of the implied volatility, by allowing the volatility coefficients to depend on the time till the next maturity date. In particular, they assumed that the volatility σ_r is of the form $\sigma_r = f(Y_r)$, where f is a real function and Y_r is a diffusion process of the form (7.6.4), being $\sqrt{\alpha(s)}$ a suitable cutoff of the function $\left(T_{n(s)} - s\right)^{-\frac{1}{2}}$, with fixed maturity dates $\{T_k\}$ (the third Friday of each month) and $n(t) = \inf\{n : T_n > s\}$.

Following this idea, we can consider Y_r to be a diffusion process of the form (7.6.4), with $\sqrt{\alpha(s)} = \left(T_{n(s)} - s\right)^{-\frac{1}{2}+\varepsilon}$, for some $\varepsilon > 0$. It is now easy to see that $Y_r \in \mathbb{L}^{1,2}$and that

$$\lim_{T\to t} \frac{1}{(T-t)^{2+\left(\frac{1}{2}-\varepsilon\right)}} \int_t^T \int_s^T E_t\left(D_s\sigma_r\right) drds$$

$$= -\rho c \left(\frac{1}{-1/2+\varepsilon}\right)\left(\frac{1}{1/2+\varepsilon}\right)\frac{f'(Y_t)}{2}.$$

Hence we deduce that, in this case,

$$\lim_{T\to t}(T-t)^{\frac{1}{2}-\epsilon}\frac{\partial I_t}{\partial k}(k_t^*) = -\frac{\rho c}{\sigma_t^2}\left(\frac{1}{-1/2+\varepsilon}\right)\left(\frac{1}{1/2+\varepsilon}\right)f'(Y_t). \quad (7.6.15)$$

That is, the short-date skew slope of the implied volatility is of the order $O(T-t)^{-\frac{1}{2}+\varepsilon}$.

Remark 7.6.4 *Notice that, among the above examples, only models based on the fractional Brownian motion and models with time-varying coefficients succeed in reproducing the short-end blow-up of the skew* $\frac{\partial I_t}{\partial k}(k_t^*)$ **for all** $t \in [0,T]$.

7.7 IS THE VOLATILITY LONG-MEMORY, SHORT-MEMORY, OR BOTH?

Long-memory volatilities have been proposed in Comte and Renault (1998) in order to capture the occurrence of fairly pronounced smile effects for long maturities. In these models, long-term heteroskedasticity is higher, and this translates into more pronounced skew and smile effects for options with long maturities. On the other hand, the analysis in Section 7.6.4.1 indicates that short-memory process as those based on the fBm/RLfBm with $H < \frac{1}{2}$ can reproduce the observed blow-up of the at-the-money skew slope. At the light of Example 5.5.1 and Section 7.6.4.1, both properties (short and long memory) are not contradictory. For example, consider a volatility process of the form $\sigma_t = f(Y_t^H + Y_t^{H'})$, where Y_t^H is a fractional Ornstein-Uhlenbeck process driven by a RLfBm with $H > \frac{1}{2}$ and $Y_t^{H'}$ is another Ornstein-Uhlenbeck process driven by a RLfBm with $H' < \frac{1}{2}$. Then, similar computations as in Section 7.6.4.1 give us that

$$\lim_{T \to t}(T-t)^{\frac{1}{2}-H}\frac{\partial I_t}{\partial k}(k_t^*) = c\sqrt{2\alpha}\frac{\rho}{\sigma_t}f'\left(Y_t^H + Y_t^{H'}\right),$$

since $D_r^W Y_s^H \to 0$ as $r, s \to t$, which implies that Y^H does not have impact on the ATM short-time limit skew. A detailed analysis of the effect of two different Hurst parameters in the implied volatility surface can be found in Funahashi and Kijima (2017).

7.8 A COMPARISON WITH JUMP-DIFFUSION MODELS: THE BATES MODEL

It is well known (see for example Cont and Tankov (2004)) that the implied volatility skew can be reproduced not only by the correlation between the volatility and the asset price but also by the presence of jumps in asset prices. In this section, we compare the effect of jumps and the effect of stochastic volatilities in the implied volatility skew. Towards this end, we consider the following extension of the model 6.1.1

$$X_t = x+(r-\lambda\alpha)t-\frac{1}{2}\int_0^t \sigma_s^2 ds+\int_0^t \sigma_s(\rho dW_s+\sqrt{1-\rho^2}dB_s)+Z_t,\ t \in [0,T]. \tag{7.8.1}$$

Here, x is the current log-price, r is the instantaneous interest rate, W and B are independent standard Brownian motions, $\rho \in (-1,1)$, Z is a compound Poisson process, independent of W and B, with intensity Φ and Lévy measure ν, and with $\alpha = \frac{1}{\lambda}\int_{\mathbb{R}}(e^y - 1)\nu(dy) < \infty$, and σ is a second-order stochastic process adapted to the filtration generated by W. This model is a generalisation of the classical Bates model introduced in Bates (1996), in the sense that we do not assume the volatility to be a diffusion process, but it can be, for example, a model based on a fBm.

For the sake of simplicity, let us focus first on the effect of jumps. That is, consider first the case where σ is deterministic (and then $\rho = 0$)

$$X_t = x + (r-\lambda\alpha)t - \frac{1}{2}\int_0^t \sigma_s^2 ds + \int_0^t \sigma_s dB_s + Z_t,\ t \in [0,T]. \tag{7.8.2}$$

Notice that, if σ is constant, (7.8.2) coincides with the Merton jump-diffusion model (see Merton (1976)). Now, the idea is to apply the classical Itô's formula to the continuous part of the process $BS(t, X_t, k, v_t)$ and to take expectations as in Section 6.2.1.2. This gives us that

$$\begin{aligned} V_t &= BS(t, X_t, k, v_t) \\ &-\lambda\alpha E_t(\int_t^T e^{-r(s-t)} \frac{\partial BS}{\partial x}(s, X_s, k, v_s)ds) \\ &+ \sum_{0\leq s\leq t} E_t[BS(s, X_{s-} + \Delta X_s, k, v_s) - BS(s, X_{s-}, k, v_s)], \end{aligned} \quad (7.8.3)$$

from where we get

$$\begin{aligned} V_t &= E_t(BS(t, X_t, k, v_t) \\ &+E_t\left(\int_t^T \int_{\mathbb{R}} e^{-r(s-t)}(BS(s, X_s + y, k, v_s) - BS(s, X_s, k, v_s))\nu(dy)ds\right) \\ &-\lambda\alpha E_t\left(\int_t^T e^{-r(s-t)} \frac{\partial BS}{\partial x}(s, X_s, k, v_s)ds\right). \end{aligned}$$

Then, the same arguments as in the proof of Proposition 7.4.1 lead to

$$\frac{\partial I_t}{\partial k}(k_t^*) = \frac{E_t\left(\int_t^T (\frac{\partial F}{\partial k}(s, X_s, k_t^*, v_s) - \frac{1}{2}F(s, X_s, k_t^*, v_s))ds\right)}{\frac{\partial BS}{\partial \sigma}(t, X_t, k_t^*, I_t(k_t^*))}, \text{ a.s.}$$

where

$$\begin{aligned} F(s, X_s, k, v_s) :&= \int_{\mathbb{R}} e^{-r(s-t)}[BS(s, X_s + y, k, v_s) - BS(s, X_s, k, v_s)]\nu(dy) \\ &- \lambda\alpha e^{-r(s-t)} \frac{\partial BS}{\partial x}(s, X_s, k, v_s). \end{aligned}$$

Then, we can prove the following result.

Proposition 7.8.1 *Assume the model (7.8.2) where σ is a deterministic and square integrable process. Then,*

$$\begin{aligned} &\frac{\partial BS}{\partial \sigma}(t, X_t, k_t^*, I_t(k_t^*)) \frac{\partial I_t}{\partial k}(k_t^*) \\ &= -\lambda\alpha E_t(\hat{G}(t, X_t, k_t^*, v_t)(T-t)) + O(T-t), \end{aligned}$$

as $T \to t$ and where $\hat{G}(t, x_t^*, v_t) = \left(\frac{\partial^2}{\partial k \partial x} - \frac{\partial}{\partial k}\right) BS(t, x_t^*, v_t)$.

Proof. The above computations give us that

$$
\begin{aligned}
&\frac{\partial BS}{\partial \sigma}(t, X_t, k_t^*, I_t(k_t^*))\frac{\partial I_t}{\partial k}(k_t^*) \\
=\; & E_t\left(\int_t^T \int_{\mathbb{R}} e^{-r(s-t)}\left(\frac{\partial}{\partial k}-\frac{1}{2}\right)\right. \\
&\times[BS(s, X_s+y, k_t^*, v_s)-BS(s, X_s, k_t^*, v_s)]\nu(dy)ds) \\
&-\lambda\alpha E_t\left(\int_t^T e^{-r(s-t)}\left(\frac{\partial}{\partial k}-\frac{1}{2}\right)\frac{\partial BS}{\partial x}(s, X_s, k_t^*, v_s)ds\right)=T_1+T_2.
\end{aligned}
\tag{7.8.4}
$$

As $|BS(t,x,\sigma)|+\left|\frac{\partial BS}{\partial k}(t,x,\sigma)\right| \leq 2e^k+e^x$ it follows that $T_1 = O(T-t)$. Now, let us see that

$$
T_2=-\lambda\alpha E_t\left(\left(\frac{\partial^2 BS}{\partial k\partial x}-\frac{\partial BS}{\partial k}\right)(t, X_t, k_t^*, v_t)(T-t)\right)+O\left(T-t\right). \tag{7.8.5}
$$

In fact,

$$
\begin{aligned}
&\left(\int_t^T e^{-r(s-t)}\left(\frac{\partial}{\partial k}-\frac{1}{2}\right)\partial_x BS(s, X_s, k_t^*, v_s)ds\right) \\
=\; & E_t\left(\int_t^T e^{-r(s-t)}\left(\frac{\partial^2 BS}{\partial k\partial x}-\frac{\partial BS}{\partial k}\right)(s, X_s, k_t^*, v_s)ds\right) \\
&+\frac{1}{2}E\left(\int_t^T e^{-r(s-t)}\frac{\partial BS}{\partial k}(s, X_s, k_t^*, v_s)ds\right).
\end{aligned}
$$

As $\left|\frac{\partial BS}{\partial k}(t,x,\sigma)\right| \leq e^k$ it follows easily that the second term in the right-hand side of this equality is $O(T-t)$. On the other hand, Itô's allows us to write

$$
\begin{aligned}
&E_t\left(e^{-r(s-t)}\hat{G}(s, X_s, k_t^*, v_s)\right) \\
&= E_t\left(\hat{G}(t, X_t, k_t^*, v_t)\right) \\
&+ E_t\left(\int_t^s \int_R e^{-r(u-t)}\left(\hat{G}(u, X_u+y, k_t^*, v_u)-\hat{G}(u, X_u, k_t^*, v_u)\right)\nu\left(dy\right)du\right) \\
&- \lambda\alpha E_t\left(\int_t^s e^{-r(u-t)}\frac{\partial \hat{G}}{\partial x}(u, X_u, k_t^*, v_u)du\right).
\end{aligned}
$$

Using again the same arguments as for the term T_1, (7.8.5) follows. Now, the proof is complete. ■

Now, we are in a position to prove the main result of this section.

Theorem 7.8.2 *Consider the model (7.8.2) where σ is a deterministic square integrable process. Then,*

$$\lim_{T\to t}\frac{\partial I_t}{\partial k}(k_t^*)=\frac{\lambda\alpha}{\sigma_t}. \tag{7.8.6}$$

Proof. Proposition 7.8.1 gives us that

$$\frac{\partial I_t}{\partial k}(k_t^*)=-\lambda\alpha\frac{E_t(\hat{G}(t,X_t,k_t^*,v_t)(T-t))}{\frac{\partial BS}{\partial\sigma}(t,X_t,k_t^*,I_t(k_t^*))}+O(T-t)^{\frac{1}{2}}.$$

Then, taking limits as $T\to t$ the result follows. ■

Remark 7.8.3 *The results in Theorems 7.5.2 and 7.8.2 can be joined to hold for general processes of the form (7.8.1). Then,*

- *If the hypotheses for the continuous volatility process in Theorem 7.5.2 hold with $H>\frac{1}{2}$, we get that*

$$\lim_{T\to t}\frac{\partial I_t}{\partial k}(x_t^*)=\frac{\lambda\alpha}{\sigma_t}. \tag{7.8.7}$$

- *If these hypotheses are satisfied with $H=\frac{1}{2}$ it follows that*

$$\lim_{T\to t}\frac{\partial I_t}{\partial k}(k_t^*)=\frac{\lambda\alpha}{\sigma_t}-\frac{\rho}{\sigma_t^2}\lim_{t\to T}\frac{1}{(T-t)^2}\int_t^T\left(D_s^W\int_s^T\sigma_r^2dr\right)ds. \tag{7.8.8}$$

- *And if these hypotheses hold for $H\leq\frac{1}{2}$ we have that*

$$\lim_{T\to t}(T-t)^{\frac{1}{2}-H}\frac{\partial I_t}{\partial k}(k_t^*)=\frac{\rho}{\sigma_t^2}\lim_{t\to T}\frac{1}{(T-t)^{\frac{3}{2}+H}}\int_t^T\left(D_s^W\int_s^T\sigma_r^2dr\right)ds. \tag{7.8.9}$$

Remark 7.8.4 *Theorem 7.8.2 proves that the skew slope due to jumps in the Bates model is $O(1)$ as in classical diffusion volatility models. What is then the main difference between these skews due to jumps and the skews in classical models with $H=\frac{1}{2}$ as Heston or SABR? Propositions 7.5.1 and 7.8.1 reveal that the corresponding term structure is different. In fact, and as pointed out in Chapter 15 in Cont*

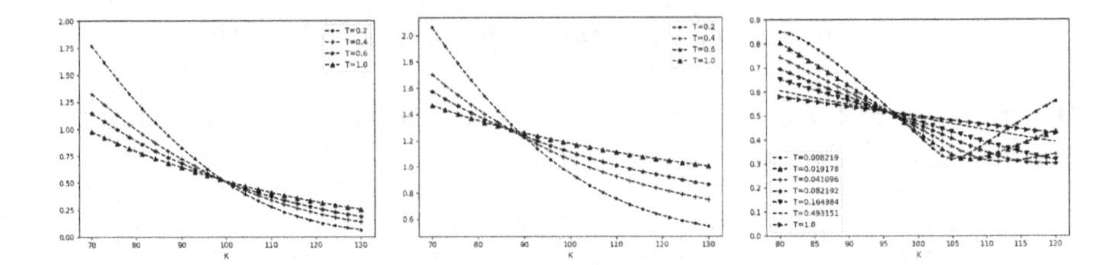

Figure 7.7 Skews for a Heston model with $\nu = 0.75, k = 0.6, \rho = -0.5, v_0 = 0.25, \theta = 0.5$ (left); a Merton jump process with intensity Φ, jump mean $\alpha = -0.4$, and jump standard deviation = 1.5 (centre); and a rBergomi model with $\nu = 0.8, \rho = -0.2$, $\sigma_0^2 = 0.05$, and $H = 0.05$ (right).

and Tankov (2004), the flattening of the skews with time to maturity is much faster for skews due to jumps than for stochastic volatility models with $H = \frac{1}{2}$. We can observe this behaviour in Figure 7.7, where we can compare the skews for a Heston model with $\nu = 0.75, k = 0.6, \rho = -0.5, v_0 = 0.25, and \theta = 0.5$; for a Merton jump-diffusion model with $S_0 = 100, \sigma = 0.4$, intensity $\Phi = 1.7$, jump mean $\alpha = -0.4$, and jump standard deviation equal to 1.5; and for a rBergomi model with $\nu = 0.8, \rho = -0.2$, $\sigma_0^2 = 0.05$, and $H = 0.05$.

7.9 CHAPTER'S DIGEST

The decomposition approach presented in Chapter 6 also allows us to write the ATMI skew slope $\frac{\partial I}{\partial k}(k_t^*)$ as the sum of this slope in the uncorrelated case (that is equal to zero), plus a term due to the correlation. This representation gives us a tool to compute explicitly the short-time limit of the ATMI skew slope for models with continuous volatilities, in terms of the Malliavin derivative of the volatility process.

This allows us to see that, in the case of models with a Hurst parameter $H \geq \frac{1}{2}$, there is a $O(\rho^2)$ relationship between the volatility swap, the ATMI, and the ATMI skew, according to some well-known results in the literature. On the other hand, we can stablish when this skew slope tends to infinity (as observed in real market data) and to compute the corresponding rate of divergence. More precisely:

- For stochastic volatility models,

$$\lim_{T\to t}\frac{\partial I_t}{\partial k}(k_t^*) = \frac{\rho}{2\sigma_t}f'(Y_t)b\left(t,Y_t\right)$$

(see (9.5.2)). This limit is finite for all classical stochastic volatility models, including models based on the OU process, Heston, SABR, or the 2-factor Bergomi model.

- For local volatility models,

$$\lim_{T\to t}\frac{\partial I_t}{\partial k}(k_t^*) = \frac{1}{2\sigma_t}\frac{\partial\sigma}{\partial x}(t,S_t)\sigma(t,S_t)S_t = \frac{1}{2}\frac{\partial\sigma}{\partial x}(t,S_t)S_t$$

(see (7.9.1)) and then, for short maturities, the skew of the local volatility is twice the skew of the implied volatility, a result that fits the well-known one-half rule. Moreover, this quantity is infinite if and only if $\frac{\partial\sigma}{\partial x}(t,S_t)$ is infinite, a property that cannot be satisfied for all (t,S_t). That is, even when local volatilities can reproduce the implied volatility surface at some fixed moment t, the same model cannot reproduce the surface for every t, and then these models are dynamically inconsistent and have to be re-calibrated from day to day.

- For models based on the fractional Brownian motion (including models based on the fractional OU model and the rough Bergomi model), $\lim_{T\to t}\frac{\partial I_t}{\partial k}(k_t^*) = 0$ if $H > \frac{1}{2}$ and $\frac{\partial I_t}{\partial k}(k_t^*) = O(T-t)^{H-\frac{1}{2}}$ if $H < \frac{1}{2}$, for all $t \in [0,T]$.

- Fixed $t \in [0,T]$, volatilities driven by Markovian approximations of the fractional Brownian motion can reproduce a skew slope of order $O((T-t)^{-\frac{1}{2}+\epsilon})$, for all $\epsilon > 0$.

- A skew slope of order $(T-t)O^{-\frac{1}{2}+\epsilon}$, for all $\epsilon > 0$, can also be described by a volatility model where the diffusion coefficient depends on the maturity time T.

- The skew slope due to jumps in the Bates model has a different behaviour than in classical stochastic volatility models. In particular, it flattens faster. Nevertheless, the corresponding skew slope is $O(1)$ as the time to maturity tends to zero.

Notice that, even when local volatilities and models based on Markovian approximations of the fBm can describe the blow-up of the skew slope at a fixed moment $t \in [0, T]$, the same model cannot describe this blow-up for all $t \in [0, T]$. This dynamic inconsistency is one of the main differences with rough volatilities, which can describe this blow-up for all $t \in [0, T]$. This is a crucial point in, for example, the pricing of forward-start options, as we see in Chapter 9.

Both from the analysis of the implied volatility surface for short and long maturities and from the study of real market data, we claim that the volatility is a complex aggregation of different components, with different Hursts parameters. The components with a Hurst parameter $H > \frac{1}{2}$ are more relevant for long-maturity options, while the components with $H < \frac{1}{2}$ have a stronger impact at short maturities.

CHAPTER 8

The ATM short-time curvature

In this chapter, following Alòs and León (2017), we study the second derivative of the implied volatility as a function of the strike price and we explicitly compute its ATM short-end limit in terms of the Malliavin derivatives of the volatility process and the correlation parameter. As a particular example, we study this limit for classical local and diffusion volatility models as well as for fractional volatilities. In particular, our results allow us to derive a condition for the at-the-money local convexity of the implied volatility, in terms of the correlation parameter and the Malliavin derivatives of the volatility process. Moreover, we see that, in the case of rough volatilities, the ATM short-time curvature blows-up at a higher rate than the ATM skew, which implies that, for short maturities, the smile effect is more relevant than the skew effect, according to real market data (see Section 1.4.1).

Through this chapter, we use the same notation as in Chapters 6 and 7. In particular, we recall that

$$\Lambda_r = E_r \left(BS \left(t, X_t, k_t^*, v_t \right) \right) \tag{8.0.1}$$

and

$$U_r = E_r \left(D_r^W BS(t, X_t, k_t^*, v_t) \right). \tag{8.0.2}$$

8.1 SOME EMPIRICAL FACTS

Let us observe some common empirical facts related to the curvature. First of all, we typically see that the smiles become more symmetric around the ATM strike when the time to maturity decreases. We can see this behaviour in Figure 8.1, corresponding to the observed implied volatilities with different times to maturity for the EURO STOXX50, as of November 17, 2020. Why does this happen? A natural explanation could be as follows: imagine that the curvature blew-up at a higher rate than the skew, then this curvature would be more relevant than the skew for short-time maturities, and this is why we see these smiles becoming more symmetric for short maturities. But, does this happen? Computing the curvature from real market data is not easy, but let us proceed as follows. Consider the above EURO STOXX50 smiles and calibrate a SABR model for every fixed maturity. According to Equation (2.3.7), the squared vol-of-vol α^2 controls the ATM curvature. Then the analysis of this estimated parameter gives us some clues about the behaviour of the curvature. In Figure 8.2 we can see the estimated values of α^2 for every fixed maturity T. We see that these estimated parameters follow a power law with an exponent equal to -0.87581. This rate is higher (in absolute value) to the one observed for the skew slope (see Figure 7.2 in Section 7.1), which was estimated to be equal to -0.42538. That is, the empirical curvature blows up at a higher rate than the empirical skew. Our objective in this chapter is to identify the class of volatility models that are consistent with these empirical properties.

8.2 THE UNCORRELATED CASE

Similarly as in the study of the ATM level and skew in Chapters 6 and 7, first step in our analysis of the at-the-money curvature $\frac{\partial^2 I_t}{\partial k^2}(k_t^*)$ is to study it in the uncorrelated case (i.e., $\rho = 0$).

8.2.1 A representation for the ATM curvature

Let us assume the following regularity conditions of the volatility process in the Malliavin calculus sense. For the sake of simplicity, we consider strict and simple hypotheses, but the results in this chapter also hold under other adequate integrability conditions (see Alòs and León (2017)).

(H1) σ^2 belongs to $\mathbb{L}_W^{2,2}$, and there exists a positive constant C, and $H \in (0,1)$ such that, for all $t < \tau < \theta < r < u < T$

$$\begin{aligned} \left| D_\theta^W \sigma_r^2 \right| &\leq C (r-\theta)^{H-\frac{1}{2}}, \\ \left| D_\theta^W D_r^W \sigma_u^2 \right| &\leq C (u-r)^{H-\frac{1}{2}} (u-\theta)^{H-\frac{1}{2}}, \end{aligned}$$

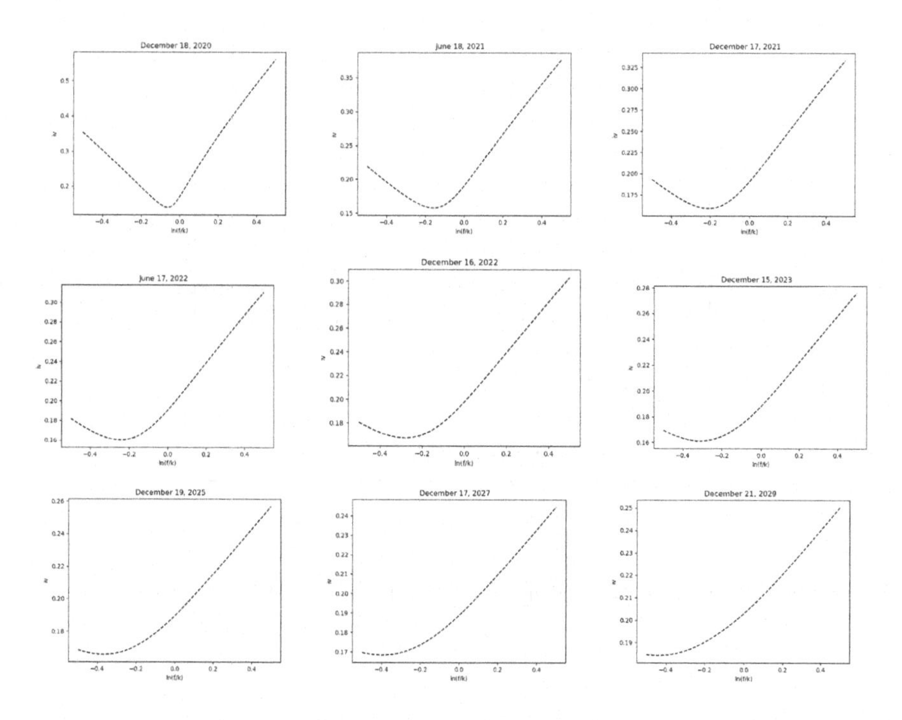

Figure 8.1 EURO STOXX50 implied volatility smiles and skews for different maturities, as of November 17, 2020.

tends to a finite limit as $T \to t$.

(H2) There exist two positive constants a, b such that

$$a \leq \sigma_t^2 \leq b, \quad t \in [0, T].$$

The starting point of our analysis is the following representation formula for the curvature in the uncorrelated case.

Theorem 8.2.1 *Assume that $\rho = 0$ in model (6.1.1), and that hypotheses (H1) and (H2) are satisfied. Then, for all $t \in [0, T]$,*

$$\frac{\partial^2 I_t}{\partial k^2}(k_t^*) = \frac{1}{2} \frac{E_t \left[\int_t^T \Psi'' \left(\Lambda_u \right) U_u^2 du \right]}{\frac{\partial BS}{\partial \sigma} \left(t, X_t, k_t^*, I_t(k_t^*) \right)}, \tag{8.2.1}$$

where

$$\Psi \left(a \right) := \frac{\partial^2 BS}{\partial k^2} \left(t, X_t, k_t^*, BS^{-1} \left(t, X_t, k_t^*, a \right) \right). \tag{8.2.2}$$

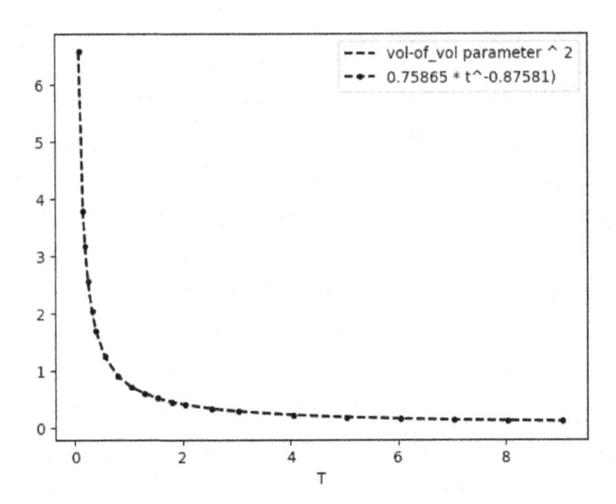

Figure 8.2 Calibrated values of α^2 for the EURO STOXX50 as of November 17, 2020.

Proof. From the definition of I, and taking implied derivatives, we get

$$\begin{aligned}\frac{\partial^2 V_t}{\partial k^2} &= \frac{\partial^2 BS}{\partial k^2}(t, X_t, k; I_t(k)) + 2\frac{\partial^2 BS}{\partial k \partial \sigma}(t, X_t, k; I_t(k))\frac{\partial I_t}{\partial k}(k) \\ &+ \frac{\partial^2 BS}{\partial \sigma^2}(t, X_t, k; I_t(k))\left(\frac{\partial I_t}{\partial k}(k)\right)^2 + \frac{\partial BS}{\partial \sigma}(t, X_t, k; I_t(k))\frac{\partial^2 I_t}{\partial k^2}(k).\end{aligned} \tag{8.2.3}$$

Now, Proposition 7.3.1 gives that $\frac{\partial I}{\partial k}(k_t^*) = 0$, from where we deduce

$$\frac{\partial BS}{\partial \sigma}(t, X_t, k_t^*, I_t(k_t^*))\frac{\partial^2 I}{\partial k^2}(t, k_t^*) = \frac{\partial^2 V_t}{\partial k^2}(k_t^*) - \frac{\partial^2 BS}{\partial k^2}(t, X_t, k_t^*, I_t(k_t^*)). \tag{8.2.4}$$

Now, the proof follows the same arguments as in the proof of Proposition 6.5.1. The Hull and White formula allows us to write

$$\frac{\partial^2 V_t}{\partial k^2}(k_t^*) = E_t\left(\frac{\partial^2 BS}{\partial k^2}(t, X_t, k_t^*, v_t)\right) \tag{8.2.5}$$

and then the term in the right-hand side of (8.2.4) reads as

$$E_t\left[\frac{\partial^2 BS}{\partial k^2}(t, X_t, k_t^*, v_t) - \frac{\partial^2 BS}{\partial k^2}(t, X_t, k_t^*, I_t(k_t^*))\right]. \tag{8.2.6}$$

Now, observe that

$$v_t = BS^{-1}\left(BS\left(t, X_t, k_t^*, v_t\right)\right) = BS^{-1}\left(\Lambda_T\right)$$

and

$$I_t(k_t^*) = BS^{-1}\left(E_t(BS\left(t, X_t, k_t^*, v_t\right))\right) = BS^{-1}\left(\Lambda_t\right),$$

where we denote $BS^{-1}(t, X_t, k_t^*, \cdot)$ by $BS^{-1}(\cdot)$ for the sake of simplicity. Notice that the term in (8.2.6) can be seen as the same function of the same process (Λ) evaluated at times T and t. That is,

$$\begin{aligned} & E_t\left[\frac{\partial^2 BS}{\partial k^2}\left(t, X_t, k_t^*, v_t\right) - \frac{\partial^2 BS}{\partial k^2}\left(t, X_t, k_t^*, I_t(k_t^*)\right)\right] \\ = \; & E_t\left[\frac{\partial^2 BS}{\partial k^2}\left(t, X_t, k_t^*, BS^{-1}(\Lambda_T)\right) - \frac{\partial^2 BS}{\partial k^2}(t, X_t, k_t^*, BS^{-1}(\Lambda_t))\right]. \end{aligned} \tag{8.2.7}$$

Now, the Clark-Ocone-Haussman formula (4.1.1), together with Hypotheses (H1) and (H2), leads to

$$\Lambda_T = \Lambda_t + \int_t^T U_r dW_r.$$

Then, applying the classical Itô's formula, we get

$$\begin{aligned} = \; & E_t\left[\frac{\partial^2 BS}{\partial k^2}\left(t, X_t, k_t^*, BS^{-1}\left(\Lambda_T\right)\right) - \frac{\partial^2 BS}{\partial k^2}(t, X_t, k_t^*, BS^{-1}\left(\Lambda_t\right))\right] \\ = \; & E_t\left[\int_t^T \Psi'\left(\Lambda_u\right) U_u dW_u + \frac{1}{2}\int_t^T \Psi''\left(\Lambda_u\right) U_u^2 du\right] \\ = \; & \frac{1}{2} E_t\left[\int_t^T \Psi''\left(\Lambda_u\right) U_u^2 du\right], \end{aligned}$$

which, jointly with (8.2.4) and (8.2.6), allows us to complete the proof.
■

Remark 8.2.2 *Let us analyse the result in Theorem 8.2.1. First of all notice that it is not an expansion but an exact formula. Moreover, a direct computation gives us that*

$$\Psi''(a) = \frac{2\sqrt{2\pi}\exp(-X_t)}{(T-t)^{\frac{3}{2}}} \frac{\exp\left(\frac{\left(BS^{-1}(a)\right)^2 (T-t)}{8}\right)}{\left(BS^{-1}\left(a\right)\right)^3} \tag{8.2.8}$$

is positive. Then, as the vega

$$\frac{\partial BS}{\partial \sigma}(t, X_t, k_t^*, I_t(k_t^*))$$

is also positive, Theorem 8.2.1 implies that, if $\rho = 0$, $\frac{\partial^2 I_t}{\partial k^2}(k_t^*)$ *is positive. As* $\frac{\partial^2 I_t}{\partial k^2}(k_t^*) = 0$ *(see Proposition 7.3.1), the implied volatility is, if* $\rho = 0$, *a locally convex function of the log-strike with a minimum at* $k = k_t^*$. *This fits the previous results by Renault and Touzi (1996) and Renault (1997). Notice that this convexity result is not a short-time result, but it is valid for every maturity.*

8.2.2 Limit results

Now our objective is to use the result in Theorem 8.2.1 to compute explicitly the short-time limit of the ATM curvature $\frac{\partial^2 I_t}{\partial k^2}(k_t^*)$. Towards this end, we need the following hypothesis on the short-time behaviour of the volatility process:

(H3) Hypothesis (H1) holds and for every $t \in [0, T]$, the limit

$$\lim_{T \to t}(T-t)^{1-2H} \frac{\left[E_t \int_t^T \left(E_r \left(\int_r^T D_r^W \sigma_\theta^2 d\theta\right)\right)^2 dr\right]}{(T-t)^{2H+2}}$$

exists and is finite.

Now, we are in a position to compute the short-end limit of the ATM curvature.

Theorem 8.2.3 *Assume the model 6.1.1 with* $\rho = 0$ *and assume that Hypotheseis (H1)–(H3) hold. Then,*

$$\lim_{T \to t}(T-t)^{1-2H} \frac{\partial^2 I_t}{\partial k^2}(k_t^*) = \frac{1}{4\sigma_t^5} \lim_{T \to t} \frac{\left[E_t \int_t^T \left(E_r \left(\int_r^T D_r^W \sigma_\theta^2 d\theta\right)\right)^2 dr\right]}{(T-t)^{2H+2}}.$$

Proof. Theorem 8.2.1 implies that

$$\lim_{T \to t}(T-t)^{1-2H} \frac{\partial^2 I}{\partial k^2}(t, k_t^*) = \frac{1}{2} \lim_{T \to t} \frac{E_t \left[\int_t^T \Psi''(\Lambda_r) U_r^2 dr\right]}{\frac{\partial BS}{\partial \sigma}(t, X_t, k_t^*, I_t(k_t^*))\,(T-t)^{2H+2}}. \tag{8.2.9}$$

Now, we apply similar ideas as in the Malliavin expansion technique in

Remark 6.5.6 to estimate $E_t\left[\int_t^T \Psi''(\Lambda_r)U_r^2 dr\right]$. Applying the anticipating Itô's formula to

$$\Psi''(\Lambda_s)\int_s^T U_r^2 dr$$

and taking expectations we get

$$\begin{aligned} 0 &= \Psi''(\Lambda_t)E_t\int_t^T U_r^2 dr - E_t\left[\int_t^T \Psi''(\Lambda_r)U_r^2 dr\right] \\ &+E\left[\int_t^T \Psi'''(\Lambda_u)\left(\int_u^T D_u^W U_r^2 dr\right)d\Lambda_u\right] \\ &+\frac{1}{2}E\left[\int_t^T \Psi^{iv}(\Lambda_u)\left(\int_u^T U_r^2 dr\right)d\langle\Lambda,\Lambda\rangle_u du\right]. \end{aligned} \tag{8.2.10}$$

Then, as

$$\Lambda_T = \Lambda_t + \int_t^T U_r dW_r$$

it follows that

$$\begin{aligned} 0 &= \Psi''(\Lambda_t)E_t\int_t^T U_r^2 dr - E_t\left[\int_t^T \Psi''(\Lambda_r)U_r^2 dr\right] \\ &+E\left[\int_t^T \Psi'''(\Lambda_u)\left(\int_u^T D_u^W U_r^2 dr\right)U_u du\right] \\ &+\frac{1}{2}E\left[\int_t^T \Psi^{iv}(\Lambda_u)\left(\int_u^T U_r^2 dr\right)U_u^2 du\right]. \end{aligned} \tag{8.2.11}$$

This, jointly with (8.2.9), implies

$$\begin{aligned} &\lim_{T\to t}(T-t)^{1-2H}\frac{\partial^2 I}{\partial k^2}(t,k_t^*) \\ &= \frac{1}{2}\lim_{T\to t}\frac{\Psi''(\Lambda_t)E_t\left[\int_t^T U_r^2 dr\right]}{\frac{\partial BS}{\partial\sigma}(t,X_t,k_t^*,I_t(k_t^*))(T-t)^{2H+2}} \\ &+\frac{1}{2}\lim_{T\to t}\frac{E\left[\int_t^T \Psi'''(\Lambda_u)\left(\int_u^T D_u^W U_r^2 dr\right)U_u du\right]}{\frac{\partial BS}{\partial\sigma}(t,X_t,k_t^*,I_t(k_t^*))(T-t)^{2H+2}} \\ &+\frac{1}{4}\lim_{T\to t}\frac{E\left[\int_t^T \Psi^{iv}(\Lambda_u)\left(\int_u^T U_r^2 dr\right)U_u^2 du\right]}{\frac{\partial BS}{\partial\sigma}(t,X_t,k_t^*,I_t(k_t^*))(T-t)^{2H+2}}. \end{aligned} \tag{8.2.12}$$

Now, we have that

$$\Psi'''(a) = \frac{\sqrt{2\pi}\exp(-2X_t)}{2(T-t)^2\left(BS^{-1}(a)\right)^4} e^{\frac{1}{4}(T-t)\left(BS^{-1}(a)\right)^2} \times \left((T-t)\left(BS^{-1}(a)\right)^2 - 12\right), \tag{8.2.13}$$

and

$$\Psi^{iv}(a) = \frac{\sqrt{2\pi}\exp(-3X_t)}{4(T-t)^{\frac{5}{2}}\left(BS^{-1}(a)\right)^5} e^{\frac{1}{2}(T-t)\left(BS^{-1}(a)\right)^2} \times \left((T-t)^2\left(BS^{-1}(a)\right)^4 - 16(T-t)\left(BS^{-1}(a)\right)^2 + 96\right). \tag{8.2.14}$$

It is easy to see (see Alòs and León (2017) for details) that $|D_u^W U_r^2| < C_t(r-u)^{2H-1}$. This, jointly with (8.2.13), (8.2.14), (H1), and (H2), gives us that the last two limits in the right-hand side of (8.2.12) are zero. Then,

$$\lim_{T\to t}(T-t)^{1-2H}\frac{\partial^2 I}{\partial k^2}(t,k_t^*) = \frac{1}{2}\lim_{T\to t}\frac{\Psi''(\Lambda_t)E_t\left[\int_t^T U_r^2 dr\right]}{\frac{\partial BS}{\partial\sigma}(t,X_t,k_t^*,I_t(k_t^*))(T-t)^{2H+2}}. \tag{8.2.15}$$

Now, taking into account (8.2.8), and that $\lim_{T\to t} I_t(k_t^*) = \sigma_t$ we get

$$\lim_{T\to t}(T-t)^{1-2H}\frac{\partial^2 I_t}{\partial k^2}(k_t^*) = \frac{1}{4\sigma_t^5}\lim_{T\to t}\frac{\left[E_t\int_t^T\left(E_r\left(\int_r^T D_r^W\sigma_\theta^2 d\theta\right)\right)^2 dr\right]}{(T-t)^{2H+2}},$$

and now the proof is complete. ■

Remark 8.2.4 *Assume that, for every $t\in[0,T]$, there exists an $\mathcal{F}_t^W$-measurable random variable $D_t^+\sigma_t^2$ such that $D_r^W\sigma_s^2$ tends (in some adequate sense) to $D_t^+\sigma_t^2$ as $r,s\to t$. Then, Theorem 8.2.3 reduces to*

$$\lim_{T\to t}\frac{\partial^2 I_t}{\partial k^2}(k_t^*) = \frac{(D^+\sigma_t^2)^2}{12\sigma_t^5} = \frac{(D^+\sigma_t)^2}{3\sigma_t^3}.$$

This is the case of classical diffusion volatilities, as we study in detail in Section 8.2.3.

Remark 8.2.5 (Multifactor volatilities) *Consider a multifactor volatility as in Remark 7.5.4, and assume the correlations $\rho_i = 0$, $i = 1, ..., d$. That is,*

$$dX_t = \left(r - \frac{\sigma_t^2}{2}\right) dt + \sigma_t dB_t,$$

where σ is a square integrable process adapted to the filtration generated by a d-dimensional Brownian motion W. Then, the same arguments as in the proof of Theorem 8.2.3 give us that

$$\lim_{T\to t}(T-t)^{1-2H}\frac{\partial^2 I_t}{\partial k^2}(k_t^*) = \frac{1}{4\sigma_t^5}\lim_{T\to t}\frac{\left[E_t\int_t^T \|E_r\left(\int_r^T D_r^W \sigma_\theta^2 d\theta\right)\|_2^2 dr\right]}{(T-t)^{2H+2}},$$

where $\|\cdot\|_2$ denotes the Euclidian norm in $\mathbb{R}^d$.

Remark 8.2.6 (The ATM level and curvature and the volatility swap) Theorems 6.5.3 and 8.2.3 give us that in the uncorrelated case

$$\lim_{T\to t}\frac{I(t,T,X_t,k_t^*) - E_t[v_t]}{(T-t)^{1+2H}} = -\frac{\sigma_t^4}{8}\lim_{T\to t}\frac{\frac{\partial^2 I}{\partial k^2}(t,T,X_t,k_t^*)}{(T-t)^{2H-\frac{1}{2}}}.$$

That is,

$$\lim_{T\to t}\frac{I(t,T,X_t,k_t^*) - E_t[v_t]}{(T-t)^{1+2H}} = -\frac{1}{8}\lim_{T\to t}\frac{I\left(t,T,X_t,k_t^*\right)^4 \frac{\partial^2 I}{\partial k^2}\left(t,T,X_t,k_t^*\right)}{(T-t)^{2H-\frac{1}{2}}}.$$

This gives us a model-free relationship between the volatility swap and the short-end ATM level and curvature in the case $\rho = 0$ (see Alòs and Shiraya (2019)).

8.2.3 Examples

This section is devoted to the study of the curvature for some popular stochastic volatility models in the uncorrelated case $\rho = 0$.

8.2.3.1 Diffusion stochastic volatilities

Assume that the volatility σ_r is of the form $\sigma_r = f(Y_r)$, where $f \in C_b^1(\mathbb{R})$ and Y_r is the solution of a stochastic differential equation

$$dY_r = a\left(r, Y_r\right) dr + b\left(r, Y_r\right) dW_r, \qquad (8.2.16)$$

for some real functions $a, b \in C_b^1(\mathbb{R})$. Then, from the results in Section 3.2.2 $D_r^W \sigma_s = f'(Y_s) D_r Y_s$, which implies that $D^+ \sigma_t^2 = 2\sigma_t f'(Y_t) b(t, Y_t)$. Then, Remark 8.2.4 gives us that

$$\lim_{T \to t} \frac{\partial I_t}{\partial k^2}(k_t^*) = \frac{1}{3\sigma_t^3}(f'(Y_t) b(t, Y_t))^2.$$

Let us see some particular examples.

Example 8.2.7 (Ornstein-Uhlenbeck volatilities) *Assume that Y_r is a stationary Ornstein-Uhlenbeck process satisfying the equation*

$$dY_r = \alpha(m - Y_r)dt + c\sqrt{2\alpha} dW_r,$$

for some positive constants m, c, and α. We know that

$$D_s^W Y_r = c\sqrt{2\alpha} \exp(-\alpha(r - s))$$

for all $t < r < s$. Then,

$$\lim_{T \to t} \frac{\partial I_t}{\partial k^2}(k_t^*) = \frac{2\alpha c^2}{3\sigma_t^3}(f'(Y_t))^2.$$

Example 8.2.8 (The SABR model) *We have seen in Section 3.3.1 that in the SABR volatility model*

$$d\sigma_t = \alpha \sigma_t dW_t, \quad t \in [0, T]$$

and its Malliavin derivative is given by

$$D_r^W \sigma_t = \alpha \sigma_t \mathbf{1}_{[0,t]}(r).$$

This implies that

$$\lim_{T \to t} \frac{\partial I_t}{\partial k^2}(k_t^*) = \frac{\alpha^2}{3\sigma_t}.$$

Example 8.2.9 (The Heston model) *The Heston model for $v_t = \sigma_t^2$ is given by*

$$dv_t = k(\theta - v_t)dt + \nu\sqrt{v_t} dW_t$$

(see Section 3.3.2). We have seen that its Malliavin derivative is given by

$$D_r^W \sigma_t = \frac{\nu}{2} \mathbf{1}_{[0,t]}(r) \exp\left(\int_r^t \left(- \left(\frac{k\theta}{2} - \frac{\nu^2}{8} \right) \frac{1}{\sigma_s^2} - \frac{k}{2} \right) ds \right),$$

from where we get

$$D_t^+ \sigma_t = \frac{\nu}{2}.$$

Then,

$$\lim_{T \to t} \frac{\partial I_t}{\partial k^2}(k_t^*) = \frac{\nu^2}{12\sigma_t^3}.$$

Example 8.2.10 (The two-factor Bergomi model) *Assume, as in Section 7.6.1.4, that the volatility can be written as*

$$\begin{aligned} v_t &= v_0 \mathcal{E}(\nu x_t) \\ &= v_0 \exp\left(\nu x_t - \frac{\nu^2}{2}(\delta^2 + (1-\delta)^2 + 2\rho_{12}\delta(1-\delta))t\right) \end{aligned} \quad (8.2.17)$$

with $\nu > 0$ *and* $x_t := \delta Y_t^1 + (1-\delta) Y_t^2$ *for some* $\delta \in (0,1)$*, where* Y^1 *and* Y^2 *are two processes of the form*

$$dY_r^1 = -\alpha_1 Y_r^1 dt + dW_r^1,$$

$$dY_r^2 = -\alpha_2 Y_r^2 dt + \rho_{12} dW_r^1 + \sqrt{1-\rho_{12}^2} dW_r^2.$$

Then, the results in Section 7.6.1.4 give us that

$$\begin{aligned} &\|D_r^W v_t\|_2^2 \\ &= v_t^2 \nu^2 \left(\delta e^{-\alpha_1(t-r)} + (1-\delta)\rho_{12} e^{-\alpha_2(t-r)}\right)^2 \\ &+ v_t^2 \nu^2 \delta^2 (1-\rho_{12}^2) e^{-2\alpha_2(t-r)}. \end{aligned} \quad (8.2.18)$$

Then, Remark 8.2.5 leads to

$$\lim_{T \to t} \frac{\partial I_t}{\partial k^2}(k_t^*) = \frac{\nu^2}{12\sigma_t}[2\delta^2(1-\rho_{12}) + 2\delta\rho_{12}(1-\rho_{12}) + \rho_{12}^2)]. \quad (8.2.19)$$

8.2.3.2 *Fractional volatility models*

Theorem 8.2.3 shows that the short-end of the ATM curvature is linked to the short-end of the Malliavin derivative $D_u^W \sigma_s^2$. If these Malliavin derivatives have a singularity as $u - s$ tends to zero (as this is the case of models based on the fBm or the RLfBm), the limit of the curvature is infinite. Then, in fractional volatility models, the curvature blows-up for short maturities. Let us see some examples.

Example 8.2.11 (Fractional Ornstein-Uhlenbeck volatilities) Assume as in Section 5.6.1 that the squared volatility σ^2 is of the form $\sigma_r = f(Y_r)$, where $f \in C_b^1(\mathbb{R})$ and Y_r is a *fractional Ornstein-Uhlenbeck* process of the form

$$Y_r = m + (Y_0 - m)\, e^{-\alpha r} + c\sqrt{2\alpha} \int_0^r e^{-\alpha(r-s)} d\hat{W}_s^H. \tag{8.2.20}$$

We proved that, in this case

$$D_r^W \sigma_t^2 = f'(Y_t) h(t,r),$$

where

- If $H > \frac{1}{2}$,
$$h(t,r) = c\sqrt{2\alpha}\left(H - \frac{1}{2}\right)\left(\int_t^r e^{-\alpha(r-u)} (u-s)^{H-\frac{3}{2}} du\right)$$
and
- If $H < \frac{1}{2}$,
$$h(t,r) = c\sqrt{2\alpha}\left(H - \frac{1}{2}\right)\left(\int_t^r e^{-\alpha(r-u)} (u-t)^{H-\frac{3}{2}} du\right).$$

This allows to prove that

$$\lim_{T\to t} \frac{\partial I_t}{\partial k^2}(k_t^*) = 0 \tag{8.2.21}$$

if $H > \frac{1}{2}$, while if $H < \frac{1}{2}$

$$\begin{aligned}
&\lim_{T\to t} (T-t)^{1-2H} \frac{\partial^2 I}{\partial k^2}(t,k^*) \\
&= \frac{1}{4\sigma_t^5} \lim_{T\to t} \frac{E_t \int_t^T \left(E_r\left(\int_r^T D_r^W \sigma_s^2 ds\right)\right)^2 dr}{(T-t)^{2H+2}} \\
&= \frac{1}{4\sigma_t^5} \lim_{T\to t} \frac{E_t \int_t^T \left(E_r\left(\int_r^T f'(Y_s) h(s,r) ds\right)\right)^2 dr}{(T-t)^{2H+2}} \\
&= \frac{(f'(Y_t))^2}{4\sigma_t^5} \lim_{T\to t} \frac{\int_t^T \left(\int_r^T h(s,r) ds\right)^2 dr}{(T-t)^{2H+2}} \\
&= \frac{1}{(2H+2)(H+1/2)^2} \frac{(\nu f'(Y_t))^2}{4\sigma_t^5}.
\end{aligned}$$

That is, fractional volatility components have no impact on the short-end curvature if $H > \frac{1}{2}$, while this curvature blows-up for $H < \frac{1}{2}$.

Example 8.2.12 (The rough Bergomi model) *We know that in the rough Bergomi model (see Section 5.6.2), $D_r^W v_s = \nu\sqrt{2H} v_s (s-r)^{H-\frac{1}{2}}$. Then,*

$$\begin{aligned}
\lim_{T\to t}(T-t)^{1-2H}\frac{\partial^2 I}{\partial k^2}(t,k^*) &= \frac{2H\nu^2}{4\sigma_t^5}\lim_{T\to t}\frac{\int_t^T\left(\int_r^T (s-r)^{H-\frac{1}{2}}ds\right)^2 dr}{(T-t)^{2H+2}} \\
&= \frac{1}{(2H+2)(H+1/2)^2}\frac{2H\nu^2}{4\sigma_t^5}.
\end{aligned}$$

8.3 THE CORRELATED CASE

In this section, we extend the above results to the correlated case. We see that this curvature depends on the correlation parameter, and we study under which conditions the short-time limit of this curvature is positive. Moreover, we study when this curvature blows-up for short maturities, and we compute the corresponding rate of divergence. Our approach follows the same ideas as in Chapters 6 and 7, where we make use of an adequate decomposition formula for option prices and Malliavin expansion techniques to extend the results we have obtained in the uncorrelated case.

8.3.1 A representation for the ATM curvature

Our first step in this section is to prove the following representation formula, which links the curvatures in the uncorrelated and the correlated cases. This result follows from an implicit derivation and the decomposition formula for option prices proved in Theorem 6.3.2.

Proposition 8.3.1 *Assume that the model (6.1.1) and hypotheses (H1) and (H2) are satisfied. Then,*

$$
\begin{aligned}
\frac{\partial^2 I_t(k_t^*)}{\partial k^2} &= \frac{\partial^2 I_t^0}{\partial k^2}(k_t^*)\frac{\frac{\partial BS}{\partial \sigma}(t, X_t, k_t^*, I_t^0(k_t^*))}{\frac{\partial BS}{\partial \sigma}(t, X_t, k_t^*, I_t(k_t^*))} \\
&+ \frac{1}{\frac{\partial BS}{\partial \sigma}(t, X_t, k_t^*, I_t(k_t^*))} \\
&\times \left[\frac{\partial^2 BS}{\partial k^2}(t, X_t, k_t^*, I_t^0(k_t^*)) - \frac{\partial^2 BS}{\partial k^2}(t, X_t, k_t^*, I_t(k_t^*)) \right. \\
&+ \frac{\rho}{2} E_t \left(\int_t^T e^{-r(s-t)} \frac{\partial^3 G}{\partial k^2 \partial x}(s, X_s, k_t^*, v_s)\Phi_s ds \right) \\
&- 2\frac{\partial^2 BS}{\partial k \partial \sigma}(t, X_t, k_t^*, I_t(k_t^*)) \frac{\partial I_t(k_t^*)}{\partial k} \\
&\left. - \frac{\partial^2 BS}{\partial \sigma^2}(t, X_t, k_t^*, I_t(k_t^*)) \left(\frac{\partial I_t(k_t^*)}{\partial k} \right)^2 \right], \qquad (8.3.1)
\end{aligned}
$$

where $I_t^0(k_t^*)$ denotes the ATM implied volatility in the uncorrelated case and $\Phi_s := \sigma_s \int_s^T D_s^W \sigma_u^2 du$.

Proof. As in the proof of Theorem 8.2.1, implicit derivation gives us that

$$
\begin{aligned}
\frac{\partial^2}{\partial k^2} V_t &= \frac{\partial^2 BS}{\partial k^2}(t, X_t, k_t^*, I_t(k_t^*)) \\
&+ 2\frac{\partial^2 BS}{\partial k \partial \sigma}(t, X_t, k_t^*, I_t(k_t^*)) \frac{\partial I_t(k_t^*)}{\partial k} \\
&+ \frac{\partial^2 BS}{\partial \sigma^2}(t, X_t, k_t^*, I_t(k_t^*)) \left(\frac{\partial I_t(k_t^*)}{\partial k} \right)^2 \\
&+ \frac{\partial BS}{\partial \sigma}(t, X_t, k_t^*, I_t(k_t^*)) \frac{\partial^2 I_t(k_t^*)}{\partial k^2}. \qquad (8.3.2)
\end{aligned}
$$

This allows us to get the following expression for the curvature

$$
\begin{aligned}
\frac{\partial^2 I_t(k_t^*)}{\partial k^2} &= \frac{1}{\frac{\partial BS}{\partial \sigma}(t, X_t, k_t^*, I_t(k_t^*))} \\
&\times \left[\frac{\partial^2 V_t}{\partial k^2} - \frac{\partial^2 BS}{\partial k^2}(t, X_t, k_t^*, I_t(k_t^*)) \right. \\
&- 2\frac{\partial^2 BS}{\partial k \partial \sigma}(t, X_t, k_t^*, I_t(k_t^*)) \frac{\partial I_t(k_t^*)}{\partial k} \\
&\left. - \frac{\partial^2 BS}{\partial \sigma^2}(t, X_t, k_t^*, I_t(k_t^*)) \left(\frac{\partial I_t(k_t^*)}{\partial k} \right)^2 \right]. \qquad (8.3.3)
\end{aligned}
$$

Now, notice that, the extension of the Hull and White formula in Theorem 6.3.2 allows us to write

$$
\begin{aligned}
&\frac{\partial^2}{\partial k^2} V_t \\
&= \frac{\partial^2}{\partial k^2} E_t(BS(t, X_t, k_t^*, v_t)) + \frac{\rho}{2} E_t\left(\int_t^T e^{-r(s-t)} \frac{\partial^3 G}{\partial k^2 \partial x}(s, X_s, k_t^*, v_s)\Phi_s ds\right) \\
&= \frac{\partial^2}{\partial k^2}\left(V_t^0\right) + \frac{\rho}{2} E_t\left(\int_t^T e^{-r(s-t)} \frac{\partial^3 G}{\partial k^2 \partial x}(s, X_s, k_t^*, v_s)\Phi_s ds\right),
\end{aligned}
$$

where V_t^0 denotes the price in the uncorrelated case $\rho = 0$. Now, from Equation (8.2.4) in the proof of Theorem 8.2.1 we get

$$
\begin{aligned}
\frac{\partial^2 V_t^0}{\partial k^2} &= \frac{\partial^2 BS}{\partial k^2}\left(t, X_t, k_t^*, I_t^0(k_t^*)\right) \\
&+ \frac{\partial BS}{\partial \sigma}\left(t, X_t, k_t^*, I_t^0(k_t^*)\right) \frac{\partial^2 I_t^0(k_t^*)}{\partial k^2}.
\end{aligned}
\tag{8.3.4}
$$

This allows us to write

$$
\begin{aligned}
&\frac{\partial^2}{\partial k^2} V_t \\
= \; &\frac{\partial^2 BS}{\partial k^2}(t, X_t, k_t^*, I_t^0(k_t^*)) \\
&+ \frac{\partial BS}{\partial \sigma}(t, X_t, k_t^*, I_t^0(k_t^*)) \frac{\partial^2 I_t^0}{\partial k^2}(k_t^*) \\
&+ \frac{\rho}{2} E_t\left(\int_t^T e^{-r(s-t)} \frac{\partial^3 G}{\partial k^2 \partial x}(s, X_s, k_t^*, v_s)\Phi_s ds\right).
\end{aligned}
\tag{8.3.5}
$$

This, jointly with (8.3.4), allows us to complete the proof. ■

Remark 8.3.2 *Notice that $I_t^0(k_t^*)$ and $I_t(k_t^*)$ tend to the spot volatility σ_t as $T \to t$. This, jointly with the classical expression for the vega, gives us that*

$$
\frac{\frac{\partial BS}{\partial \sigma}\left(t, X_t, k_t^*; I_t^0(k_t^*)\right)}{\frac{\partial BS}{\partial \sigma}\left(t, X_t, k_t^*, I_t(k_t^*)\right)} \to 1
$$

as $T \to t$. Then, the first term in the right-hand side of (8.3.1) tends to the curvature in the uncorrelated case $\frac{\partial^2 I_t^0(k_t^*)}{\partial k^2}$. The sum of the other four terms can be interpreted as the effect of the correlation on the curvature.

8.3.2 Limit results

Now our objective is to use the result in Proposition 8.3.1 to compute the short-time limit of the ATM curvature. Towards this end, we need the following hypotheses:

(H1') σ^2 belongs to $\mathbb{L}_W^{3,2}$, and there exists a positive constant C, and $H \in (0,1)$ such that, for all $t < \tau < \theta < r < u < T$

$$\begin{aligned}\left|D_\theta^W \sigma_r^2\right| &\leq C\,(r-\theta)^{H-\frac{1}{2}}, \\ \left|D_\theta^W D_r^W \sigma_u^2\right| &\leq C\,(u-r)^{H-\frac{1}{2}}\,(u-\theta)^{H-\frac{1}{2}}, \\ \left|D_\tau^W D_\theta^W D_r^W \sigma_u\right| &\leq C\,(u-r)^{H-\frac{1}{2}}\,(u-\theta)^{H-\frac{1}{2}}\,(u-\tau)^{H-\frac{1}{2}}.\end{aligned}$$

(H3') Hypotheses (H1') and (H2) hold and, for every $t \in [0,T]$,

$$\frac{1}{(T-t)^{1+2H}}\left(\int_t^T \left(\int_s^T D_s^W \sigma_\theta^2 d\theta\right) ds\right)$$

and

$$\frac{1}{(T-t)^{2+2H}}\int_t^T (\int_s^T D_s^W \left(\sigma_r \left(\int_s^T D_s^W \sigma_\theta^2 d\theta\right)\right) dr) ds$$

have a finite limit as $T \to t$.

Theorem 8.3.3 *Assume that the model (6.1.1), and hypotheses (H1'), (H2'), and (H3) are satisfied. Then,*

$$\begin{aligned}&\lim_{T\to t}(T-t)^{1-2H}\frac{\partial^2 I}{\partial k^2}(t,k_t^*) \\ &= \frac{1}{4\sigma_t^5}\lim_{T\to t}\frac{\left[E_t\int_t^T\left(E_r\left(\int_r^T D_r^W\sigma_\theta^2 d\theta\right)\right)^2 dr\right]}{(T-t)^{2+2H}} \\ &+\rho^2\lim_{T\to t}E_t\left[-\frac{3}{2\sigma_t^5\,(T-t)^{5-2H}}\left(\int_t^T\left(\int_s^T D_s^W\sigma_\theta^2 d\theta\right)ds\right)^2\right. \\ &\left.+\frac{1}{\sigma_t^4\,(T-t)^{2+2H}}\int_t^T\int_s^T D_s^W\left(\sigma_r\left(\int_s^T D_s^W\sigma_\theta^2 d\theta\right)\right)dr)ds\right].\end{aligned}$$

Proof. Proposition 8.3.1 allows us to write

$$
\begin{aligned}
&(T-t)^{1-2H}\frac{\partial^2 I_t}{\partial k^2}(k_t^*)\\
&\quad=(T-t)^{1-2H}\left[\frac{\frac{\partial BS}{\partial\sigma}(t,X_t,k_t^*;I_t^0(k_t^*))}{\frac{\partial BS}{\partial\sigma}(t,X_t,k_t^*,I_t(k_t^*))}\frac{\partial^2 I_t^0}{\partial k^2}(k_t^*)\right.\\
&\qquad+\frac{\frac{\partial^2 BS}{\partial k^2}(t,X_t,k_t^*,I_t^0(k_t^*))-\frac{\partial^2 BS}{\partial k^2}(t,X_t,k_t^*;I_t(k_t^*))}{\frac{\partial BS}{\partial\sigma}(t,X_t,k_t^*,I_t(k_t^*))}\\
&\qquad+\frac{\rho}{2}\frac{E_t\left(\int_t^T e^{-r(s-t)}\frac{\partial^3 G}{\partial k^2\partial x}(s,X_s,k_t^*,v_s)\Phi_s ds\right)}{\frac{\partial BS}{\partial\sigma}(t,X_t,k_t^*,I_t(k_t^*))}\\
&\qquad-2\frac{\frac{\partial^2 BS}{\partial k\partial\sigma}(t,X_t,k_t^*;I_t(k_t^*))}{\frac{\partial BS}{\partial\sigma}(t,X_t,k_t^*,I_t(k_t^*))}\frac{\partial I_t}{\partial k}(k_t^*)\\
&\qquad\left.-\frac{\frac{\partial^2 BS}{\partial\sigma^2}(t,X_t,k_t^*;I_t(k_t^*))}{\frac{\partial BS}{\partial\sigma}(t,X_t,k_t^*,I_t(k_t^*))}\left(\frac{\partial I_t}{\partial k}(k_t^*)\right)^2\right]\\
&:=\quad(T-t)^{1-2H}\frac{\frac{\partial BS}{\partial\sigma}(t,X_t,k_t^*;I_t^0(k_t^*))}{\frac{\partial BS}{\partial\sigma}(t,X_t,k_t^*,I_t(k_t^*))}\frac{\partial^2 I_t^0}{\partial k^2}(k_t^*)\\
&\quad+T_1+T_2+T_3+T_4.
\end{aligned}
\tag{8.3.6}
$$

From 8.3.2 we know that

$$
\frac{\frac{\partial BS}{\partial\sigma}(t,X_t,k_t^*;I_t^0(k_t^*))}{\frac{\partial BS}{\partial\sigma}(t,X_t,k_t^*,I_t(k_t^*))}\to 1
$$

as $T\to t$. On the other hand, as

$$
\frac{\partial^2 BS}{\partial k\partial\sigma}(t,X_t,k_t^*;I_t(k_t^*))=\frac{1}{2}\frac{\sqrt{T-t}}{\sqrt{2\pi}}e^{X_t}\exp\left(-\frac{I^2(t,k_t^*)(T-s)}{8}\right),
$$

we get

$$
\frac{\frac{\partial^2 BS}{\partial k\partial\sigma}(t,X_t,k_t^*;I_t(k_t^*))}{\frac{\partial BS}{\partial\sigma}(t,X_t,k_t^*,I_t(k_t^*))}=\frac{1}{2},
$$

which implies that $\lim_{T\to t}T_3=-\lim_{T\to t}(T-t)^{1-2H}\frac{\partial I_t}{\partial k}(k_t^*)$. Moreover, the classical relationship between the ATM vega $\frac{\partial BS}{\partial\sigma}$ and vomma $\frac{\partial^2 BS}{\partial\sigma^2}$

$$
\frac{\partial^2 BS}{\partial\sigma^2}(t,X_t,k_t^*;I_t(k_t^*))=\frac{I_t(k_t^*)}{4}\frac{\partial BS}{\partial\sigma}(t,X_t,k_t^*,I_t(k_t^*))
$$

leads to

$$T_4 = -(T-t)^{1-2H} \frac{\frac{\partial^2 BS}{\partial \sigma^2}(t, X_t, k_t^*; I_t(k_t^*))}{\frac{\partial BS}{\partial \sigma}(t, X_t, k_t^*, I_t(k_t^*))} \left(\frac{\partial I_t}{\partial k}(k_t^*) \right)^2$$
$$= -\frac{(T-t)^{2-2H} I_t(k_t^*)}{4} \left(\frac{\partial I_t}{\partial k}(k_t^*) \right)^2.$$

Then, as $\frac{\partial I}{\partial k}(t, k_t^*) = O(T-t)^{H-\frac{1}{2}}$ (see Theorem 7.5.2) it follows that $T_4 \to 0$ as $T \to t$. Then, we have proved that

$$\lim_{T\to t} (T-t)^{1-2H} \frac{\partial^2 I_t}{\partial k^2}(k_t^*) = \lim_{T\to t} (T-t)^{1-2H} \left[\frac{\partial^2 I_t^0}{\partial k^2}(k_t^*) - \frac{\partial I_t}{\partial k}(k_t^*) + T_1 + T_2 \right]. \tag{8.3.7}$$

Then, we only need to study the terms T_1 and T_2. We are going to expand these terms using the Malliavin expansion techniques as in Remark 6.5.6. Now the proof decomposes into two steps.

Step 1. Let us study the term T_1. We have that

$$\frac{\partial^2 BS}{\partial k^2}(t, X_t, k_t^*, I_t^0(k_t^*)) - \frac{\partial^2 BS}{\partial k^2}(t, X_t, k_t^*, I_t(k_t^*))$$
$$= \frac{\partial^2 BS}{\partial k^2}(t, X_t, k_t^*, BS^{-1}(V_t^0)) - \frac{\partial^2 BS}{\partial k^2}(t, X_t, k_t^*, BS^{-1}(V_t)).$$

We know, from the extension of the Hull and White formula in Theorem 6.3.2, that

$$V_t = V_t^0 + E_t \left(\int_t^T e^{-r(s-t)} \frac{\partial G}{\partial x}(s, X_s, k_t^*, v_s) \Phi_s ds \right).$$

Then,

$$\frac{\partial^2 BS}{\partial k^2}(t, X_t, k_t^*, I_t^0(k_t^*)) - \frac{\partial^2 BS}{\partial k^2}(t, X_t, k_t^*, I_t(k_t^*))$$
$$= -\Psi'(\mu(T,t)) \left(\frac{\rho}{2} E_t \left(\int_t^T e^{-r(s-t)} \frac{\partial G}{\partial x}(s, X_s, k_t^*, v_s) \Phi_s ds \right) \right),$$

where Ψ is defined in Theorem 8.2.1 and $\mu(T,t)$ is a positive value between V_t^0 and V_t. Now, in order to estimate

$$E_t \left(\int_t^T e^{-r(s-t)} \frac{\partial G}{\partial x}(s, X_s, k_t^*, v_s) \Phi_s ds \right),$$

we apply the Malliavin expansion technique explained in Remark 6.5.6. Then, we get

$$
\begin{aligned}
T_1 &= -\frac{\rho}{2}(T-t)^{1-2H}\frac{\Psi'(\mu(T,t))E_t\left(\int_t^T e^{-r(s-t)}\frac{\partial G}{\partial x}(s,X_s,k_t^*,v_s)\Phi_s ds\right)}{\frac{\partial BS}{\partial\sigma}(t,X_t,k_t^*,I(t,k_t^*))} \\
&= -\frac{\rho}{2}(T-t)^{1-2H}\frac{\Psi'(\mu(T,t))}{\frac{\partial BS}{\partial\sigma}(t,X_t,k_t^*,I_t(k_t^*))}\left[E_t\left(\frac{\partial G}{\partial x}(t,X_t,k_t^*,v_t)\int_t^T\Phi_s ds\right)\right. \\
&+\frac{\rho}{2}E_t\left(\int_t^T\int_s^T\left(\frac{\partial^3}{\partial x^3}-\frac{\partial^2}{\partial x^2}\right)\frac{\partial G}{\partial x}(s,X_s,k_t^*,v_s)\Phi_r dr)\Lambda_s ds\right) \\
&\left.+\rho E_t\left(\int_t^T\frac{\partial^2 G}{\partial x^2}(s,X_s,k_t^*,v_s)(\int_s^T D_s^W\Phi_r dr)\sigma_s ds\right)\right].
\end{aligned}
$$

We can easily see that the last two terms do not tend to zero. Then, we apply again the Malliavin expansion technique to these terms and we get

$$
\begin{aligned}
T_1 &= -\frac{\rho}{2}(T-t)^{1-2H}\frac{\Psi'(\mu(T,t))E_t\left(\int_t^T e^{-r(s-t)}\frac{\partial G}{\partial x}(s,X_s,k_t^*,v_s)\Phi_s ds\right)}{\frac{\partial BS}{\partial\sigma}(t,X_t,k_t^*,I(t,k_t^*))} \\
&= -\frac{\rho}{2}(T-t)^{1-2H}\frac{\Psi'(\mu(T,t))}{\frac{\partial BS}{\partial\sigma}(t,X_t,k_t^*,I_t(k_t^*))}\left[E_t\left(\frac{\partial G}{\partial x}(t,X_t,k_t^*,v_t)\int_t^T\Phi_s ds\right)\right. \\
&+\frac{\rho}{2}E_t\left(\left(\frac{\partial^3}{\partial x^3}-\frac{\partial^2}{\partial x^2}\right)\frac{\partial G}{\partial x}(t,X_t,k_t^*,v_t)\int_t^T(\int_s^T\Phi_r dr)\Lambda_s ds\right) \\
&\left.+\rho E_t\left(\frac{\partial^2 G}{\partial x^2}(t,X_t,k_t^*,v_t)\int_t^T(\int_s^T D_s^W\Phi_r dr)\sigma_s ds\right)\right]+T_1^4 \\
&= T_1^1+T_1^2+T_1^3+T_1^4,
\end{aligned}
$$

where T_1^4 is a term such that $T_1^4\to 0$ as $T\to t$ (see Alòs and León (2017) for further details). Now, straightforward computations allow us to explicitly compute the limits of T_1^1, T_1^2, and T_1^3:

- Let us study the term T_1^1. A direct computation gives us that

$$
\frac{\partial G}{\partial x}(t,X_t,k_t^*,v_t)=\frac{\exp(X_t)\exp\left(-\frac{v_t^2(T-t)}{2}\right)}{\sqrt{2\pi}v_t\sqrt{t-t}}.
$$

Then, it follows that

$$\begin{aligned}
&\lim_{T\to t} T_1^1 \\
&= -\frac{\rho}{2}\lim_{T\to t}(T-t)^{1-2H}\frac{\Psi'(\mu(T,t))}{\frac{\partial BS}{\partial\sigma}(t,X_t,k_t^*,I(t,k_t^*))} \\
&\qquad \times E_t\left(\frac{\partial G}{\partial x}(t,X_t,k_t^*,v_t)\int_t^T \Phi_s ds\right) \\
&= \frac{\rho}{4}\lim_{T\to t}(T-t)^{1-2H}\frac{1}{BS^{-1}(\mu(T,t))^2 v_t(T-t)^2}E_t\left(\int_t^T \Phi_s ds\right).
\end{aligned}$$

Notice that $BS^{-1}(\mu(T,t))$ is an intermediate value between $I_t^0(k_t^*)$ and $I_t(k_t^*)$. Then, $BS^{-1}(\mu(T,t)) \to \sigma_t$ as $T \to t$. Moreover, due to Theorem 7.5.2, the above quantity is linked to the skew. More precisely, we get

$$\lim_{T\to t} T_1^1 = \frac{1}{2}\lim_{T\to t}(T-t)^{1-2H}\frac{\partial I_t}{\partial k}(t,k_t^*). \tag{8.3.8}$$

- Now let us study the term T_1^2. As

$$\begin{aligned}
&\left(\frac{\partial^3}{\partial x^3}-\frac{\partial^2}{\partial x^2}\right)\frac{\partial G}{\partial x}(t,X_t,k_t^*,v_t) \\
&= -\frac{1}{16}\frac{\exp(X_t)\exp\left(-\frac{v_t^2(T-t)}{2}\right)}{\sqrt{2\pi}\left(v_t\sqrt{t-t}\right)^5}\left(3v_t^4(T-t)^2+8v_t^2(T-t)-48\right),
\end{aligned}$$

it follows that

$$\begin{aligned}
\lim_{T\to t} T_1^2 &= \frac{\rho^2}{4\sigma_t^7}\lim_{T\to t}\frac{3}{(T-t)^{5-2H}}\left(\int_t^T\left(\int_s^T \Phi_r dr\right)\Phi_s ds\right) \\
&= \frac{\rho^2}{8\sigma_t^7}\lim_{T\to t}\frac{3}{(T-t)^{5-2H}}\left(\int_t^T \Phi_s ds\right)^2 \\
&= \frac{\rho^2}{8\sigma_t^5}\lim_{T\to t}\frac{3}{(T-t)^{5-2H}}\left(\int_t^T\left(\int_s^T D_s^W\sigma_\theta^2 d\theta\right)ds\right)^2.
\end{aligned} \tag{8.3.9}$$

- Finally, let us study T_1^3. As

$$\frac{\partial^2 G}{\partial x^2}(t,X_t,k_t^*,v_t) = \frac{1}{8}\frac{\exp(X_t)\exp\left(-\frac{v_t^2(T-t)}{2}\right)}{\sqrt{2\pi}\left(v_t\sqrt{t-t}\right)^3}\left(v_t^2(T-t)+4\right)$$

we get

$$T_1^3 = -\frac{\rho^2}{2\sigma_t^5} \lim_{T\to t} \frac{1}{(T-t)^{4-2H}} \times E_t\left(\int_t^T (\int_s^T D_s^W \left(\sigma_r \left(\int_s^T D_s^W \sigma_\theta^2 d\theta \right)\right) dr) \sigma_s ds \right). \quad (8.3.10)$$

Then (8.3.8), (8.3.9), and (8.3.10) give us that

$$\lim_{T\to t} T_1 = \frac{1}{2} \lim_{T\to t} (T-t)^{1-2H} \frac{\partial I_t}{\partial k}(k_t^*) + \frac{\rho^2}{\sigma_t^5} \lim_{T\to t} \left[\frac{3}{8(T-t)^{5-2H}} \left(\int_t^T \left(\int_s^T D_s^W \sigma_\theta^2 d\theta \right) ds \right)^2 - \frac{1}{2(T-t)^{4-2H}} E_t \left(\int_t^T (\int_s^T D_s^W \left(\sigma_r \left(\int_s^T D_s^W \sigma_\theta^2 d\theta \right)\right) dr) \sigma_s ds \right) \right]. \quad (8.3.11)$$

Step 2. Again, the Malliavin expansion technique allows us to study the term T_2. First of all, we apply the anticipating Itô's formula to the process

$$e^{-r(s-t)} \frac{\partial^3 G}{\partial k^2 \partial x}(s, X_s, k_t^*, v_s) \int_s^T \Phi_r dr$$

and we take conditional expectations. Then, we get

$$\begin{aligned} T_2 &= \frac{\rho}{2}(T-t)^{1-2H} \frac{E_t\left(\int_t^T e^{-r(s-t)} \frac{\partial^3 G}{\partial k^2 \partial x}(s, X_s, k_t^*, v_s) \Phi_s ds \right)}{\frac{\partial BS}{\partial \sigma}(t, X_t, k_t^*, I_t(k_t^*))} \\ &= \frac{\rho}{2}(T-t)^{1-2H} \frac{1}{\frac{\partial BS}{\partial \sigma}(t, X_t, k_t^*, I_t(k_t^*))} \left[E_t \left(\frac{\partial^3 G}{\partial k^2 \partial x}(t, X_t, k_t^*, v_t) \int_t^T \Phi_s ds \right) \right. \\ &+ \frac{\rho^2}{4} E_t \left(\int_t^T e^{-r(s-t)} \left(\frac{\partial^3}{\partial x^3} - \frac{\partial^2}{\partial x^2} \right) \frac{\partial^3 G}{\partial k^2 \partial x}(s, X_s, k_t^*, v_s) (\int_s^T \Phi_r dr) \Phi_s ds \right) \\ &+ \left. \frac{\rho^2}{2} E_t \left(\int_t^T e^{-r(s-t)} \frac{\partial^4 G}{\partial k^2 \partial x^2}(s, X_s, k_t^*, v_s) (\int_s^T D_s^W \Phi_r dr) \sigma_s ds \right) \right]. \end{aligned} \quad (8.3.12)$$

Again, the last two terms do not tend to zero. Applying again the Malliavin expansion technique to the last two terms we get

$$
\begin{aligned}
T_2 &= \frac{\rho}{2}(T-t)^{1-2H}\frac{E_t\left(\int_t^T e^{-r(s-t)}\frac{\partial^3 G}{\partial k^2\partial x}(s,X_s,k_t^*,v_s)\Phi_s ds\right)}{\frac{\partial BS}{\partial\sigma}(t,X_t,k_t^*,I_t(k_t^*))} \\
&= \frac{\rho}{2}(T-t)^{1-2H}\frac{1}{\frac{\partial BS}{\partial\sigma}(t,X_t,k_t^*,I_t(k_t^*))}\left[E_t\left(\frac{\partial^3 G}{\partial k^2\partial x}(t,X_t,k_t^*,v_t)\int_t^T\Phi_s ds\right)\right. \\
&+\frac{\rho^2}{4}E_t\left(\left(\frac{\partial^3}{\partial x^3}-\frac{\partial^2}{\partial x^2}\right)\frac{\partial^3 G}{\partial k^2\partial x}(t,X_t,k_t^*,v_t)\int_t^T e^{-r(s-t)}(\int_s^T\Phi_r dr)\Phi_s ds\right) \\
&+\left.\frac{\rho^2}{2}E_t\left(\frac{\partial^4 G}{\partial k^2\partial x^2}(t,X_t,k_t^*,v_t)\int_t^T e^{-r(s-t)}(\int_s^T D_s^W\Phi_r dr)\sigma_s ds\right)\right]+T_2^4 \\
&=: T_2^1+T_2^2+T_2^3+T_2^4,
\end{aligned}
\tag{8.3.13}
$$

where T_2^4 is a term that tends to zero (see again Alòs and León (2017) for details). Now, we compute the limits of T_2^1, T_2^2 , and T_2^3.

- Let us compute the limit of T_2^1. As

$$
\frac{\partial^3 G}{\partial k^2\partial x}(t,X_t,k_t^*,v_t)=\frac{\exp(X_t)\exp\left(-\frac{v_t^2(T-t)}{2}\right)}{\sqrt{2\pi}v_t\sqrt{T-t}}\left[\frac{3}{8}+\frac{1}{2v_t^2(T-t)}\right]
$$

and this, jointly with Theorem 7.5.2, implies that

$$
\begin{aligned}
\lim_{T\to t}T_2^1 &= \frac{\rho}{4\sigma_t^2}\lim_{T\to t}\frac{E_t\left[\int_t^T\left(\int_s^T D_s^W\sigma_\theta^2 d\theta\right)ds\right]}{(T-t)^{2H+1}} \\
&= \frac{1}{2}\lim_{T\to t}(T-t)^{1-2H}\frac{\partial I_t}{\partial k}(k_t^*).
\end{aligned}
\tag{8.3.14}
$$

- Let us compute the limit of T_2^2. Taking into account that

$$
\begin{aligned}
&\left(\frac{\partial^3}{\partial x^3}-\frac{\partial^2}{\partial x^2}\right)\frac{\partial^3 G}{\partial k^2\partial x}(t,X_t,k_t^*,v_t) \\
&= \frac{3}{64}\frac{\exp(X_t)\exp\left(-\frac{v_t^2(T-t)}{2}\right)}{\sqrt{2\pi}\left(v_t\sqrt{T-t}\right)^7} \\
&\quad\times\left(-v_t^6(T-t)^3+12v_t^4(T-t)^2+176v_t^2(T-t)-320\right)
\end{aligned}
$$

we get

$$
\begin{aligned}
&\lim_{T\to t} T_2^2 \\
&= \lim_{T\to t}(T-t)^{1-2H}\frac{\frac{\rho^2}{4}E_t\left(\left(\frac{\partial^3}{\partial x^3}-\frac{\partial^2}{\partial x^2}\right)\frac{\partial^3 G}{\partial k^2\partial x}(t,X_t,k_t^*,v_t)\int_t^T(\int_s^T\Phi_r dr)\Phi_s ds\right)}{\frac{\partial BS}{\partial\sigma}\left(t,X_t,k_t^*,I(t,k_t^*)\right)} \\
&= \frac{\rho^2}{4}\lim_{T\to t}E_t\left(-\frac{15}{v_t^7(T-t)^{3+2H}}\int_t^T(\int_s^T\Phi_r dr)\Phi_s ds\right) \\
&= \frac{\rho^2}{8}\lim_{T\to t}E_t\left(-\frac{15}{v_t^7(T-t)^{3+2H}}\left(\int_t^T\Phi_s ds\right)^2\right) \\
&= \frac{-15\rho^2}{8\sigma_t^5}\lim_{T\to t}E_t\left(\frac{1}{(T-t)^{3+2H}}\left(\int_t^T\left(\int_s^T D_s^W\sigma_\theta^2 d\theta\right)ds\right)^2\right). \qquad (8.3.15)
\end{aligned}
$$

- Finally, let us study the term T_2^3. As

$$
\frac{\partial^4 G}{\partial k^2\partial x^2}(t,X_t,k_t^*,v_t) = -\frac{1}{16}\frac{1}{\sqrt{2\pi}\,(v_t(T-t))^5}\exp(X_t)
\exp\left(-\frac{v_t^2(T-t)}{2}\right)\left(v_t^4(T-t)^2-48\right),
$$

we can write

$$
\lim_{T\to t}T_2^3 = \frac{3\rho^2}{2\sigma_t^4}\lim_{T\to t}E_t
\left(\frac{1}{(T-t)^{4+2H}}\int_t^T(\int_s^T D_s^W\left(\sigma_r\left(\int_s^T D_s^W\sigma_\theta^2 d\theta\right)\right)dr)ds\right).
$$

This, jointly with (8.3.14) and (8.3.15), implies that

$$
\begin{aligned}
\lim_{T\to t}T_2 &= \frac{1}{2}\lim_{T\to t}(T-t)^{1-2H}\frac{\partial I_t}{\partial k}(k_t^*) \\
&+ \rho^2\lim_{T\to t}E_t\left[-\frac{15}{8\sigma_t^5(T-t)^{5-2H}}\left(\int_t^T\left(\int_s^T D_s^W\sigma_\theta^2 d\theta\right)ds\right)^2\right. \\
&\left.+\frac{3}{2\sigma_t^4(T-t)^{4-2H}}\int_t^T(\int_s^T D_s^W\left(\sigma_r\left(\int_s^T D_s^W\sigma_\theta^2 d\theta\right)\right)dr)ds\right].
\end{aligned}
\qquad (8.3.16)
$$

Now, (8.3.7), (8.3.11), (8.3.16), and Theorem 8.2.3 allow us to complete the proof. ■

Remark 8.3.4 *The above results show that, in fractional volatility models, the at-the-money curvature is $O(T-t)^{2H-1}$. As the skew slope is $O(T-t)^{H-\frac{1}{2}}$, this means that for short maturities, the curvature is higher than the skew, according to the empirical data in Figure 8.1.*

Remark 8.3.5 *Theorem 8.3.3 proves that*

$$\lim_{T-t}(T-t)^{1-2H}\frac{\partial^2 I_t}{\partial k^2}(k_t^*) = \lim_{T-t}(T-t)^{1-2H}\frac{\partial^2 I_t^0}{\partial k^2}(k_t^*) + O(\rho^2),$$

where I_t^0 denotes the implied volatility in the uncorrelated case. That is, the difference between the curvature in the uncorrelated and the correlated case is $O(\rho^2)$.

Remark 8.3.6 *Assume that, for every $t \in [0,T]$, there exist two $\mathcal{F}_t^W$-measurable random variables $D_t^+\sigma_t^2$ and $(D_t^+)^2\sigma_t^2$ such that $D_r^W\sigma_s^2$ tends (in some adequate sense) to $D_t^+\sigma_t^2$ and $D_u^W D_r^W \sigma_s^2$ tends to $(D_t^+)^2\sigma_t^2$, as $r,s \to t$. Then, straightforward computations allow us to see that Theorem 8.3.3 reduces to*

$$\lim_{T\to t}\frac{\partial^2 I}{\partial k^2}(t,k_t^*) = \left(\frac{1}{12}-\frac{7}{24}\rho^2\right)\frac{\left(D_t^+\sigma_t^2\right)^2}{\sigma_t^5} + \frac{\rho^2}{6\sigma_t^3}\left(D_t^+\right)^2\sigma_t^2. \quad (8.3.17)$$

Remark 8.3.7 *Notice that, if the volatility is given by a semimartingale process, (8.3.17) agrees with the results in Section 5 of Durrleman (2010).*

8.3.3 The convexity of the short-time implied volatility

We know that, in the uncorrelated case, the implied volatility is a locally convex function around the at-the-money strike (see Renault and Touzi (1996) and Remark 8.2.2). In the uncorrelated case, Theorem 8.3.3 gives us that

$$\begin{aligned}
&\lim_{T\to t}(T-t)^{1-2H}\frac{\partial^2 I_t}{\partial k^2}(k_t^*) \\
&= \lim_{T-t}(T-t)^{1-2H}\frac{\partial^2 I_t^0}{\partial k^2}(k_t^*) \\
&+\rho^2\lim_{T\to t}E_t\left[-\frac{3}{2\sigma_t^5(T-t)^{3+2H}}\left(\int_t^T\left(\int_s^T D_s^W\sigma_\theta^2 d\theta\right)ds\right)^2\right. \\
&\left.+\frac{1}{\sigma_t^4(T-t)^{2+2H}}\int_t^T\left(\int_s^T D_s^W\left(\sigma_r\left(\int_s^T D_s^W\sigma_\theta^2 d\theta\right)\right)dr\right)ds\right].
\end{aligned}$$

Then, if

$$U_t := \lim_{T\to t} E_t \left[-\frac{3}{2\sigma_t^5 (T-t)^{3+2H}} \left(\int_t^T \left(\int_s^T D_s^W \sigma_\theta^2 d\theta \right) ds \right)^2 + \frac{1}{\sigma_t^4 (T-t)^{2+2H}} \int_t^T (\int_s^T D_s^W \left(\sigma_r \left(\int_s^T D_s^W \sigma_\theta^2 d\theta \right) \right) dr) ds \right] \geq 0$$

the limit of $(T-t)^{\frac{1}{2}-H} \frac{\partial^2 I_t}{\partial k^2}(k_t^*)$ as $T \to t$ is positive for all the values of ρ. If U_t is negative, $(T-t)^{\frac{1}{2}-H} \frac{\partial^2 I_t}{\partial k^2}(k_t^*)$ is positive if and only if

$$\rho^2 \leq -\frac{\lim_{T\to t}(T-t)^{1-2H} \frac{\partial^2 I_t^0}{\partial k^2}(k_t^*)}{U_t}.$$

Notice that, under the hypotheses of Remark 8.3.6,

$$\lim_{T\to t}(T-t)^{1-2H} \frac{\partial^2 I_t^0}{\partial k^2}(k_t^*) = \frac{1}{12\sigma_t^5}(D_t^+ \sigma_t^2)^2$$

and

$$U_t = \frac{7}{24\sigma_t^5}(D_t^+ \sigma_t^2)^2 - \frac{1}{6\sigma_t^3}(D_t^+)^2 \sigma_t^2.$$

Then, if $\frac{(D_t^+)^2 \sigma_t^2}{6\sigma_t^3} - \frac{7}{24} \frac{(D_t^+ \sigma_t^2)^2}{\sigma_t^5} \geq 0$ this is satisfied independently of the correlation parameter ρ. If $\frac{(D_t^+)^2 \sigma_t^2}{6\sigma_t^3} - \frac{7}{24} \frac{(D_t^+ \sigma_t^2)^2}{\sigma_t^5} < 0$ this condition holds if

$$\rho^2 \leq \frac{\frac{1}{12\sigma_t^5}(D_t^+ \sigma_t^2)^2}{\frac{7}{24\sigma_t^5}(D_t^+ \sigma_t^2)^2 - \frac{1}{6\sigma_t^3}(D_t^+)^2 \sigma_t^2}.$$

8.4 EXAMPLES

8.4.1 Local volatility models

Assume a local volatility model $\sigma_t := \sigma(t, S_t)$, where $\sigma(t, \cdot) \in \mathcal{C}_b^1$. We have seen in Section 7.6.2 that

$$D_t^+ \sigma_t = \frac{\partial \sigma}{\partial S}(t, S_t) \sigma(t, S_t) S_t,$$

which implies that

$$D_t^+ \sigma_t^2 = 2\sigma_t \frac{\partial \sigma}{\partial S}(t, S_t) \sigma(t, S_t) S_t = 2 \frac{\partial \sigma}{\partial S}(t, S_t) \sigma^2(t, S_t) S_t.$$

In a similar way,

$$
\begin{aligned}
&(D_t^+)^2\sigma_t^2 \\
&= 2\frac{\partial^2\sigma}{\partial S^2}(t,S_t)\sigma^3(t,S_t)S_t^2 + 4\left(\frac{\partial\sigma}{\partial S}(t,S_t)\right)^2\sigma^2(t,S_t)S_t^2 \\
&\quad + 2\frac{\partial\sigma}{\partial S}(t,S_t)\sigma^3(t,S_t)S_t. \qquad (8.4.1)
\end{aligned}
$$

Then,

$$
\begin{aligned}
&\lim_{T\to t}\frac{\partial^2 I}{\partial k^2}(t,k_t^*) \\
&= \left(\frac{1}{12}-\frac{7}{24}\right)\frac{4\left(\frac{\partial\sigma}{\partial S}(t,S_t)S_t\right)^2}{\sigma(t,S_t)} \\
&+ \frac{1}{3}\frac{\partial^2\sigma}{\partial S^2}(t,S_t)S_t^2 + \frac{2}{3\sigma(t,S_t)}\left(\frac{\partial\sigma}{\partial S}(t,S_t)\right)^2 S_t^2 + \frac{1}{3}\frac{\partial\sigma}{\partial S}(t,S_t)S_t \\
&= \frac{1}{3}\frac{\partial^2\sigma}{\partial S^2}(t,S_t)S_t^2 - \frac{1}{6\sigma(t,S_t)}\left(\frac{\partial\sigma}{\partial S}(t,S_t)\right)^2 S_t^2 + \frac{1}{3}\frac{\partial\sigma}{\partial S}(t,S_t)S_t. \quad (8.4.2)
\end{aligned}
$$

Now, define $\hat{\sigma}(t,z)=\sigma(t,e^z)$, and notice that (8.4.2) can be written as

$$
\begin{aligned}
&\lim_{T\to t}\frac{\partial^2 I}{\partial k^2}(t,k_t^*) \\
&= \frac{1}{3}\frac{\partial^2\hat{\sigma}}{\partial z^2}(t,e^{X_t}) - \frac{1}{6\hat{\sigma}(t,e^{X_t})}\left(\frac{\partial\hat{\sigma}}{\partial z}(t,e^{X_t})\right)^2. \qquad (8.4.3)
\end{aligned}
$$

In particular, the implied volatility is (in the short end) a locally convex function around the at-the-money strike if

$$
\frac{\partial^2\hat{\sigma}}{\partial z^2}(t,e^{X_t}) \geq \frac{1}{2\hat{\sigma}(t,e^{X_t})}\left(\frac{\partial\hat{\sigma}}{\partial z}(t,e^{X_t})\right)^2. \qquad (8.4.4)
$$

On the other hand, notice that the one-half slope rule (7.9.1), $\lim_{T\to t}\left(\frac{\partial\hat{\sigma}}{\partial z}(t,e^{X_t})\right)^2 = 4\left(\frac{\partial I}{\partial k}(k_t^*)\right)^2$. Then, as the ATM implied volatility tends to the spot volatility (see Durrleman (2008)) we get

$$
\lim_{T\to t}\left(3\frac{\partial^2 I}{\partial k^2}(t,k_t^*) + \frac{2}{I(k_t^*)}\left(\frac{\partial I}{\partial k}(k_t^*)\right)^2\right) = \frac{\partial^2\hat{\sigma}}{\partial z^2}(t,e^{X_t}). \qquad (8.4.5)
$$

Both (8.4.3) and (8.4.5) are model-free relationships showing that the short-time ATM curvature of the local volatility is higher or equal than three times the curvature of the implied volatility. This fits real market data (see Figure 2.5), where the local volatility smiles are more pronounced than the implied volatility smiles.

When the implied volatility is a perfect smile (that is, $\frac{\partial I}{\partial k}(k_t^*) = 0$), we get

$$3 \lim_{T \to t} \frac{\partial^2 I}{\partial k^2}(t, k_t^*) = \frac{\partial^2 \hat{\sigma}}{\partial z^2}(t, e^{X_t}), \tag{8.4.6}$$

according to the results in Hagan, Kumar, Lesniewski, and Woodward (2002). In the next example, we study the particular case of the CEV model.

Example 8.4.1 *Assume the CEV model defined in Example 2.2.6, where* $\sigma(t, S_t) = \sigma S_t^{\gamma-1}$ *and* $\hat{\sigma}(t, X_t) = \sigma e^{(\gamma-1)X_t}$. *Then,* $\frac{\partial \hat{\sigma}}{\partial z}(t, e^{X_t}) = \sigma(\gamma - 1)e^{(\gamma-1)X_t}$ *and* $\frac{\partial^2 \hat{\sigma}^2}{\partial z}(t, e^{X_t}) = \sigma(\gamma - 1)^2 e^{(\gamma-1)X_t}$. *Then, (8.4.3) reads as*

$$\begin{aligned} \lim_{T \to t} \frac{\partial^2 I}{\partial k^2}(t, k_t^*) &= (\gamma - 1)^2 e^{(\gamma-1)X_t} \left(\frac{\sigma}{3} - \frac{\sigma}{6} \right) \\ &= \frac{\sigma(\gamma - 1)^2}{6} e^{(\gamma-1)X_t}, \end{aligned} \tag{8.4.7}$$

according to Hagan, Kumar, Lesniewski, and Woodward (2014). In particular, in the short end, the CEV implied volatility is locally convex around the ATM strike.

8.4.2 Diffusion volatility models

In this subsection we assume that $\sigma = f(Y)$, for some positive function f, and where Y is the solution of a stochastic differential equation

$$dY_r = a(r, Y_r)\, dr + b(r, Y_r)\, dW_r, \; r \in [0, T], \tag{8.4.8}$$

for some real functions $a, b \in C_b^2$. Then, under some general regularity hypotheses on f, a, and b, classical computations allow us to see that the conditions of Remark 8.3.6 are satisfied with

$$D_t^+ \sigma_t^2 = 2\sigma_t f'(Y_t) b(t, Y_t),$$

and

$$\left(D_t^+\right)^2 \sigma_t^2 := 2\left[\left((f'(Y_t))^2 + \sigma_t f''(Y_t) \right) b^2(t, Y_t) + \sigma_t f'(Y_t) \frac{\partial b}{\partial x}(t, Y_t) b(t, Y_t) \right].$$

Then,

$$
\begin{aligned}
&\lim_{T\to t}\frac{\partial^2 I}{\partial k^2}(t,k_t^*)\\
&=\left(\frac{1}{12}-\frac{7}{24}\rho^2\right)\frac{(D_t^+\sigma_t^2)^2}{\sigma_t^5}+\frac{\rho^2}{6\sigma_t^3}\left(\left(D_t^+\right)^2\sigma_t^2\right)\\
&=\left(\frac{1}{12}-\frac{7}{24}\rho^2\right)\frac{\left(2\sigma_t f'(Y_t)b(t,Y_t)\right)^2}{\sigma_t^5}\\
&+\frac{\rho^2}{3\sigma_t^3}\left(\left((f'(Y_t))^2+\sigma_t f''(Y_t)\right)b^2(t,Y_t)+\sigma_t f'(Y_t)\frac{\partial b}{\partial x}(t,Y_t)b(t,Y_t)\right)\\
&=\frac{1}{3\sigma_t^3}\left(f'(Y_t)b(t,Y_t)\right)^2\\
&+\frac{\rho^2}{3\sigma_t^3}\left(b^2(t,Y_t)\left(-\frac{5}{2}\left(f'(Y_t)\right)^2+\sigma_t f''(Y_t)\right)+\sigma_t f'(Y_t)\frac{\partial b}{\partial x}(t,Y_t)b(t,Y_t)\right).
\end{aligned}
\tag{8.4.9}
$$

Remark 8.4.2 *The limit in Equation (8.4.9) fits the implied volatility expansion of Medvedev and Scaillet (2007).*

Remark 8.4.3 *Notice that the above limit depends on the derivatives of f and the function b, but it does not depend on the function a.*

Example 8.4.4 *Assume the SABR model as in 3.3.1. We have seen that $D_t^+\sigma_t^2 = 2\alpha\sigma_t^2$. In a similar way, we can see that $(D_t^+)^2\sigma_t^2 = 4\alpha^2\sigma_t^2$. Then,*

$$
\begin{aligned}
\lim_{T\to t}\frac{\partial^2 I}{\partial k^2}(t,k_t^*) &= \left(\frac{1}{12}-\frac{7}{24}\rho^2\right)\frac{4\alpha^2}{\sigma_t}+\frac{\rho^2}{6\sigma_t}4\alpha^2\\
&= \left(\frac{1}{3}-\frac{1}{2}\rho^2\right)\frac{\alpha^2}{\sigma_t},
\end{aligned}
$$

according to the results in Section 2.3.2. This expression allows us to deduce that the short-maturity SABR implied volatility is convex in the at-the-money strike if $\rho^2 \le \frac{2}{3}$.

Example 8.4.5 *Assume the Heston model as in Section 3.3.2. We have seen that $D_t^+\sigma_t^2 = \nu\sigma_t$. In a similar way, we can see that $(D_t^+)^2\sigma_t^2 = \frac{\nu^2}{2}$*

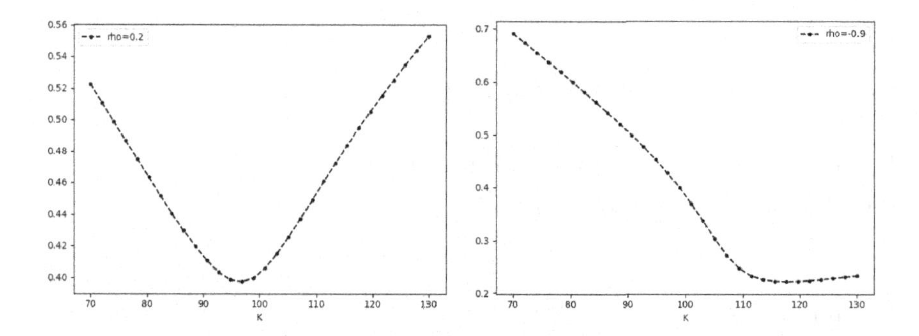

Figure 8.3 Heston simile and skew with parameters $\nu = 2.5, k = 0.3, v_0 = 0.2, \theta = 0.4, T = 0.1$, and $\rho = 0$ (left), and 0.9 (right).

(see Alòs and Ewald (2008)). Then,

$$\begin{aligned}\lim_{T \to t} \frac{\partial^2 I}{\partial k^2}(t, k_t^*) &= \left(\frac{1}{12} - \frac{7}{24}\rho^2\right) \frac{\left(D_t^+ \sigma_t^2\right)^2}{\sigma_t^5} + \frac{\rho^2}{6\sigma_t^3} \left(D_t^+\right)^2 \sigma_t^2 \\ &= \left(\frac{1}{12} - \frac{7}{24}\rho^2\right) \frac{\nu^2 \sigma_t^2}{\sigma_t^5} + \frac{\rho^2}{6\sigma_t^3} \frac{\nu^4}{2} \\ &= \left(1 - \rho^2 \frac{5}{2}\right) \frac{\nu^2}{12\sigma_t^3}.\end{aligned}$$

This implies that the short-time Heston implied volatility is convex in the at-the-money strike if $\rho^2 \leq \frac{2}{5}$. We can observe this behaviour in Figure 8.3, corresponding to the parameters $\nu = 2.5, k = 0.3, v_0 = 0.2, \theta = 0.4, T = 0.1$, and $\rho = 0, 0.9$.

8.4.3 Fractional volatilities

8.4.3.1 Models based on fractional Ornstein-Uhlenbeck processes

Assume as in Section 5.6.1 that the squared volatility σ^2 is of the form $\sigma_r^2 = f(Y_r)$, where $f \in C_b^1(\mathbb{R})$ and Y_r is a *fractional Ornstein-Uhlenbeck* process of the form

$$Y_r = m + (Y_0 - m)\, e^{-\alpha r} + c\sqrt{2\alpha} \int_0^r e^{-\alpha(r-s)} d\hat{W}_s^H. \tag{8.4.10}$$

We proved that, in this case

$$D_r^W \sigma_t^2 = f'(Y_t) h(t, r),$$

where $h(t,r)$ is defined as in Example 8.2.11. In a similar way,

$$D_s^W D_r^W \sigma_t^2 = f''(Y_t)h(t,r)h(t,s).$$

In the case $H > \frac{1}{2}$, the above Malliavin derivatives tend to zero, and then $\frac{\partial^2 I}{\partial k^2}(t,k_t^*) \to 0$. If $H < \frac{1}{2}$ Theorem 8.3.3 implies that

$$\begin{aligned}
&\lim_{T\to t}(T-t)^{1-2H}\frac{\partial^2 I}{\partial k^2}(t,k_t^*) \\
&= \frac{(\nu f'(Y_t))^2}{4\sigma^5}\lim_{T\to t}\frac{\left[E_t\int_t^T\left(\int_r^T h(u,r)du\right)^2 dr\right]}{(T-t)^{2+2H}} \\
&+\rho^2\lim_{T\to t}E_t\left[-\frac{3\,(\nu f'(Y_t))^2}{2\sigma_t^5\,(T-t)^{5-2H}}\left(\int_t^T\left(\int_s^T h(u,s)du\right)ds\right)^2\right. \\
&+\frac{(\nu f'(Y_t))^2}{2\sigma_t^5\,(T-t)^{2+2H}}\int_t^T\int_s^T h(r,s)\left(\int_r^T h(u,r)du\right)drds \\
&\left.+\frac{\nu^2 f''(Y_t)}{\sigma_t^3\,(T-t)^{2+2H}}\int_t^T\int_s^T\left(\int_r^T h(u,r)h(u,s)du\right)drds\right] \\
&= \frac{(\nu f'(Y_t))^2}{4\sigma^5}\frac{1}{(1/2+H)^2(2+2H)} \\
&+\rho^2\left[\frac{(\nu f'(Y_t))^2}{2\sigma_t^5}\left(\frac{-3}{(1/2+H)^2(3/2+H)^2}+\frac{B(3/2+H,3/2+H)}{(1/2+H)^2}\right)\right. \\
&\left.+\frac{\nu^2 f''(Y_t)}{\sigma_t^3}\frac{B(3/2+H,3/2+H)}{(1/2+H)^2}\right].
\end{aligned}$$

That is, the curvature of the at-the-money implied volatility for a fractional volatility model with Hurst parameter $H < \frac{1}{2}$ is $O(T-t)^{2H-1}$.

Example 8.4.6 (The rough Bergomi model) *In the particular case of the rough Bergomi model (see Section 5.6.2),* $D_r^W\sigma_t^2 = \nu\sqrt{2H}\sigma_t^2(r-s)^{H-\frac{1}{2}}$ *and then* $D_s^W D_r^W\sigma_t^2 = \nu^2 2H\sigma_t^2(r-t)^{H-\frac{1}{2}}(s-t)^{H-\frac{1}{2}}$. *This allows us to write*

$$
\begin{aligned}
&\lim_{T\to t}(T-t)^{1-2H}\frac{\partial^2 I}{\partial k^2}(t,k_t^*)\\
&=\frac{\nu^2}{4\sigma_t}\lim_{T\to t}\frac{\left[E_t\int_t^T\left(\int_r^T(u-r)^{H-\frac{1}{2}}du\right)^2dr\right]}{(T-t)^{2+2H}}\\
&+\rho^2\lim_{T\to t}E_t\left[-\frac{3\nu^2}{2\sigma_t(T-t)^{5-2H}}\left(\int_t^T\left(\int_s^T(u-s)^{H-\frac{1}{2}}du\right)ds\right)^2\right.\\
&+\frac{\nu^2}{2\sigma_t^3(T-t)^{2+2H}}\int_t^T\int_s^T(r-s)^{H-\frac{1}{2}}\left(\int_r^T(u-r)^{H-\frac{1}{2}}du\right)drds\\
&\left.+\frac{\nu^2}{\sigma_t(T-t)^{2+2H}}\int_t^T\int_s^T\left(\int_r^T(u-r)^{H-\frac{1}{2}}(u-s)^{H-\frac{1}{2}}du\right)drds\right]\\
&=\frac{\nu^2}{4\sigma_t}\frac{1}{(1/2+H)^2(2+2H)}\\
&+\rho^2\left[\frac{\nu^2}{2\sigma_t}\left(\frac{-3}{(1/2+H)^2(3/2+H)^2}+\frac{B(3/2+H,3/2+H)}{(1/2+H)^2}\right)\right.\\
&\left.+\frac{\nu^2}{\sigma_t}\frac{B(3/2+H,3/2+H)}{(1/2+H)^2}\right]\\
&=\frac{\nu^2}{2\sigma_t(1/2+H)^2}\left[\frac{1}{(1+H)}\right.\\
&\left.+3\rho^2\left(\frac{-1}{(3/2+H)^2}+B(3/2+H,3/2+H)\right)\right].
\end{aligned}
$$

A numerical analysis shows us that $\frac{-1}{(3/2+H)^2}+B(3/2+H,3/2+H)$ *is negative for all* $H\in(0,1)$*. Then, the above quantity is greater than*

$$
\frac{1}{(1+H)}+3\left(\frac{-1}{(3/2+H)^2}+B(3/2+H,3/2+H)\right),
$$

which is always positive. That is, the ATM curvature of the implied volatility of a rBergomi model with $H<\frac{1}{2}$ *is positive and* $O((T-t)^{2H-1})$*. In Figure 8.4 we can see the smiles corresponding to a rBergomi model with* $H=0.3,\nu=0.5,\rho=-0.6$, $\sigma=\sqrt{0.05}$, $t=0$, *and varying values of* T*. Observe that, as the curvature is* $O(T^{2H-1})$ *while the skew is* $O(T^{H-\frac{1}{2}})$*, the smile effect becomes more relevant than the skew effect for short maturities. This fits real market data, as observed in Figure 8.1.*

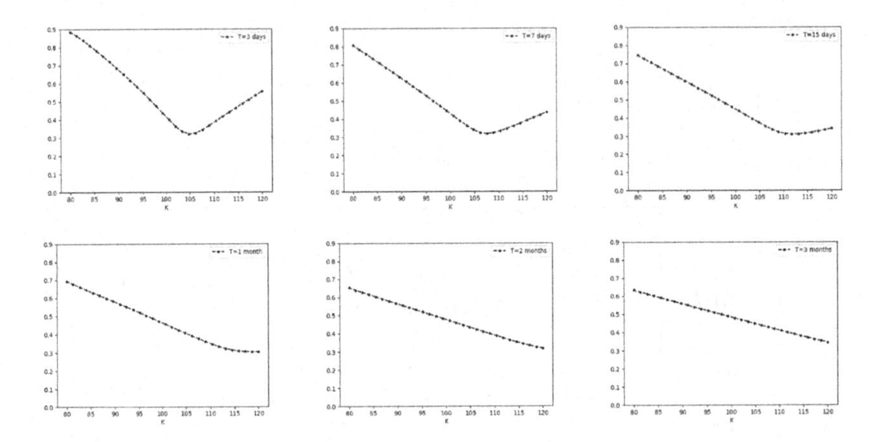

Figure 8.4 rBergomi smiles with $H = 0.3, \nu = 0.5, \rho = -0.6$, and $\sigma = \sqrt{0.05}$.

8.5 CHAPTER'S DIGEST

Malliavin calculus techniques and the decomposition formulas in Chapter 6 give us an analytic expression for the short-time limit of the second derivative of the implied volatility in terms of the Malliavin derivatives of the volatility process. This expression allows us to study in detail the short-end behaviour of the ATM curvature for local, stochastic, and rough volatilities that can be summarised as follows.

Our limit results show that this curvature is finite for classical models where the volatility is a diffusion. Moreover, it allows us to compute explicitly the short-end curvature, recovering previous results for local and stochastic volatilities. In the case of fractional volatility models, the at-the-money curvature is $O(T-t)^{2H-1}$. As the skew slope is $O(T-t)^{H-\frac{1}{2}}$ in the case $H < \frac{1}{2}$ and for short maturities, the curvature is higher than the skew, a behaviour that fits real market data.

On the other hand, we can study when this short-end curvature is positive or negative, which gives us a condition for the ATM local convexity of the implied volatility. In particular, this condition holds for the rBergomi and the CEV models, for the SABR model if the correlation parameter $\rho^2 \leq \frac{2}{3}$, and for the Heston model if $\rho^2 \leq \frac{2}{5}$.

The short-end ATM curvature of the local volatility is greater than three times the implied volatility curvature, being approximately three times the implied volatility curvature in the uncorrelated case $\rho = 0$, according to previous results in the literature.

IV

The implied volatility of non-vanilla options

CHAPTER 9

Options with random strikes and the forward smile

Now, our objective is to study European options with a payoff of the form

$$(S_T - A)_+, \tag{9.0.1}$$

where A is some random variable. As these derivatives can be seen as an extension of classical European calls, where we allow the strike to be a random variable, we refer to them as *random strike options.* This class of derivatives includes, among others:

- **Forward -start options**, where A depends on the price of the underlying asset S at some time before maturity.
- **Spread options**, where A is a random variable that depends on the prices of other assets.
- **Floating strike Asian options**, where A depends on the average asset price before maturity T.

A natural approximation of the corresponding option price is given by the price of a call with strike equal to the expected value of A. In the following sections, following the ideas in Alòs and León (2013) and Section 7.3 in Alòs and León (2016), we see how Malliavin calculus can allow us to decompose prices around this leading term, as well as to use this decomposition to analyse the properties of the corresponding implied volatility.

The first step in this chapter is to present the general approach for random strike options in Section 9.1. In the following sections, we apply these techniques to the study of the ATM short-end level, skew, and curvature of forward-start options.

9.1 A DECOMPOSITION FORMULA FOR RANDOM STRIKE OPTIONS

Consider a stochastic volatility model of the form

$$dX_t = \left(r - \frac{\sigma_t^2}{2}\right) dt + \sigma_t \left(\sum_{i=1}^{d} \rho_i dW_t^i + \sqrt{1 - \sum_{i=1}^{d} \rho_i^2 dB_t} \right), \tag{9.1.1}$$

for some interest rate r, $d \geq 1$, correlation parameters $\rho_i \in [-1, 1]$, $i = 1, ..., d$, and where $W = (W^1, ..., W^d)$ is a d-dimensional Brownian motion and σ is a square integrable process adapted to the filtration generated by W. That is, (9.1.1) is a model similar to (6.1.1), but where we allow the Brownian motion W to be d-dimensional. This extension of (6.1.1) allows us not only to describe the correlation between asset prices and the volatility but also with the random strike A.

Given a random strike option with payoff (9.0.1), we consider the following random variables:

- $M_t := E_t(A)$. M is a martingale adapted to W, it admits a representation of the form

$$M_t = M_0 + \int_0^t m_s dZ_s = \sum_{i=1}^{d} \int_0^t m_s^i dW_s^i,$$

  for some square integrable process m adapted to the filtration generated by W.

- $v_t := \sqrt{\frac{1}{T-t}\int_t^T a_u^2 du}$, where

$$\begin{aligned} a_u^2 du &= d\left(\langle X, X\rangle_u - 2\frac{\langle X, M\rangle_u}{M_u} + \frac{\langle M, M\rangle_u}{M_u^2}\right) \\ &= \left(\sigma_u^2 - \frac{2\sigma_u \sum_{i=1}^d \rho_i m_u^i}{M_u^2} + \frac{\sum_{i=1}^d (m_u^i)^2}{M_u^2}\right) du \\ &= \sigma_u^2\left(1 - \sum_{i=1}^d \rho_i^2\right) + \sum_{i=1}^d \left(\sigma_u \rho_i - \frac{m_u^i}{M_u}\right)^2. \end{aligned} \quad (9.1.2)$$

Now, notice that

$$\begin{aligned} V_t &= e^{-r(T-t)} E_t\left(e^{X_T} - A\right)_+ = e^{-r(T-t)} E_t\left(e^{X_T} - M_T\right)_+ \\ &= e^{-r(T-t)} E_t(BS(T, X_T, M_T, v_T)). \end{aligned}$$

Therefore, the anticipating Itô's formula (4.3.1) gives us that (notice that both X and M are adapted, while v is anticipating)

$$\begin{aligned} &E_t\left(e^{-rT} BS(T, X_T, M_T, v_T)\right) \\ &= E_t\Bigg[e^{-rt} BS(t, X_t, M_t, v_t) - r\int_t^T e^{-ru} BS(u, X_u, M_u, v_u) du \\ &\quad + \int_t^T e^{-ru} \frac{\partial BS}{\partial u}(u, X_u, M_u, v_u) du \\ &\quad + \int_t^T e^{-ru} \frac{\partial BS}{\partial x}(u, X_u, M_u, v_u)\left(r - \frac{\sigma_u^2}{2}\right) du \\ &\quad + \frac{1}{2}\int_t^T e^{-ru} \frac{\partial^2 BS}{\partial x^2}(u, X_u, M_u, v_u)\sigma_u^2 du \\ &\quad + \int_t^T e^{-ru} \frac{\partial^2 BS}{\partial x \partial K}(u, X_u, M_u, v_u) d\langle X, M\rangle_u \\ &\quad + \frac{1}{2}\int_t^T e^{-ru} \frac{\partial^2 BS}{\partial K^2}(u, X_u, M_u, v_u) d\langle M, M\rangle_u \\ &\quad + \frac{1}{2}\int_t^T e^{-ru}\left(\frac{\partial^2}{\partial x^2} - \frac{\partial}{\partial x}\right) BS(u, X_u, M_u, v_u)\left(v_u^2 du - a_u^2 du\right) \\ &\quad + \sum_{i=1}^d \frac{\rho^i}{2}\int_t^T e^{-ru} \frac{\partial}{\partial x}\left(\frac{\partial^2}{\partial x^2} - \frac{\partial}{\partial x}\right) BS(u, X_u, M_u, v_u)\sigma_u\left(\int_u^T D_u^{W^i} a_\theta^2 d\theta\right) du \\ &\quad + \sum_{i=1}^d \frac{1}{2}\int_t^T e^{-ru} \frac{\partial}{\partial K}\left(\frac{\partial^2}{\partial x^2} - \frac{\partial}{\partial x}\right) BS(u, X_u, M_u, v_u) m_u^i\left(\int_u^T D_u^{W^i} a_\theta^2 d\theta\right) du\Bigg]. \end{aligned}$$

Then, grouping terms we get

$$
\begin{aligned}
V_t &= E_t \Big[BS(t, X_t, M_t, v_t) \\
&+ \int_t^T e^{-r(u-t)} \mathcal{L}_{BS}(v_u)\, BS(u, X_u, M_u, v_u) du \\
&+ \frac{1}{2} \int_t^T e^{-r(u-t)} \left(\frac{\partial^2}{\partial x^2} - \frac{\partial}{\partial x} \right) BS(u, X_u, M_u, v_u) \left(\sigma_u^2 du - m_u^2 du \right) du \\
&+ \int_t^T e^{-r(u-t)} \frac{\partial^2 BS}{\partial x \partial K}(u, X_u, M_u, v_u) d \langle X, M \rangle_u \\
&+ \frac{1}{2} \int_t^T e^{-r(u-t)} \frac{\partial^2 BS}{\partial K^2}(u, X_u, M_u, v_u) d \langle M, M \rangle_u \\
&+ \sum_{i=1}^d \frac{\rho^i}{2} \int_t^T e^{-ru} \frac{\partial}{\partial x} \left(\frac{\partial^2}{\partial x^2} - \frac{\partial}{\partial x} \right) BS(u, X_u, M_u, v_u) \sigma_u \left(\int_u^T D_u^{W^i} a_\theta^2 d\theta \right) du \\
&+ \sum_{i=1}^d \frac{1}{2} \int_t^T e^{-ru} \frac{\partial}{\partial K} \left(\frac{\partial^2}{\partial x^2} - \frac{\partial}{\partial x} \right) BS(u, X_u, M_u, v_u) m_u^i \left(\int_u^T D_u^{W^i} a_\theta^2 d\theta \right) du \Big].
\end{aligned}
$$

Now, as

$$
\mathcal{L}_{BS}(v_u)(BS)(u, X_u, M_u, v_u) = 0,
$$

$$
\frac{\partial^2 BS}{\partial x \partial K}(t, x, K, \sigma) = -\frac{1}{K} \left(\frac{\partial^2 BS}{\partial x^2} - \frac{\partial BS}{\partial x} \right)(t, x, K, \sigma),
$$

and

$$
\frac{\partial^2 BS}{\partial K^2}(t, x, K, \sigma) = \frac{1}{K^2} \left(\frac{\partial^2 BS}{\partial x^2} - \frac{\partial BS}{\partial x} \right)(t, x, K, \sigma),
$$

the above equality reduces to

$$
\begin{aligned}
V_t &= E_t \Bigg[BS(t, X_t, M_t, v_t) \\
&+ \frac{1}{2} \int_t^T e^{-r(u-t)} \left(\frac{\partial^2}{\partial x^2} - \frac{\partial}{\partial x} \right) BS(u, X_u, M_u, v_u) \\
&\times \left(\sigma_u^2 du - a_u du - 2 \frac{d \langle X, M \rangle_t}{M_t} + \frac{d \langle M, M \rangle_t}{M_t^2} \right) du \\
&+ \frac{\rho}{2} \int_t^T e^{-r(u-t)} \frac{\partial}{\partial x} \left(\frac{\partial^2}{\partial x^2} - \frac{\partial}{\partial x} \right) BS(u, X_u, M_u, v_u) \sigma_u \rho \left(\int_u^T D_u^W a_\theta^2 d\theta \right) du \\
&+ \frac{\rho_{Z,W}}{2} \int_t^s e^{-r(u-t)} \frac{\partial}{\partial K} \left(\frac{\partial^2}{\partial x^2} - \frac{\partial}{\partial x} \right) BS(u, X_u, M_u, v_u) m_u \left(\int_u^T D_u^W a_\theta^2 d\theta \right) \Bigg].
\end{aligned}
$$

Then, as $a_u^2 = \sigma_u^2 - 2\frac{d\langle X,M\rangle_t}{M_t} + \frac{d\langle M,M\rangle_t}{M_t^2}$, we get

$$
\begin{aligned}
V_t = E_t \Bigg[& BS(t, X_t, M_t, v_t) \\
& + \sum_{i=1}^{d} \frac{\rho^i}{2} \int_t^T e^{-ru} \frac{\partial}{\partial x} \left(\frac{\partial^2}{\partial x^2} - \frac{\partial}{\partial x} \right) BS(u, X_u, M_u, v_u) \sigma_u \left(\int_u^T D_u^{W^i} a_\theta^2 d\theta \right) du \\
& + \sum_{i=1}^{d} \frac{1}{2} \int_t^T e^{-ru} \frac{\partial}{\partial K} \left(\frac{\partial^2}{\partial x^2} - \frac{\partial}{\partial x} \right) BS(u, X_u, M_u, v_u) m_u^i \left(\int_u^T D_u^{W^i} a_\theta^2 d\theta \right) du \Bigg].
\end{aligned}
\tag{9.1.3}
$$

The interpretation of (9.1.3) is simple. From one side, $a_u du = d\langle X - \ln M \rangle_u$ and then a can be seen as the volatility of $S - M$. Then, the first term on the right-hand side in the above equality is the Hull-White term and coincides with the price of the option if this process a is independent of X and M. The second term is given by the covariance between X and a, while the last one comes from the correlation between the random strike and a. When the strike is constant this term cancels, and we recover the decomposition formula for vanilla prices in Theorem (6.3.2). In the next chapter, we see how to adapt this procedure to the study of forward-starting options.

9.2 FORWARD-START OPTIONS AS RANDOM STRIKE OPTIONS

Forward-start options are defined as options whose strike depends on the price of the underlying at some time $s \in (0, T)$. That is, the corresponding payoff is given by

$$\left(e^{X_T} - e^{\alpha} e^{X_s} \right)_+, \tag{9.2.1}$$

where $s \in [0, T]$ is the forward-start date and $\alpha \in \mathbb{R}$ is the log-forward moneyness. Another version of forward-start options has a payoff of the form

$$\left(e^{X_T - X_s} - e^{\alpha} \right)_+$$

(see for example Lucic (2003)). The non-arbitrage price at inception $t \leq s$ of an option with payoff (9.2.1) is given as

$$V_t = e^{-r(T-t)} E_t \left(e^{X_T} - e^{\alpha} e^{X_s} \right)_+. \tag{9.2.2}$$

If $t \geq s$, this is simply a standard European call option evaluated during the life of the contract.

Under the Black-Scholes formula with volatility σ, it is easy to see, taking conditional expectations at time s, that the price of a forward-start option is determined by the price of a vanilla call with time to maturity $T - s$. That is,

$$\begin{aligned} V_t &= e^{-r(T-t)} E_t[E_s \left(e^{X_T} - e^{\alpha} e^{X_s}\right)_+] \\ &= e^{-r(s-t)} E_t(BS(s, X_s, \alpha + X_s, \sigma)) \\ &= e^{-r(s-t)} E_t(e^{X_s} BS(s, 0, \alpha), \sigma)) \\ &= e^{X_t} BS(s, 0, \alpha, \sigma). \end{aligned} \tag{9.2.3}$$

In the stochastic volatility case, a classical pricing approach is to apply a change-of-measure, which links the price of the forward option with the price of a plain vanilla (see for example Lucic (2003) and Musiela and Rutkowski (2005)). Some other works are based on knowledge of the characteristic function, as in Kruse and Nögel (2005) and Guo and Hung (2008).

For $t \in [0, s]$, we can define the **forward implied volatility** $I(t, s; \alpha)$ as the $\mathcal{F}_t$-adapted process satisfying

$$V_t = \exp(X_t) BS\left(s, 0, e^{\alpha}, I(t, s; \alpha)\right). \tag{9.2.4}$$

Notice that, in the constant volatility case, $\sigma_u = \sigma$. When the volatility is random, the implied volatility surface for forward-starting options exhibits substantial differences to the classical vanilla case, as studied in Jacquier and Roome (2013, 2015, 2016). For example, in the Heston model, the short-time Heston forward smile blows up to infinity (except ATM) and it is symmetric, reproducing a small-maturity 'U-shaped' effect (see Jaquier and Roome (2015) and Bühler (2002)). Approximation formulas for the Heston forward implied volatility can be found in the quoted papers by Jacquier and Roome (2013, 2015, 2916). The out-of-the-money forward smile has been studied in the case of stochastic local volatility models in Pascucci and Mazzon (2017).

Our objective is to study the ATM short-time level, skew, and curvature of forward-starting options. Following Alòs, Jacquier, and León (2019), our analysis is based on their representation as random strike options.

Let us adapt our problem to the scheme presented in Section 9.1. From (9.2.1) we have that the random strike A is given by $e^{\alpha}e^{X_s}$, for some $s \in (0,T)$. Notice that this random strike is driven by the same Brownian motion as asset prices (it does not have a random component on its own). Then, we only need a second Brownian motion to model the volatility process.That is, we can consider model 9.1.1 with $d=1$, which coincides with the model 6.1.1 considered in Chapters 6, 7, and 8. Now,

$$M_t := E_t\left(e^{\alpha}e^{X_s}\right).$$

Notice that, if $t<s$, $M_t = e^{\alpha}e^{r(s-t)}e^{X_t}$, while if $t>s$, $M_t = e^{\alpha}e^{X_s}$. Then, straightforward computations allow us to write

$$M_t = e^{\alpha}e^{rs}e^{X_0} + \int_0^t \sigma_u 1_{[0,s]}(u)e^{\alpha}e^{r(s-u)}e^{X_u}\left(\rho dW_u + \sqrt{1-\rho^2}dB_u\right).$$

That is, $M_0 = e^{\alpha}e^{rs}e^{X_0}$,

$$m_u = \rho\sigma_u 1_{[0,s]}(u)e^{\alpha}e^{r(s-u)}e^{X_u},$$

and

$$m^B = \sqrt{1-\rho^2}\sigma_u 1_{[0,s]}(u)e^{\alpha}e^{r(s-u)}e^{X_u}.$$

On the other hand,

$$d\langle M,X\rangle_u = \sigma_u^2 e^{\alpha}e^{r(s-u)}e^{X_u}1_{[0,s]}(u)du,$$

$$d\langle M,M\rangle_u = \sigma_u^2 e^{2\alpha}e^{2r(s-u)}e^{2X_u}1_{[0,s]}(u)du,$$

and then $v_t := \left(\frac{Y_t}{T-t}\right)^{\frac{1}{2}}$, with

$$Y_t := \int_t^T \sigma_u^2 1_{[s,T]}(u)du = \int_{t\vee s}^T \sigma_u^2 du$$

Notice that, if $t<s$, $v_t\sqrt{T-t} = v_s\sqrt{T-s}$.

Now, a direct application of (9.1.3) allows us to prove a decomposition result for forward-start option prices, as we see in the following section.

9.3 FORWARD-START OPTIONS AND THE DECOMPOSITION FORMULA

The first step in our analysis is to write the decomposition 9.1.3 in the specific case of forward start options. Towards this end, we consider the following hypotheses:

(H1) There exist two constants $0 < a < b$ such that $a < \sigma_t < b$, for all $t \in [0, T]$, with probability one.

(H2) $\sigma, \sigma e^X \in \mathbb{L}^{2,2}$.

We remark that these hypotheses have been chosen for the sake of simplicity and that they can be substituted by other adequate integrability conditions.

Now, we are in a position to prove the main result of this section.

Theorem 9.3.1 *Consider the model (6.1.1) and assume that hypotheses (H1) and (H2) hold. Then, for all* $0 \le t \le s \le T$,

$$\begin{aligned} V_t &= E_t \Big[\exp(X_t) BS\left(s, 0, e^{\alpha}, v_s\right) \\ &\quad + \frac{\rho}{2} \int_s^T e^{-r(u-t)} H(u, X_u, M_u, v_u) \sigma_u \left(\int_u^T D_u^W \sigma_\theta^2 d\theta \right) du \\ &\quad + \frac{\rho}{2} G(s, 0, e^{\alpha}, v_s) \int_t^s e^{-r(u-t)} e^{X_u} \sigma_u \left(\int_s^T D_u^W \sigma_\theta^2 d\theta \right) du \Big]. \end{aligned}$$

Proof. A direct application of (9.1.3) allows us to write

$$\begin{aligned} V_t = E_t \Big[& BS(t, X_t, M_t, v_t) \\ & + \frac{1}{2} \int_t^T e^{-r(u-t)} \frac{\partial}{\partial x} \left(\frac{\partial^2}{\partial x^2} - \frac{\partial}{\partial x} \right) BS(u, X_u, M_u, v_u) \sigma_u \rho \\ & \left(\int_{u \vee s}^T D_u^W \sigma_\theta^2 d\theta \right) du \\ & + \frac{1}{2} \int_t^s e^{-r(u-t)} \frac{\partial}{\partial K} \left(\frac{\partial^2}{\partial x^2} - \frac{\partial}{\partial x} \right) BS(u, X_u, M_u, v_u) \\ & \times \sigma_u e^{\alpha} e^{r(s-u)} e^{X_u} \rho \left(\int_{u \vee s}^T D_u^W \sigma_\theta^2 d\theta \right) du \Big] \end{aligned}$$

$$
\begin{aligned}
= \; & E_t \Bigg[BS(t, X_t, M_t, v_t) \\
& + \frac{1}{2} \int_s^T e^{-r(u-t)} \frac{\partial}{\partial x} \left(\frac{\partial^2}{\partial x^2} - \frac{\partial}{\partial x} \right) BS(u, X_u, M_u, v_u) \sigma_u \rho \\
& \left(\int_u^T D_u^W \sigma_\theta^2 d\theta \right) du \\
& + \frac{1}{2} \int_t^s e^{-r(u-t)} \frac{\partial}{\partial x} \left(\frac{\partial^2}{\partial x^2} - \frac{\partial}{\partial x} \right) BS(u, X_u, M_u, v_u) \sigma_u \rho \\
& \left(\int_s^T D_u^W \sigma_\theta^2 d\theta \right) du \\
& + \frac{1}{2} \int_t^s e^{-r(u-t)} \frac{\partial}{\partial K} \left(\frac{\partial^2}{\partial x^2} - \frac{\partial}{\partial x} \right) BS(u, X_u, M_u, v_u) \sigma_u M_u \rho \\
& \left(\int_s^T D_u^W \sigma_\theta^2 d\theta \right) du \Bigg].
\end{aligned}
$$

Now, using the relationships

$$
\frac{\partial}{\partial x} \left(\frac{\partial^2 BS}{\partial x^2} - \frac{\partial BS}{\partial x} \right) (t, x, k, \sigma) = \frac{e^x N'(d_+)}{\sigma \sqrt{T-t}} \left(1 - \frac{d_+}{\sigma \sqrt{T-t}} \right)
$$

and

$$
\frac{\partial}{\partial K} \left(\frac{\partial^2 BS}{\partial x^2} - \frac{\partial BS}{\partial x} \right) (t, x, k, \sigma) = \frac{e^x N'(d_+)}{K \sigma \sqrt{T-t}} \left(\frac{d_+}{\sigma \sqrt{T-t}} \right),
$$

we can write

$$
\begin{aligned}
V_t = \; & E_t \Bigg[BS(t, X_t, M_t, v_t) \\
& + \frac{\rho}{2} \int_s^T e^{-r(u-t)} H(u, X_u, M_u, v_u) \sigma_u \left(\int_u^T D_u^W \sigma_\theta^2 d\theta \right) du \\
& + \frac{\rho}{2} \int_t^s e^{-r(u-t)} G(u, X_u, M_u, v_u) \sigma_u \left(\int_s^T D_u^W \sigma_\theta^2 d\theta \right) du \Bigg].
\end{aligned}
$$

Finally, notice that

$$\begin{aligned}
&BS(t, X_t, M_t, v_t) \\
&= \exp(X_t) N\left(\frac{-\alpha + r(T-s)}{v_s\sqrt{T-s}} + \frac{v_s\sqrt{T-s}}{2}\right) \\
&- e^{\alpha}\exp(X_t)e^{-r(T-s)} N\left(\frac{-\alpha + r(T-s)}{v_s\sqrt{T-s}} - \frac{v_s\sqrt{T-s}}{2}\right) \\
&= \exp(X_t) BS\left(s, 0, e^{\alpha}, v_s\right)
\end{aligned}$$

and, for all $u < s$,

$$\begin{aligned}
G(u, X_u, M_u, v_u) &= \frac{e^{X_u} N'\left(\frac{-\alpha+r(T-s)}{v_s\sqrt{T-s}} + \frac{v_s\sqrt{T-s}}{2}\right)}{v_s\sqrt{T-s}} \\
&= e^{X_u} G(s, 0, e^{\alpha}, v_s).
\end{aligned}$$

This allows us to complete the proof. ■

Remark 9.3.2 *If the volatility process is a constant σ, then $v_s = \sigma$ and $D_u^W \sigma_\theta^2 = 0$ for all $u < \theta$, and then we recover the option pricing formula for forward-start options under the Black-Scholes model (9.2.3).*

Now, our first objective is to compute the short-end limit of the ATM implied forward volatility.

9.4 THE ATM SHORT-TIME LIMIT OF THE IMPLIED VOLATILITY

This section is devoted to the study of the limit

$$\lim_{T \to s} I(t, s; \alpha^*),$$

where the implied volatility $I(t, s; \alpha)$ is defined as in 9.2.4 and $\alpha^* := r(T-s)$ denotes the ATM log-forward moneyness. In the classical vanilla case (see Chapter 6), this limit is equal to the spot volatility σ_t. In the forward case, we will see that this limit also depends on the correlation parameter.

For the sake of simplicity, we will consider first the uncorrelated case $\rho = 0$. In this case, the short-time limit is simply the expectation of the spot volatility, as we see in the following result.

Lemma 9.4.1 *Consider the model (6.1.1) with $\rho = 0$ and assume that hypotheses in Theorem 9.3.1 hold. Then, for all $t < s$,*

$$\lim_{T \to s} I(t, s; \alpha^*) = E_t(\sigma_s).$$

Proof. Here we follow the same ideas as in the proof of Proposition 6.5.1. We know that

$$\begin{aligned} &I(t, s; \alpha^*) \\ &= BS^{-1}(E_t(BS(s, 0, \exp(\alpha^*), v_s))) \\ &= E_t\Big[BS^{-1}(BS(s, 0, \exp(\alpha^*), v_s)) \\ &\quad + BS^{-1}(E_t(BS(s, 0, \exp(\alpha^*), v_s))) - BS^{-1}(BS(s, 0, \exp(\alpha^*), v_s))\Big] \\ &= E_t\Big[v_s + BS^{-1}(E_t(BS(s, 0, \exp(\alpha^*), v_s))) \\ &\quad - BS^{-1}(BS(s, 0, \exp(\alpha^*), v_s))\Big]. \end{aligned}$$

Notice that a direct application of Clark-Ocone-Haussman formula gives us that

$$BS(s, 0, \exp(\alpha^*), v_s) = E_t(BS(s, 0, \exp(\alpha^*), v_s)) + \int_t^T U_r dW_r,$$

with

$$U_u = E_u\left(\left(\frac{\partial BS}{\partial \sigma}(s, 0, \exp(\alpha^*), v_s)\right) \frac{D_u^W \int_s^T \sigma_r^2 dr}{2(T-s)v_s}\right). \tag{9.4.1}$$

Then, applying Itô's formula to the process $BS^{-1}(A_u)$, where

$$A_u := E_t(BS(s, 0, \exp(\alpha^*), v_s)) + \int_t^u U_r dW_r$$

and taking expectations, we get

$$\begin{aligned} &E_t\Big[BS^{-1}(E_t(BS(s, 0, \exp(\alpha^*), v_s))) - BS^{-1}(BS(s, 0, \exp(\alpha^*), v_s))\Big] \\ &= -\frac{1}{2}E_t \int_t^T (BS^{-1})''(E_r(BS(s, 0, \exp(\alpha^*), v_s))U_r^2 dr. \end{aligned}$$

Hence

$$\begin{aligned} &\lim_{T \to s} I(t, s, \alpha^*) \\ &= E_t(\sigma_s) \\ &\quad - \lim_{T \to s} \frac{1}{2} E_t \int_t^T (BS^{-1})''(E_r(BS(s, 0, \exp(\alpha^*), v_s))U_r^2 dr. \end{aligned}$$

Now, considering that

$$
\begin{aligned}
&(BS^{-1})''(E_r(BS\,(s,0,\exp(\alpha^*),v_s)) \\
&= \frac{1}{\left(N'(d_1)\sqrt{T-s}\right)^2}\left(\frac{BS^{-1}(E_r(BS\,(s,0,\exp(\alpha^*),v_s))(T-s)}{4}\right),
\end{aligned}
$$

where $d_1 := \frac{BS^{-1}(E_r(BS(s,0,\exp(\alpha^*),v_s))\sqrt{T-s}}{2}$, we get that

$$
\lim_{T\to s}\frac{1}{2}E_t\int_t^T (BS^{-1})''(E_r(BS\,(s,0,\exp(\alpha^*),v_s))U_r^2 dr = 0.
$$

Therefore, the proof is complete. ■

As a consequence of the previous result, we have the following limit in the correlated case.

Theorem 9.4.2 *Consider the model (6.1.1) and assume that hypotheses in Theorem 9.3.1 hold. Then, for all $t < s$,*

$$
\lim_{T\to s} I(t,s;\alpha^*) = E_t\left(\sigma_s\right) + \rho e^{-X_t} E_t\left(\int_t^s e^{-r(u-t)} e^{X_u}\sigma_u D_u^W \sigma_s du\right).
$$

Proof. We know, by Theorem 9.3.1, that

$$
\begin{aligned}
&I(t,s;\alpha^*) \\
&= BS^{-1}\left[E_t(BS\,(s,0,\exp(\alpha^*),v_s))\right. \\
&\quad + \frac{\rho}{2}\exp(-X_t)E_t\int_s^T e^{-r(u-t)}H(u,X_u,M_u,v_u)\sigma_u\left(\int_u^T D_u^W\sigma_\theta^2 d\theta\right)du \\
&\quad + \frac{\rho}{2}\exp(-X_t)E_t \\
&\left.\left(G(s,0,\exp(\alpha^*),v_s)\int_t^s e^{-r(u-t)}e^{X_u}\sigma_u\left(\int_s^T D_u^W\sigma_\theta^2 d\theta\right)du\right)\right].
\end{aligned}
$$

By the mean value theorem, and taking into account the classical expression for the Back-Scholes vega, we can find θ between

$E_t\left(BS(s,0,\exp(\alpha^*),v_s)\right)$ such that

$$\begin{aligned}
&I(t,s;\alpha^*) - BS^{-1}\left(E_t\left(BS(s,0,\exp(\alpha^*),v_s)\right)\right)\\
&= \frac{\sqrt{2\pi}\exp(\frac{BS^{-1}(\theta)}{8}(T-s))}{\sqrt{T-s}}\left[\frac{\rho}{2}\exp(-X_t)\right.\\
&\qquad \times E_t\left(\int_s^T e^{-r(u-t)}H(u,X_u,M_u,v_u)\sigma_u\left(\int_u^T D_u^W\sigma_\theta^2 d\theta\right)du\right)\\
&\qquad \left. + \frac{\rho}{2}E_t\left(G(s,0,\exp(\alpha^*),v_s)\int_t^s e^{-r(u-t)}e^{X_u}\sigma_u\left(\int_s^T D_u^W\sigma_\theta^2 d\theta\right)du\right)\right]\\
&= I_1 + I_2. \qquad (9.4.2)
\end{aligned}$$

From Lemma 6.3.1, we have

$$\begin{aligned}
\lim_{T\to s}|I_1| &\le C\lim_{T\to s}\frac{\exp(-X_t)}{\sqrt{T-s}}\int_s^T E\left(e^{X_s}|\mathcal{G}_t\right)(T-s)^{-1}\left|\left(\int_u^T D_u^W\sigma_\theta^2 d\theta\right)\right|du\\
&\le C\exp(-X_t)\lim_{T\to s}(T-s)^H\left(E\left(e^{2X_s}|\mathcal{G}_t\right)\right)^{1/2} = 0,
\end{aligned}$$

where $\mathcal{G}_t := \mathcal{F}_t \vee \mathcal{F}_T^W$. Finally, (9.4.2) and Lemma 9.4.1 imply

$$\begin{aligned}
&\lim_{T\to s} I(t,s;\alpha^*)\\
&= E_t(\sigma_s)\\
&+ \frac{\rho\exp(X_t)}{2}\lim_{T\to s}E_t\left(\frac{\exp(\frac{v_s^2(T-s)}{8})}{v_s(T-s)}\int_t^s e^{-r(u-t)}e^{X_u}\sigma_u\left(\int_s^T D_u^W\sigma_\theta^2 d\theta\right)du\right)\\
&= E_t(\sigma_s) + \frac{\rho\exp(X_t)}{2}E_t\left(\frac{1}{\sigma_s}\int_t^s e^{-r(u-t)}e^{X_u}\sigma_u D_u^W\sigma_s^2 du\right).
\end{aligned}$$

Now, as $D_u^W\sigma_s^2 = 2\sigma_s D_u^W\sigma_s$, the proof is complete. ■

Remark 9.4.3 *Notice that, in the case $s = t$, we recover the results by Durrleman (2008) for classical vanilla options. Also, contrary to the plain vanilla case, the ATM short-time limit depends on the correlation parameter.*

Example 9.4.4 *Consider a SABR model as in Section 2.3.2 where $\beta = 1$ and $d\sigma_t = \alpha\sigma_t dW_t$ for some $\alpha > 0$. We know from Example*

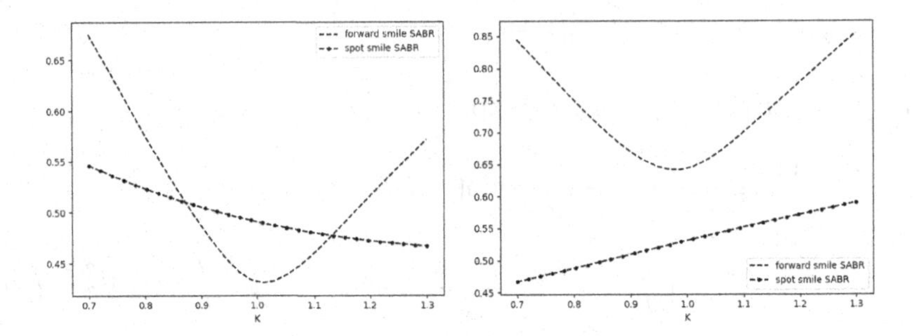

Figure 9.1 Spot and forward implied volatilities corresponding to a SABR model with $\alpha = 0.7$, $\sigma_0 = 0.5$, $T = 1$, $s = 0.5$, and $\rho = -0.4$ (left) and $\rho = 0.6$ (right).

3.3.1 that $D_u^W \sigma_s = \alpha \sigma_s$, *for all* $u < s$. *Then,*

$$\begin{aligned}
&\lim_{T \to s} I(t, s; \alpha^*) \\
&= E_t(\sigma_s) + \rho e^{-X_t} E_t \left(\int_t^s e^{-r(u-t)} e^{X_u} \sigma_u D_u^W \sigma_s du \right) \\
&= E_t(\sigma_s) + \rho e^{-X_t} E_t \left(\int_t^s e^{-r(u-t)} e^{X_u} \sigma_u E_u(D_u^W \sigma_s) du \right) \\
&= E_t(\sigma_s) + \alpha \rho e^{-X_t} E_t \left(\int_t^s e^{-r(u-t)} e^{X_u} \sigma_u E_u(\sigma_s) du \right) \\
&= \sigma_t + \alpha \rho e^{-X_t} E_t \left(\int_t^s e^{-r(u-t)} e^{X_u} \sigma_u^2 du \right).
\end{aligned}$$

Then, if $\rho > 0$, *this limit is higher than the spot volatility* σ_t, *while if* $\rho < 0$, *this limit is less, as we can see in Figure 9.1, corresponding to a SABR model with* $\alpha = 0.7$, $\sigma_0 = 0.5$, $T = 1$, $s = 0.9$, *and* $\rho = -0.4, 0.6$.

Example 9.4.5 *Consider a rough Bergomi model as in Section 2.5.1. That is,*

$$\sigma_t^2 = \sigma_0^2 \exp\left(\nu \sqrt{2H} Z_t - \frac{1}{2} \nu^2 t^{2H} \right), r \in [0, T], \tag{9.4.3}$$

for some positive ν *and* $H < \frac{1}{2}$, *and where*

$$Z_t := \int_0^t (t-u)^{H-\frac{1}{2}} dW_u.$$

We proved in Section 5.6.2 that

$$D_u^W \sigma_s^2 = \nu\sqrt{2H}\sigma_s^2(s-u)^{H-\frac{1}{2}}. \tag{9.4.4}$$

Then, Theorem 10.2.6 leads to

$$\lim_{T\to s} I(t,s;\alpha^*)$$
$$= E_t(\sigma_s) + \nu\sqrt{2H}\rho e^{-X_t} E_t\left(\int_t^s e^{-r(u-t)} e^{X_u}\sigma_u\sigma_s(s-u)^{H-\frac{1}{2}}du\right). \tag{9.4.5}$$

Again, this limit is less in the case of negative correlation. Moreover,

$$\lim_{T\to s} I(t,s;\alpha^*) = E_t(\sigma_s) + O(s-t)^{H+\frac{1}{2}}. \tag{9.4.6}$$

That is, for delays $s-t<1$, the difference between the short-end limit of the ATM forward implied volatility $I(t,s;\alpha^*)$ and the expected volatility $E_t(\sigma_s)$ depends on the Hurst parameter H.

Example 9.4.6 *Consider a CEV model where $\sigma_t = \sigma S_t^{\gamma-1}$, for some positive constants σ,γ. The results in Example 3.2.4 give us that*

$$D_u^W\sigma_t = \sigma(\gamma-1)S_t^{\gamma-2}D_u^W S_t$$
$$= (\gamma-1)\sigma_t^2 \mathbf{1}_{[0,t]}(u)\exp\left((1-\gamma)\int_u^t\left(r+\frac{\gamma}{2}\sigma_t^2\right)d\theta\right). \tag{9.4.7}$$

Then, Theorem 10.2.6 reads as

$$\lim_{T\to s} I(t,s;\alpha^*) = E_t(\sigma_s)$$
$$+(\gamma-1)e^{-X_t}E_t\left(\sigma_s^2\int_t^s e^{-r(u-t)}\sigma_u\exp\left((1-\gamma)\int_u^t\left(r+\frac{\gamma}{2}\sigma_\theta^2\right)d\theta\right)du\right).$$

Notice that in the case $\gamma=1$ (the Black-Scholes case) the above limit reduces to σ.

We will need the next result in Section 9.5.

Lemma 9.4.7 *Consider the model (6.1.1) and assume that $E\left(\exp\left(\frac{\int_0^T\sigma_u^2du}{2}\right)\right)<\infty$. Then,*

$$I(t,s;\alpha^*)\sqrt{T-s}\to 0$$

as $T\to s$.

Proof. We know that, in this case,

$$V_t = e^{-r(T-t)} E_t \left[\left(e^{X_T} - e^{r(T-s)} e^{X_s} \right)_+ \right],$$

which converges to zero as $T \to s$. Indeed, we have

$$\left(e^{X_T} - e^{r(T-s)} e^{X_s} \right)_+ \to 0,$$

as $T \to s$ with probability 1. Moreover, by Novikov theorem (see for example Karatzas and Shreve (1991)), we have

$$\left(e^{X_T} - e^{r(T-s)} e^{X_s} \right)_+ \leq e^{X_T} + e^{r(T-s)} e^{X_s}$$

and

$$E^* \left(e^{X_T} + e^{r(T-s)} e^{X_s} \right) = 2e^{rT} \to 2e^{rs}$$

as $T \to s$. Thus, the dominated convergence theorem implies that $V_t \to 0$ as $T \to s$, with probability 1. On the other hand,

$$\begin{aligned} V_t &= \exp(X_t) \left(N \left(\frac{I(t,s;\alpha^*)\sqrt{T-s}}{2} \right) - N \left(-\frac{I(t,s;\alpha^*)\sqrt{T-s}}{2} \right) \right) \\ &= \exp(X_t) \left(1 - 2N \left(-\frac{I(t,s;\alpha^*)\sqrt{T-s}}{2} \right) \right), \end{aligned}$$

which allows us to complete the proof. ■

9.5 AT-THE-MONEY SKEW

For plain vanilla options, we know that the ATM skew depends on the correlation parameter and that it is equal to zero in the uncorrelated case. The following result shows that this also holds for forward implied volatilities. Towards this end, we consider the following additional hypothesis regarding the regularity of the volatility process, similar to Hypothesis (H3) in Chapter 7.

> **(H3)** There exist $C > 0$, $H \in (0,1)$, such that, for all $t \leq \theta < u < r$,
>
> $$E_t \left[\left(D_u^W \sigma_r \right)^2 \right] \leq C(r-u)^{2H-1}$$
>
> $$\text{and} \quad E_t \left[\left(D_\theta^W D_u^W \sigma_r \right)^2 \right] \leq C[(r-u)(r-\theta)]^{2H-1}.$$

The small-maturity at-the-money forward skew in the uncorrelated case $\rho = 0$ is provided in the following proposition.

Proposition 9.5.1 *Consider the model (6.1.1) with $\rho = 0$. Then, under* ***(H1)–(H3)***,

$$\frac{\partial I}{\partial \alpha}(t, s; \alpha^*) = 0, \tag{9.5.1}$$

for all $s > t$.

Proof. Taking implicit derivatives in the definition of the forward implied volatility, we get

$$\begin{aligned}\frac{\partial V_t}{\partial \alpha} &= \exp(X_t)\frac{\partial BS}{\partial k}\left(s, 0, e^{\alpha}, I(t, s; \alpha)\right) \\ &\quad + \exp(X_t)\frac{\partial BS}{\partial \sigma}\left(s, 0, e^{\alpha}, I(t, s; \alpha)\right)\frac{\partial I}{\partial \alpha}(t, s; \alpha),\end{aligned}$$

where $\frac{\partial BS}{\partial k} := \frac{\partial BS}{\partial (\ln K)}$. Then, from Theorem 10.2.6, we are able to write

$$\begin{aligned}&\frac{\partial I}{\partial \alpha}(t, s; \alpha) \\ &= \frac{\frac{\partial V_t}{\partial \alpha} - \exp(X_t)\frac{\partial BS}{\partial k}\left(s, 0, e^{\alpha}, I(t, s; \alpha)\right)}{\exp(X_t)\frac{\partial BS}{\partial \sigma}\left(s, 0, e^{\alpha}, I(t, s; \alpha)\right)} \\ &= \frac{E_t\left[\frac{\partial BS}{\partial k}\left(s, 0, e^{\alpha}, v_s\right)\right] - \frac{\partial BS}{\partial k}\left(s, 0, e^{\alpha}, I(t, s; \alpha)\right)}{\frac{\partial BS}{\partial \sigma}\left(s, 0, e^{\alpha}, I(t, s; \alpha)\right)} \\ &\quad + \frac{\rho}{2}\frac{E_t\left[\int_s^T e^{-r(u-t)}\frac{\partial H}{\partial k}(u, X_u, e^{\alpha}e^{X_s}, v_u)\sigma_u\left(\int_u^T D_u^W\sigma_\theta^2 d\theta\right)du\right]}{\exp(X_t)\frac{\partial BS}{\partial \sigma}\left(s, 0, e^{\alpha}, I(t, s; \alpha)\right)} \\ &\quad + \frac{\rho}{2}\frac{E_t\left[\frac{\partial G}{\partial k}(s, 0, e^{\alpha}, v_s)\int_t^s e^{X_u}e^{-r(u-t)}\sigma_u\left(\int_u^T D_u^W\sigma_\theta^2 d\theta\right)du\right]}{\exp(X_t)\frac{\partial BS}{\partial \sigma}\left(s, 0, e^{\alpha}, I(t, s; \alpha)\right)}.\end{aligned} \tag{9.5.2}$$

Now, if $\rho = 0$, equality (9.5.2) implies

$$\frac{\partial I}{\partial \alpha}(t, s; \alpha) = \frac{E_t\left[\frac{\partial BS}{\partial k}\left(s, 0, e^{\alpha}, v_s\right)\right] - \frac{\partial BS}{\partial k}\left(s, 0, e^{\alpha}, I(t, s; \alpha)\right)}{\frac{\partial BS}{\partial \sigma}\left(s, 0, e^{\alpha}, I(t, s; \alpha)\right)}.$$

Then, for $\alpha = \alpha^*$, Theorem 9.3.1 gives us that

$$
\begin{aligned}
& E_t\left[\frac{\partial BS}{\partial k}(s,0,\exp(\alpha^*),v_s)\right] \\
&= E_t\left[-N\left(-\frac{v_s\sqrt{T-s}}{2}\right)\right] \\
&= E_t\left[\frac{N\left(\frac{v_s\sqrt{T-s}}{2}\right)-N\left(-\frac{v_s\sqrt{T-s}}{2}\right)-1}{2}\right] \\
&= E_t\left[\frac{BS\left(s,0,\exp(\alpha^*),v_s\right)-1}{2}\right] \\
&= \frac{V_t\exp\left(-X_t\right)-1}{2}
\end{aligned}
$$

and

$$
\begin{aligned}
& \frac{\partial BS}{\partial k}(s,0,\exp(\alpha^*),I(t,s;\alpha^*)) \\
&= -N\left(-\frac{I(t,s;\alpha^*)\sqrt{T-s}}{2}\right) \\
&= \frac{V_t\exp\left(-X_t\right)-1}{2}.
\end{aligned}
$$

So, we conclude that

$$
E_t\left[\frac{\partial BS}{\partial k}(s,0,\exp(\alpha^*),v_s)\right]-\frac{\partial BS}{\partial k}(s,0,\exp(\alpha^*),I(t,s;\alpha^*))=0
$$

and, consequently, $\frac{\partial I}{\partial \alpha}(t,s;\alpha^*)=0$. Now, the proof is complete. ■

Now, let us extend Proposition 9.5.1 to the correlated case.

Theorem 9.5.2 *Consider the model (6.1.1) . Then, under* ***(H1)–(H3)****,*

$$
\begin{aligned}
& \lim_{T\downarrow s}(T-s)^{\frac{1}{2}-H}\frac{\partial I}{\partial \alpha}(t,s;\alpha^*) \\
&= \frac{\rho\exp\left(-r(s-t)\right)}{2\exp(X_t)}\lim_{T\to s}E_t\left[\frac{\exp\left(X_s\right)}{\sigma_s^3(T-s)^2}\int_s^T\sigma_u\left(D_u^W\int_u^T\sigma_\theta^2 d\theta\right)du\right],
\end{aligned}
$$

for all $s>t$.

Proof. From (9.5.2) we obtain

$$\begin{aligned}
&\frac{\partial I}{\partial \alpha}(t,s;\alpha^*) \\
&= \frac{E_t\left[\frac{\partial BS}{\partial k}(s,0,\exp(\alpha^*),v_s)\right] - \frac{\partial BS}{\partial k}(s,0,\exp(\alpha^*),I(t,s;\alpha^*))}{\frac{\partial BS}{\partial \sigma}(s,0,\exp(\alpha^*),I(t,s;\alpha^*))} \\
&\quad + \frac{\frac{\rho}{2}E_t\left[\int_s^T e^{-r(u-t)}\frac{\partial H}{\partial k}(u,X_u,M_u,v_u)\sigma_u\left(\int_u^T D_u^W\sigma_\theta^2 d\theta\right)\right]}{\exp(X_t)\frac{\partial BS}{\partial \sigma}(s,0,\exp(\alpha^*),I(t,s;\alpha^*))} \\
&\quad + \frac{\frac{\rho}{2}E_t\left[\frac{\partial G}{\partial k}(s,0,\exp(\alpha^*),v_s)\int_t^s e^{X_u}e^{-r(u-t)}\sigma_u\left(\int_u^T D_u^W\sigma_\theta^2 d\theta\right)\right]}{\exp(X_t)\frac{\partial BS}{\partial \sigma}(s,0,\exp(\alpha^*),I(t,s;\alpha^*))} \\
&= T_1+T_2+T_3.
\end{aligned}$$

Similar arguments as in the proof of Proposition 9.5.1 and Theorem 9.3.1 allow us to write

$$\begin{aligned}
&E_t\left[\frac{\partial BS}{\partial k}(s,0,\exp(\alpha^*),v_s)\right] - \frac{\partial BS}{\partial k}(s,0,\exp(\alpha^*),I(t,s;\alpha^*)) \\
&= E_t\left[\frac{BS(s,0,\exp(\alpha^*),v_s)-1}{2}\right] - \frac{V_t\exp(-X_t)-1}{2} \\
&= -\frac{\exp(-X_t)}{2}E_t\left\{\frac{\rho}{2}\int_s^T e^{-r(u-t)}H(u,X_u,M_u,v_u)\sigma_u\left(\int_u^T D_u^W\sigma_\theta^2 d\theta\right)du|_{\alpha=\alpha^*}\right. \\
&\left. -\frac{\exp(-X_t)\rho}{4}G(s,0,\exp(\alpha^*),v_s)\int_t^s e^{-r(u-t)}\sigma_u e^{X_u}\left(\int_u^T D_u^W\sigma_\theta^2 d\theta\right)du\right\}.
\end{aligned}$$

Then,

$$\lim_{T\to s} T_1 = \lim_{T\to s} I_1 + \lim_{T\to s} I_2,$$

where

$$I_1 = -\frac{\exp(-X_t)\rho E_t\left(\int_s^T e^{-r(u-t)}H(u,X_u,M_u,v_u)\sigma_u\left(\int_u^T D_u^W\sigma_\theta^2 d\theta\right)du\right)|_{\alpha=\alpha^*}}{4\frac{\partial BS}{\partial \sigma}(s,0,\exp(\alpha^*),I(t,s;\alpha^*))}$$

and

$$I_2 = -\frac{\exp(-X_t)\rho E_t\left(G(s,0,\exp(\alpha^*),v_s)\int_t^s e^{-r(u-t)}\sigma_u e^{X_u}\left(\int_s^T D_u^W\sigma_\theta^2 d\theta\right)du\right)}{4\frac{\partial BS}{\partial \sigma}(s,0,\exp(\alpha^*),I(t,s;\alpha^*))}.$$

It is easy to see that, under (H3), $\lim_{T\to s} I_1 = 0$ due to Lemma 6.3.1. On the other hand,

$$G(s,0,\exp(\alpha^*),v_s) = 2\frac{\partial G}{\partial k}(s,0,\exp(\alpha^*),v_s),$$

which yields

$$
\begin{aligned}
&I_2 \\
&= -\frac{\rho}{2}\frac{\exp(-X_t)\, E_t\left(\frac{\partial G}{\partial k}(s,0,\exp(\alpha^*),v_s)\int_t^s e^{-r(u-t)}e^{X_u}\sigma_u\left(\int_u^T D_u^W\sigma_\theta^2 d\theta\right)du\right)}{\frac{\partial BS}{\partial\sigma}(s,0,\exp(\alpha^*),I(t,s;\alpha^*))} \\
&= -T_3.
\end{aligned}
$$

This gives us that $I_2 + T_3 = 0$. On the other hand, using Lemmas 6.3.1 and 9.4.7, and the anticipating Itô's formula again, it follows that

$$
\begin{aligned}
\lim_{T\to s} T_2 &= \lim_{T\to s}\frac{\frac{\rho}{2}E_t\left[\int_s^T e^{-r(u-t)}\frac{\partial H}{\partial k}(u,X_u,M_u,v_u)\sigma_u\left(\int_u^T D_u^W\sigma_\theta^2 d\theta\right)du\right]}{\exp(X_t)\frac{\partial BS}{\partial\sigma}(s,0,\exp(\alpha^*),I(t,s;\alpha^*))} \\
&= \frac{\rho}{2}\lim_{T\to s}\frac{E_t\left[\frac{\partial H}{\partial k}(s,X_s,M_s,v_s)e^{-r(s-t)}\int_s^T\sigma_u\left(\int_u^T D_u^W\sigma_\theta^2 d\theta\right)du\right]}{\exp(X_t)\frac{\partial BS}{\partial\sigma}(s,0,\exp(\alpha^*),I(t,s;\alpha^*))} \\
&= \frac{\rho}{2}\lim_{T\to s}E_t\left[\frac{\exp(X_s)\exp(-r(s-t))}{\exp(X_t)v_s^3(T-s)^2}\int_s^T\sigma_u\left(\int_u^T D_u^W\sigma_\theta^2 d\theta\right)du\right] \\
&= \frac{\rho}{2\exp(X_t)}\lim_{T\to s}E_t\left[\frac{\exp(X_s)\exp(-r(s-t))}{\sigma_s^3(T-s)^2}\int_s^T\sigma_u\left(\int_u^T D_u^W\sigma_\theta^2 d\theta\right)du\right],
\end{aligned}
$$

which allows us to complete the proof. ■

Remark 9.5.3 *In the vanilla case $t = s$, this formula reduces to the limit result in Theorem 7.5.2.*

Remark 9.5.4 *If the volatility process σ is right-continuous, and under some regularity conditions that allow us interchanging the expectation and the limit, the result in Theorem 9.5.2 reads as*

$$
\begin{aligned}
&\lim_{T\downarrow s}(T-s)^{\frac{1}{2}-H}\frac{\partial I}{\partial\alpha}(t,s;\alpha^*) \\
&= \frac{\rho\exp(-r(s-t))}{2\exp(X_t)}E_t\left[\frac{\exp(X_s)}{\sigma_s^2}\lim_{T\to s}\frac{1}{(T-s)^2}\int_s^T\left(D_u^W\int_u^T\sigma_\theta^2 d\theta\right)du\right], \\
&= \frac{\exp(-r(s-t))}{\exp(X_t)}E_t\left[\exp(X_s)\lim_{T\to s}(T-s)^{\frac{1}{2}-H}\frac{\partial I_s}{\partial k}(k_s^*)\right], \qquad (9.5.3)
\end{aligned}
$$

where $\frac{\partial I_s}{\partial k}(k_s^*)$ denotes the derivative of the implied volatility at time s, evaluated at the ATM strike k_s^* (see Section (7.6.2)). That is, the ATM forward skew is a mean of the future vanilla skews. Both quantities (vanilla and forward skews) are not necessarily similar, as we can see in the following examples.

9.5.1 Local volatility models

Consider a local volatility model $\sigma_t = \sigma(t, e^{X_t})$. Then, from Section 7.6.2, we have now that

$$\lim_{T \to s} \frac{\partial I_s}{\partial k}(k_s^*)) = \frac{1}{2}\frac{\partial \sigma}{\partial X}(s, e^{X_s}).$$

Then, in order to reproduce the blow-up of the short-time skew slope for vanilla options (the case $s = t$), it is enough to take σ with a singularity at (t, e^{X_t}), in such a way that $\frac{\partial \sigma}{\partial x}(t, e^{X_t}) = \infty$. But it is not possible to construct a function σ with a singularity at every (s, e^{X_s}). In general, these latest derivatives are finite, and this explains why the local volatility forward skew tends to be flatter.

Example 9.5.5 *Consider a CEV model where $\sigma_t = \sigma S_t^{\gamma - 1}$, for some positive constants σ, γ. The results in 7.6.2 give us that*

$$\lim_{T \to t} \frac{\partial I_t}{\partial k}(k_t^*) = \frac{\sigma(\gamma - 1)}{2} S_t^{\gamma - 1} = \frac{(\gamma - 1)}{2} \sigma_t. \tag{9.5.4}$$

Then

$$\begin{aligned} \lim_{T \downarrow s} \frac{\partial I}{\partial \alpha}(t, s; \alpha^*) &= \frac{\exp(-r(s-t))}{\exp(X_t)} E_t \left[\exp(X_s) \frac{(\gamma - 1)}{2} \sigma_s \right] \\ &= \frac{\sigma(\gamma - 1) \exp(-r(s-t))}{2 \exp(X_t)} E_t \left[\exp(\gamma X_s) \right]. \end{aligned} \tag{9.5.5}$$

9.5.2 Stochastic volatility models

Assume that the volatility σ_r is of the form $\sigma_r = f(Y_r)$, where $f \in C_b^1(\mathbb{R})$ and Y_r is the solution of a stochastic differential equation

$$dY_r = a(r, Y_r)\, dr + b(r, Y_r)\, dW_r, \tag{9.5.6}$$

for some real functions $a, b \in C_b^1(\mathbb{R})$. Then, the results in Section 7.6.1 give us that

$$\lim_{T \to t} \frac{\partial I_t}{\partial k}(k_t^*) = \frac{\rho}{2\sigma_t} f'(Y_t) b(t, Y_t)$$

and this implies

$$\lim_{T \downarrow s} \frac{\partial I}{\partial \alpha}(t, s; \alpha^*) = \frac{\rho \exp(-r(s-t))}{2\sigma_t \exp(X_t)} E_t \left[\exp(X_s) f'(Y_s) b(s, Y_s) \right]. \tag{9.5.7}$$

Example 9.5.6 *Assume, for example, that* $b(s, Y_s) = (s-t)^{H-\frac{1}{2}}$, *for some* $H < \frac{1}{2}$, *in a similar way as in Carr and Itkin (2020). Then, in the vanilla case* $s = t$, *the above limit is infinite, while in the forward case* $s \neq t$; *we have*

$$\lim_{T \downarrow s} \frac{\partial I}{\partial \alpha}(t, s; \alpha^*) = \frac{\rho \exp(-r(s-t))}{2\sigma_t \exp(X_t)}(s-t)^{H-\frac{1}{2}} E_t \left[\exp(X_s) f'(Y_s)\right],$$

which is finite for all $s \neq t$.

Example 9.5.7 (The SABR model) *A particular example of stochastic volatility models is the SABR, defined in Section 2.3.2, where* $\beta = 1$ *and* $d\sigma_t = \alpha\sigma_t dW_t$ *for some* $\alpha > 0$. *We know from Section 7.6.1.2 that*

$$\lim_{T \to t} \frac{\partial I_t}{\partial k}(k_t^*) = \frac{\rho\alpha}{2}.$$

Then, 9.5.11 gives us that

$$\lim_{T \downarrow s} \frac{\partial I}{\partial \alpha}(t, s; \alpha^*) = \frac{\alpha\rho \exp(-r(s-t))}{2 \exp(X_t)} E_t \left[\exp(X_s)\right] = \frac{\alpha\rho}{2}, \tag{9.5.8}$$

which coincides with the same limit as in the vanilla case. We remark that the same argument applies to the two-factor Bergomi model in Example 7.6.1.4. In general, when the limit $\lim_{T \to t} \frac{\partial I_t}{\partial k}(k_t^*)$ *is a constant non-depending on the value of the spot volatility, the short-end ATM skew slope is the same for vanilla and forward-start options.*

Example 9.5.8 (The Heston model) *Consider a Heston model of the form*

$$dv_t = k(\theta - v_t)dt + \nu\sqrt{v_t}dW_t,$$

where $v_t := \sigma_t^2$ *(see Section 3.3.2). We know, from Section 7.6.1.3, that*

$$\lim_{T \to t} \frac{\partial I_t}{\partial k}(k_t^*) = \frac{\rho\nu}{4\sigma_t}.$$

Then,

$$\lim_{T \downarrow s} \frac{\partial I}{\partial \alpha}(t, s; \alpha^*) = \frac{\nu\rho \exp(-r(s-t))}{4 \exp(X_t)} E_t \left[\frac{\exp(X_s)}{\sigma_s}\right]. \tag{9.5.9}$$

9.5.3 Fractional volatility models

Assume that the volatility σ is of the form $\sigma_r = f(Y_r)$, where $f \in C_b^1(\mathbb{R})$ and Y_r is a *fractional Ornstein-Uhlenbeck* process of the form

$$Y_r = m + (Y_0 - m)\, e^{-\alpha r} + c\sqrt{2\alpha} \int_0^r e^{-\alpha(r-s)} d\hat{W}_s^H, \qquad (9.5.10)$$

where $\hat{W}_s^H := \int_0^s (s-u)^{H-\frac{1}{2}} dW_u$ and $Y_0, m, c,$ and α are positive constants. We proved (see Section 7.6.4.1) that, for vanilla options,

- if $H > \frac{1}{2}$,

$$\lim_{T \to t} \frac{\partial I_t}{\partial k}(k_t^*) = 0,$$

  and,

- if $H < \frac{1}{2}$,

$$\lim_{T \to t} (T-t)^{\frac{1}{2}-H} \frac{\partial I_t}{\partial k}(k_t^*) = c\sqrt{2\alpha} \frac{\rho}{\sigma_t} f'(Y_t)\,.$$

  Then, a direct application of (9.5.11) gives us that, if $H > \frac{1}{2}$, the ATM forward skew is zero, while in the case $H < \frac{1}{2}$

$$\lim_{T \downarrow s} (T-s)^{\frac{1}{2}-H} \frac{\partial I}{\partial \alpha}(t, s; \alpha^*) = \frac{\rho\sqrt{2\alpha}\exp(-r(s-t))}{\exp(X_t)} E_t\left[\exp(X_s)\frac{f'(Y_s)}{\sigma_s}\right], \qquad (9.5.11)$$

  which is $O(T-s)^{H-\frac{1}{2}}$ as for vanilla options. A similar argument proves that, in general, models based on the fBm with $H < \frac{1}{2}$, as the rough Bergomi model (see Section 5.6.2), have a forward skew of order $O(T-s)^{H-\frac{1}{2}}$.

9.5.4 Time-depending coefficients

Let Y_r be a diffusion process of the form (7.6.4), with $\sqrt{\alpha(s)} = \left(T_{n(s)} - s\right)^{-\frac{1}{2}+\varepsilon}$, for some $\varepsilon > 0$. We have now, from 7.6.15, that

$$\lim_{T \to t} (T-t)^{\frac{1}{2}-\epsilon} \frac{\partial I_t}{\partial k}(k_t^*) = -\frac{\rho c}{\sigma_t^2}\left(\frac{1}{-1/2+\varepsilon}\right)\left(\frac{1}{1/2+\varepsilon}\right) f'(Y_t). \quad (9.5.12)$$

Then,

$$\lim_{T\downarrow s}(T-s)^{\frac{1}{2}-H}\frac{\partial I}{\partial\alpha}(t,s;\alpha^*)$$
$$=\frac{-\rho c\sqrt{2\alpha}\left(\frac{1}{-1/2+\varepsilon}\right)\left(\frac{1}{1/2+\varepsilon}\right)\exp\left(-r(s-t)\right)}{\exp(X_t)}E_t\left[\exp\left(X_s\right)\frac{f'(Y_s)}{\sigma_s^2}\right],$$

which implies that this class of models can reproduce the blow-up for both the vanilla and forward skews.

9.6 AT-THE-MONEY CURVATURE

In this section, we figure out $\frac{\partial^2 I}{\partial\alpha^2}(t,s;\alpha^*)$. For the sake of simplicity, we start analysing the uncorrelated case $\rho=0$.

9.6.1 The uncorrelated case

Assume that $\rho=0$. Then, we can prove the following technical lemma.

Lemma 9.6.1 *Consider the model (6.1.1) with $\rho=0$ and assume that hypotheses in Theorem 9.3.1 hold. Then,*

$$\frac{\partial BS}{\partial\sigma}(s,0,\exp(\alpha^*),I(t,s;\alpha^*))\frac{\partial^2 I}{\partial\alpha^2}(t,s;\alpha^*)$$
$$=\frac{1}{2}E_t\left[\int_t^T\frac{\partial^2\Psi}{\partial a^2}\left(E_u\left(BS\left(s,0,\exp(\alpha^*),v_s\right)\right)\right)U_u^2du\right],$$

where

$$\Psi(a):=\frac{\partial^2 BS}{\partial k^2}\left(s,0,\exp(\alpha^*),BS^{-1}\left(a\right)\right)$$

and

$$U_u=E_u\left(\left(\frac{\partial BS}{\partial\sigma}(s,0,\exp(\alpha^*),v_s)\right)\frac{D_u^W\int_s^T\sigma_r^2dr}{2(T-s)v_s}\right).\tag{9.6.1}$$

Proof. Notice that, as

$$\begin{aligned}\frac{\partial V_t}{\partial\alpha}&=\exp(X_t)\frac{\partial BS}{\partial k}(s,0,\exp(\alpha),I(t,s;\alpha))\\&\quad+\exp(X_t)\frac{\partial BS}{\partial\sigma}(s,0,\exp(\alpha),I(t,s;\alpha))\frac{\partial I}{\partial\alpha}(t,s;\alpha)\end{aligned}$$

we get

$$\begin{aligned}\frac{\partial^2 V_t}{\partial \alpha^2} &= \exp(X_t)\frac{\partial^2 BS}{\partial k^2}(s,0,\exp(\alpha),I(t,s;\alpha)) \\ &+2\exp(X_t)\frac{\partial^2 BS}{\partial\sigma\partial k}(s,0,\exp(\alpha),I(t,s;\alpha))\frac{\partial I}{\partial \alpha}(t,s;\alpha) \\ &+\exp(X_t)\frac{\partial^2 BS}{\partial \sigma^2}(s,0,\exp(\alpha),I(t,s;\alpha))\left(\frac{\partial I}{\partial \alpha}(t,s;\alpha)\right)^2 \\ &+\exp(X_t)\frac{\partial BS}{\partial \sigma}(s,0,\exp(\alpha),I(t,s;\alpha))\frac{\partial^2 I}{\partial \alpha^2}(t,s;\alpha). \quad (9.6.2)\end{aligned}$$

Then, Proposition 9.5.1 implies that

$$\begin{aligned}&\frac{\partial^2 V_t}{\partial \alpha^2}|_{\alpha=\alpha^*} \\ &= \exp(X_t)\frac{\partial^2 BS}{\partial k^2}(s,0,\exp(\alpha^*),I(t,s;\alpha^*)) \\ &+\exp(X_t)\frac{\partial BS}{\partial \sigma}(s,0,\exp(\alpha^*),I(t,s;\alpha^*))\frac{\partial^2 I}{\partial \alpha^2}(t,s;\alpha^*).\end{aligned}$$

Then, taking into account Theorem 9.3.1 and the fact that $I(t,s;\alpha^*) = BS^{-1}(\exp(-X_t)V_t)$, we are able to write

$$\begin{aligned}&\exp(X_t)\frac{\partial BS}{\partial \sigma}(s,0,\exp(\alpha^*),I(t,s;\alpha^*))\frac{\partial^2 I}{\partial \alpha^2}(t,s;\alpha^*) \\ &= \frac{\partial^2 V_t}{\partial \alpha^2}|_{\alpha=\alpha^*} - \exp(X_t)\frac{\partial^2 BS}{\partial k^2}(s,0,\exp(\alpha^*),I(t,s;\alpha^*)) \\ &= \exp(X_t)E_t\left(\frac{\partial^2 BS}{\partial k^2}(s,0,\exp(\alpha^*),v_s) - \frac{\partial^2 BS}{\partial k^2}(s,0,\exp(\alpha^*),I(t,s;\alpha^*))\right) \\ &= \exp(X_t)E_t\left[\frac{\partial^2 BS}{\partial k^2}\left(s,0,\exp(\alpha^*),BS^{-1}\left(BS(s,0,\exp(\alpha^*),v_s)\right)\right)\right. \\ &\quad\left. -\frac{\partial^2 BS}{\partial k^2}\left(s,0,\exp(\alpha^*),BS^{-1}\left(E_t\left(BS(s,0,\exp(\alpha^*),v_s)\right)\right)\right)\right].\end{aligned}$$

Now, the Clark-Ocone-Haussman formula and the chain rule for the Malliavin derivative operator give us that

$$\begin{aligned}&BS(s,0,\exp(\alpha^*)),v_s) \\ &= E_t(BS(s,0,\exp(\alpha^*),v_s)) + \int_t^T E_u(D_u^W BS(s,0,\exp(\alpha^*),v_s)))dW_u \\ &= E_t(BS(s,0,\exp(\alpha^*),v_s)) + \int_t^T U_s dW_u. \quad (9.6.3)\end{aligned}$$

Then, applying Itô's formula to the process

$$A_u := \frac{\partial^2 BS}{\partial k^2}\left(s, 0, \exp(\alpha^*), BS^{-1}\left(E_u\left(BS(s, 0, \exp(\alpha^*), v_s)\right)\right)\right)$$

and taking expectations, the result follows. ■

Now, we are in a position to compute the ATM short-end curvature in the uncorrelated case.

Theorem 9.6.2 *Consider the model (6.1.1) with $\rho = 0$ and assume that hypotheses (H1), (H2), and (H3) are satisfied. Then,*

$$\lim_{T\to s}(T-s)\frac{\partial^2 I}{\partial \alpha^2}(t, s; \alpha^*) = \frac{1}{4}E_t\left(\int_t^s \left(E_u\left(\frac{D_u^W \sigma_s^2}{\sigma_s}\right)\right)^2 \left(E_u\left(\sigma_s\right)\right)^{-3} du\right).$$

Proof. Lemma 9.6.1 yields

$$\begin{aligned}
&\lim_{T\to s}(T-s)\frac{\partial^2 I}{\partial \alpha^2}(t, s; \alpha^*) \\
&= \lim_{T\to s}(T-s)\left(\frac{E_t\left[\int_t^s \frac{\partial^2 \Psi}{\partial a^2}\left(E_u\left(BS\left(s, 0, \exp(\alpha^*), v_s\right)\right)\right) U_u^2 du\right]}{2(\frac{\partial BS}{\partial \sigma}\left(s, 0, \exp(\alpha^*), I(t, s; \alpha^*)\right)}\right. \\
&\quad \left. + \lim_{T\to s}(T-s)\left(\frac{E_t\left[\int_s^T \frac{\partial^2 \Psi}{\partial a^2}\left(E_u\left(BS\left(s, 0, \alpha^*, v_s\right)\right)\right) U_u^2 du\right]}{2(\frac{\partial BS}{\partial \sigma}\left(s, 0, \exp(\alpha^*), I(t, s; \alpha^*)\right)}\right)\right) \\
&= \lim_{T\to s}(T_1 + T_2).
\end{aligned}$$

From (H1), there exist two constants $a, b > 0$ such that $a \le \sigma \le b$. Thus, the fact that $BS(s, 0, e^{r(T-s)}, \cdot)$ is an increasing function, together with (H3) and

$$\begin{aligned}
&\frac{\partial^2 \Psi}{\partial a^2}\left(E_u\left(BS\left(s, 0, \alpha, v_s\right)\right)\right) \\
&= \frac{2\sqrt{2\pi}\exp\left(\frac{\left(BS^{-1}(E*_u(BS(s,0,e^{r(T-s)}, v_s)))\right)^2 (T-s)}{8}\right)}{\left(BS^{-1}(E *_u \left(BS(s, 0, e^{r(T-s)}, v_s)\right))\right)^3 (T-s)^{3/2}},
\end{aligned}$$

allows us to deduce that, considering that C is a constant that may change from line to line,

$$\begin{aligned} 0 &< T_2 \leq C E_t \left(\int_s^T \frac{\exp\left(\frac{C(T-s)}{8}\right)}{c^3(T-s)} U_u^2 du \right) \leq \frac{C}{(T-s)} \int_s^T E_t\left(U_u^2\right) du \\ &\leq C(T-s)^{1/2}, \end{aligned}$$

which implies that $\lim_{T\to s} T_2 = 0$.

Finally, the dominated convergence theorem and Lemma 9.4.1 give us that

$$\begin{aligned} \lim_{T\to s} T_1 &= \pi \lim_{T\to s} \int_t^s \frac{\exp\left(\frac{I(u,s;\alpha^*)^2(T-s)}{8}\right)}{I(u,s;\alpha^*)^3(T-s)} E_t\left(U_u^2\right) du \\ &= \frac{1}{4} \lim_{T\to s} E_t \int_t^s \frac{1}{I(u,s;\alpha^*)^3(T-s)} \left(E_u \left(\frac{\int_s^T (D_u^W \sigma_r^2) dr}{v_s \sqrt{T-s}} \right) \right)^2 du \\ &= \frac{1}{4} E_t \left(\int_t^s \left(E_u \left(\frac{D_s^W \sigma_r^2}{\sigma_s} \right) \right)^2 \left(E_u\left(\sigma_s\right)\right)^{-3} du \right) \end{aligned}$$

and this allows us to complete the proof. ■

Remark 9.6.3 *Theorem 9.6.2 proves that the ATM short-end curvature of the forward smile is $O(T-t)^{-1}$ in the uncorrelated case, both for stochastic and rough volatilities, according to Jaquier and Roome (2015) and Bühler (2002). Notice that this rate does not depend on the Hurst parameter or the model.*

9.6.2 The correlated case

Now, we are ready to analyse the curvature in the correlated case. So we assume $\rho \neq 0$.

Theorem 9.6.4 *Under the assumptions of Theorem 9.3.1, we have*

$$\begin{aligned} &\lim_{T\to s} (T-s) \frac{\partial^2 I}{\partial \alpha^2}(t,s;\alpha^*) \\ &= \frac{1}{4} E_t \left(\int_t^s \left(E_u \left(\frac{D_u^W \sigma_s^2}{\sigma_s} \right) \right)^2 \left(E_u\left(\sigma_s\right)\right)^{-3} du \right) \\ &\quad + \frac{1}{E_t\left(\sigma_s\right)} - \frac{1}{E_t\left(\sigma_s\right) + \frac{\rho}{2}\exp(-X_t) E_t\left(\frac{1}{\sigma_s} \int_t^s e^{-r(u-t)} e^{X_u} \sigma_u D_u^W \sigma_s^2 du \right)} \\ &\quad - \frac{\rho}{2}\exp(-X_t) E_t \left(\frac{1}{\sigma_s^3} \int_t^s e^{-r(u-t)} e^{X_u} \sigma_u \left(D_u^W \sigma_s^2 \right) du \right). \end{aligned}$$

Proof. By Theorem 9.3.1 and (9.6.2), we deduce that

$$\begin{aligned}
&\exp(X_t)\frac{\partial BS}{\partial\sigma}(s,0,\exp(\alpha^*),I(t,s;\alpha))\frac{\partial^2 I}{\partial\alpha^2}(t,s;\alpha)\\
&= \exp(X_t)E_t\left(\frac{\partial^2 BS}{\partial k^2}(s,0,\exp(\alpha^*),v_s)-\frac{\partial^2 BS}{\partial k^2}(s,0,\exp(\alpha^*),I(t,s;\alpha))\right)\\
&\quad -2\exp(X_t)\frac{\partial^2 BS}{\partial\sigma\partial k}(s,0,\exp(\alpha^*),I(t,s;\alpha))\frac{\partial I}{\partial\alpha}(t,s;\alpha)\\
&\quad -\exp(X_t)\frac{\partial^2 BS}{\partial\sigma^2}(s,0,\exp(\alpha^*),I(t,s;\alpha^*))\left(\frac{\partial I}{\partial\alpha}(t,s;\alpha^*)\right)^2\\
&\quad +\frac{\rho}{2}E_t\left[\int_s^T e^{-r(u-t)}\frac{\partial^2 H}{\partial k^2}(u,X_u,M_u,v_u)\sigma_u\left(\int_u^T D_u^W\sigma_\theta^2 d\theta\right)du\right|_{\alpha=\alpha^*}\\
&\quad +\left.\frac{\partial^2 G}{\partial k^2}(s,0,\exp(\alpha^*),v_s)\int_t^s e^{-r(u-t)}e^{X_u}\sigma_u\left(\int_s^T D_u^W\sigma_\theta^2 d\theta\right)du\right]. \qquad (9.6.4)
\end{aligned}$$

Thus, we can write

$$\begin{aligned}
&(T-s)\frac{\partial^2 I}{\partial\alpha^2}(t,s;\alpha^*)\\
&=(T-s)\frac{E_t\left(\frac{\partial^2 BS}{\partial k^2}(s,0,\exp(\alpha^*),v_s)-\frac{\partial^2 BS}{\partial k^2}(s,0,\exp(\alpha^*),I^0(t,s;\alpha^*))\right)}{\frac{\partial BS}{\partial\sigma}(s,0,\exp(\alpha^*),I(t,s;\alpha^*))}\\
&+(T-s)\frac{E_t\left(\frac{\partial^2 BS}{\partial k^2}(s,0,\exp(\alpha^*),I^0(t,s;\alpha^*))-\frac{\partial^2 BS}{\partial k^2}(s,0,\exp(\alpha^*),I(t,s;\alpha^*))\right)}{\frac{\partial BS}{\partial\sigma}(s,0,\exp(\alpha^*),I(t,s;\alpha^*))}\\
&-(T-s)\frac{2\frac{\partial^2 BS}{\partial\sigma\partial k}(s,0,\exp(\alpha^*),I(t,s;\alpha^*))\frac{\partial I}{\partial\alpha}(t,s;\alpha^*)}{\frac{\partial BS}{\partial\sigma}(s,0,\exp(\alpha^*),I(t,s;\alpha^*))}\\
&-(T-s)\frac{\frac{\partial^2 BS}{\partial\sigma^2}(s,0,\exp(\alpha^*),I(t,s;\alpha^*))\left(\frac{\partial I}{\partial\alpha}(t,s;\alpha^*)\right)^2}{\frac{\partial BS}{\partial\sigma}(s,0,\exp(\alpha^*),I(t,s;\alpha^*))}\\
&+(T-s)\left.\frac{\frac{\rho}{2}\exp(-X_t)E_t\left[\int_s^T e^{-r(u-t)}\frac{\partial^2 H}{\partial k^2}(u,X_u,M_u,v_u)\sigma_u\left(\int_u^T D_u^W\sigma_\theta^2 d\theta\right)du\right.}{\frac{\partial BS}{\partial\sigma}(s,0,\exp(\alpha^*),I(t,s;\alpha^*))}\right|_{\alpha=\alpha^*}\\
&+(T-s)\\
&\times\frac{\frac{\rho}{2}\exp(-X_t)E_t\left(\frac{\partial^2 G}{\partial k^2}(s,0,\exp(\alpha^*),v_s)\int_t^s e^{-r(u-t)}e^{X_u}\sigma_u\left(\int_s^T D_u^W\sigma_\theta^2 d\theta\right)du\right)}{\frac{\partial BS}{\partial\sigma}(s,0,\exp(\alpha^*),I(t,s;\alpha^*))}\\
&=T_1+T_2+T_3+T_4+T_5+T_6.
\end{aligned}$$

Here, $I^0(t,s;\alpha)$ denotes the implied volatility in the uncorrelated case $\rho = 0$. It is easy to see that the definition of BS leads us to $\lim_{T\to s}(T_3+T_4+T_5) = 0$. Moreover, we can easily see that

$$\lim_{T\to s}\frac{\frac{\partial BS}{\partial\sigma}(s,0,\exp(\alpha^*),I(t,s;\alpha^*))}{\frac{\partial BS}{\partial\sigma}(s,0,\exp(\alpha^*),I^0(t,s;\alpha^*))}=1.$$

Then, the proof of Lemma 9.6.1 yields

$$\lim_{T\to s} T_1 = \lim_{T\to s} (T-s)\frac{\partial^2 I^0(t,s,\alpha^*)}{\partial k^2}.$$

By computing $\frac{\partial^2 BS}{\partial k^2}$ it is easy to see that

$$\lim_{T\to s} T_2 = \lim_{T\to s}\left(\frac{1}{I^0(t,s;\alpha^*)} - \frac{1}{I(t,s;\alpha^*)}\right).$$

Therefore, we can use Lemma 9.4.1 and Theorem 10.2.6 to figure out this limit.

On the other hand,

$$\begin{aligned}
&\lim_{T\to s} T_6 \\
&= \frac{\rho}{2}\exp(-X_t)\lim_{T\to s}\ (T-s) \\
&\quad\times \frac{E_t\left(\frac{\partial^2 G}{\partial k^2}(s,0,\exp(\alpha^*),v_s)\int_t^s e^{-r(u-t)}e^{X_u}\sigma_u\left(\int_s^T D_u^W\sigma_\theta^2 d\theta\right)du\right)}{\frac{\partial BS}{\partial\sigma}(s,0,\exp(\alpha^*),I(t,s;\alpha^*))} \\
&= -\frac{\rho}{2}\exp(-X_t)E_t\left(\frac{1}{\sigma_s^3}\int_t^s e^{-r(u-t)}e^{X_u}\sigma_u\left(D_u^W\sigma_s^2\right)du\right).
\end{aligned}$$

Finally, the result is a consequence of Lemma 9.4.1 and Theorems 10.2.6 and 9.6.2. ■

Remark 9.6.5 *The above limit result proves that, except if*

$$\begin{aligned}
&\frac{1}{4}E_t\left(\int_t^s\left(E_u\left(\frac{D_u^W\sigma_s^2}{\sigma_s}\right)\right)^2(E_u(\sigma_s))^{-3}du\right) \\
&+\frac{1}{E_t(\sigma_s)} - \frac{1}{E_t(\sigma_s)+\frac{\rho}{2}\exp(-X_t)E_t\left(\frac{1}{\sigma_s}\int_t^s e^{-r(u-t)}e^{X_u}\sigma_u D_u^W\sigma_s^2 du\right)} \\
&-\frac{\rho}{2}\exp(-X_t)E_t\left(\frac{1}{\sigma_s^3}\int_t^s e^{-r(u-t)}e^{X_u}\sigma_u\left(D_u^W\sigma_s^2\right)du\right) = 0,
\end{aligned}$$

the forward curvature is $O(T-s)^{-1}$, and this rate is the same for classical diffusions and for rough volatilities, according again to Jaquier and Roome (2015) and Bühler (2002). This means that, when s is closer to T, the smile effect is more pronounced than the skew effect, as we see in the following examples.

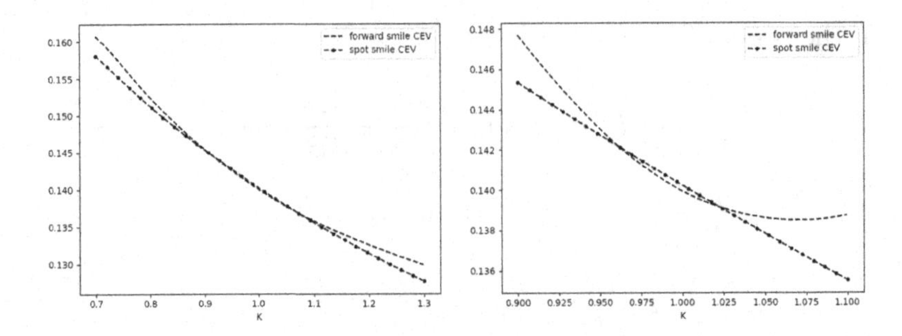

Figure 9.2 Spot and forward implied volatilities corresponding to a CEV model with $T = 1Y, \sigma = \gamma = 0.3$, and $s = 0.5$ (left), and $s = 0.9$ (right).

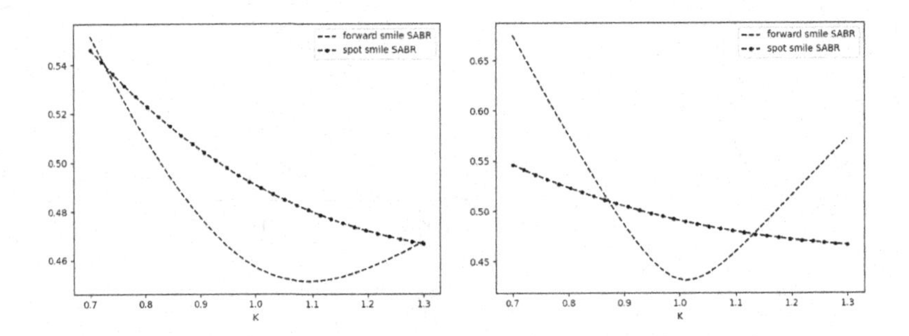

Figure 9.3 Spot and forward implied volatilities corresponding to a SABR model with $\sigma_0 = 0.5, \alpha = 0.7, \rho = -0.4, t = 0, T = 1Y$, and $s = 0.5$ (left), and $s = 0.9$ (right).

Example 9.6.6 *Consider a CEV model where $\sigma_t = \sigma S_t^{\gamma-1}$, where we take $\sigma = \gamma = 0.3$. In Figure 9.2 we can observe the corresponding forward smiles with $T = 1Y$, and $s = 0.5, 0.9$. Notice that the smile effect increases for small values of $T - s$.*

Example 9.6.7 *Consider a SABR model as in Section 2.3.2 where $\beta = 1$ and $d\sigma_t = \alpha\sigma_t dW_t$ for some $\alpha > 0$. In Figure 9.3 we can observe the forward smiles for a SABR model with $\sigma_0 = 0.5, \alpha = 0.7, \rho = -0.4, t = 0, T = 1Y$, and $s = 0.5, 0.9$. Notice that, as $T - s$ decreases, the smile effect becomes more relevant than the skew effect.*

Example 9.6.8 *Consider a rough Bergomi model as in Section 5.6.2,*

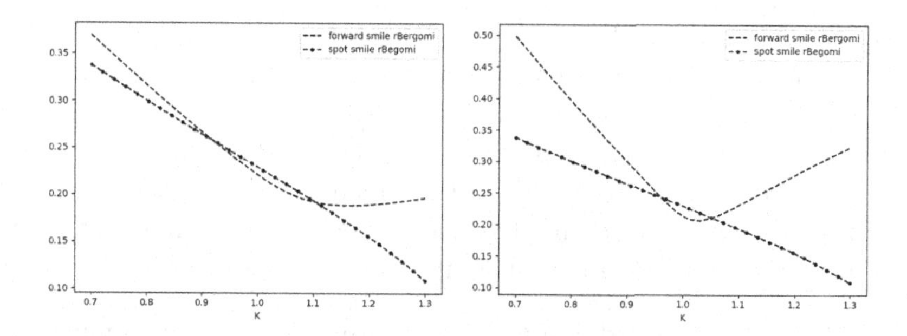

Figure 9.4 Spot and forward implied volatilities corresponding to a rBergomi model with $\sigma_0 = \sqrt{0.5}, H = 0.7, \rho = -0.6, \nu = 0.5, t = 0, T = 1Y$, and $s = 0.5$ (left), and $s = 0.9$ (right).

where

$$\sigma_t^2 = E(\sigma_t^2) \exp\left(\nu\sqrt{2H} Z_t - \frac{1}{2}\nu^2 t^{2H}\right), r \in [0, T], \tag{9.6.5}$$

for some positive ν *and* $H < \frac{1}{2}$, *with*

$$Z_t := \int_0^t (t-u)^{H-\frac{1}{2}} dW_u.$$

In Figure 9.4 we can observe the forward smiles for a rBergomi model with $\sigma_0 = \sqrt{0.5}, H = 0.7, \rho = -0.6, \nu = 0.5, t = 0, T = 1Y$, *and* $s = 0.5, 0.9$. *Notice again that, as* $T - s$ *decreases, the smile effect becomes more relevant than the skew effect.*

9.7 CHAPTER'S DIGEST

The same arguments as in Chapter 6 can be applied to get an adequate decomposition formula for options with random strikes, as forward-start option. The obtained formulas can be applied to the cases of local, stochastic, and rough volatilities. Based on this representation, we can see that

- The short-end limit of the ATM forward implied volatility (in contrast to the vanilla case) depends on the correlation parameter ρ. For example, in the SABR model, this limit is an increasing function of ρ. In the case of fractional volatilities, the effect of correlation is $O(s-t)^{H+\frac{1}{2}}$.

- The ATM forward skew is the expectation of the product of the future asset price and the future vanilla skew. Then, for local, stochastic, and local stochastic volatilities, as well as for models based on Markovian approximations of the fBm, the forward skew slope tends to a finite limit as the time to maturity decreases. But rough volatilities and models with diffusions depending on time to maturity can describe a blow-up of the ATM forward skew slope.

- The ATM forward implied curvature is $O(T-t)^{-1}$ for local, stochastic, and rough volatilities. That is, independently of the model (in contrast to the vanilla case) this curvature blows up, at a faster speed than for vanilla options.

CHAPTER 10

Options on the VIX

In this chapter, following Alòs, García-Lorite, and Muguruza (2018) we study European options with maturity T with a payoff of the form

$$(VIX_T - K)_+,$$

where VIX_T denotes the CBOE volatility index defined in Section 4.1.4 and K denotes the strike price. Options on the volatility are becoming increasingly popular (see Carr, Geman, Madan, and Yor (2005), Carr and Madan (2014), Carr, Lee, and Lorig (2019), and Horvath, Jacquier, and Tankov (2019), for instance). One of the most interesting modelling problems in this field is to construct a model reproducing at the same time the properties of the implied volatility surfaces for the S&P index and for the VIX. In particular, the empirical ATM VIX skew is positive, as we can see in Figure 10.1, corresponding to the market VIX smiles as of April 4, 2019, with maturities April 17, 2019, and May 22, 2019 (data courtesy of BBVA). This property is not easily replicated by classical stochastic volatility models (see for example Fouque and Saporito (2015), Kokholm and Stisen (2015), Goutte, Ismail, and Pham (2017), de Marco (2018), Guyon (2018), or Gatheral, Jusselin, and Rosenbaum (2020)). For example, it is well-known that the Heston model tends to reproduce a negative VIX skew, while the SABR model generates a flat VIX skew (see Baldeaux and Badran (2014) and Fouque and Saporito (2015)). In this chapter, we study analytically the VIX implied volatility, and we identify the class of processes that can reproduce this positive skew.

From the technical point of view, options on the VIX are a particular example of options where the underlying asset does not follow a log-normal dynamics as in model (6.1.1). As noticed in Section 4.1.4 that, given a random variable A adapted to the σ-algebra generated by a

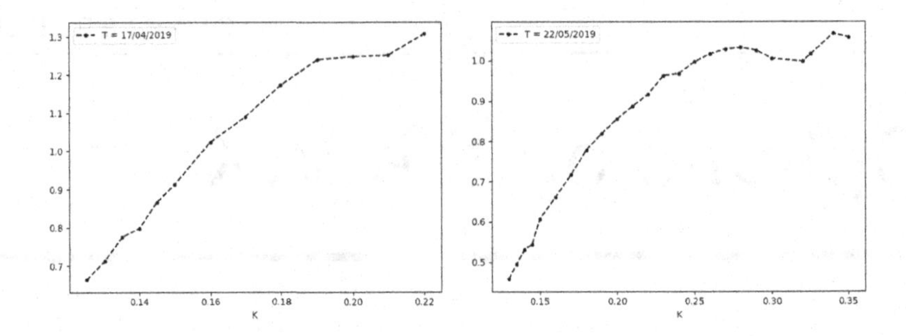

Figure 10.1 Market VIX smiles as of April 4, 2019, with maturities April 17, 2019 (left), and May 22, 2019 (right). Data courtesy of BBVA.

Brownian motion in a time interval $[0,T]$, we can proceed as follows: consider the process $M_t^T := E_t(A)$, $t \in [0,T]$. Then, the corresponding option price V_t can be written as

$$V_t = e^{-r(T-t)} E_t(A-K)_+ = e^{-r(T-t)} E_t(M_T^T - K)_+.$$

Now, because of the martingale representation theorem

$$M_t^T = E(M_t^T) + \int_0^t m_s^T dW_s, \tag{10.0.1}$$

for some adapted and square integrable process m^T. Then, we get that M_t^T admits the following dynamics

$$M_t^T = E(M_t^T) + \int_t^T \phi_s M_s^T dW_s, \tag{10.0.2}$$

where

$$\phi_s := \frac{m_s^T}{M_s^T}. \tag{10.0.3}$$

That is, given a random variable A, we have constructed a martingale M^T such that $M_T^T = A$ and M^T follows a stochastic volatility model (Equation (10.0.2)) with a stochastic volatility ϕ. Is it straightforward to apply the results in the previous chapters? Not entirely. Decomposition formulas and Malliavin expansions can be obtained in a similar way, but in the computation of the limits we have to take into account that ϕ may depend on T.

As we pointed out in Section 4.1.4, the process m_t^T can be computed by means of the Clark-Ocone-Haussman formula. In particular, if A is the VIX index defined as

$$VIX_T = \sqrt{\frac{1}{\Delta} E_T \int_T^{T+\Delta} \sigma_s^2 ds}, \tag{10.0.4}$$

(where Δ is 30 days) and σ denotes the spot volatility, we have that

$$m_t^{T,VIX} = \frac{1}{2\Delta} E_t \left(\frac{1}{VIX_T} \int_T^{T+\Delta} (D_t \sigma_s^2) ds \right), \tag{10.0.5}$$

where D denotes the Malliavin derivative with respect to the Brownian motion W.

Then, using Equality (10.0.3) we identify the maturity-dependent stochastic volatility as

$$\phi_t^{VIX} = \frac{1}{2\Delta M_t^{T,VIX}} E_t \left(\frac{1}{VIX_T} \int_T^{T+\Delta} (D_t \sigma_s^2) ds \right), \tag{10.0.6}$$

where $M_t^{T,VIX} := E_t(VIX_T)$.

In the following sections, we analyse the short-end behaviour of the ATM level and skew for options with a general underlying asset A. For the sake of simplicity, we take $r = 0$.

10.1 THE ATM SHORT-TIME LEVEL AND SKEW OF THE IMPLIED VOLATILITY

From (10.0.2), all the techniques in the previous chapters can be adapted to options on a general underlying asset A, with the only difference that now the volatility process ϕ may depend on the maturity time T. Our purpose in this section is the analysis of the short-time level and skew of the ATMI for call options on a general random variable A. In the following sections, we will apply these results to the particular case of VIX options.

Let us define the implied volatility as the quantity $I_t(k)$ such that

$$V_t = BS(t, \ln(E_t(A)), k, I_t(k)).$$

Denote by $k_0^* := \ln(E_t(A))$ the ATM log-strike. Moreover, in the sequel we will write $BS^{-1}(t, u) = BS^{-1}(t, \ln(E_t(A)), k_t^*, u)$.

10.1.1 The ATMI short-time limit

Now our first objective is to study the short-time behaviour of the implied volatility level. Towards this end, we will need the following technical conditions:

(H1) $A \in \mathbb{D}^{2,2} \cap L^p(\Omega)$, $DA \in L^p(\Omega; L^2([0,T]))$, and $D^2 A \in L^p(\Omega; L^2([0,T]^2))$, for all $p > 1$.

(H2) $\frac{1}{M_t^T} \in L^p(\Omega)$, for all $p > 1$.

(H3) The term

$$\frac{1}{T^{\frac{1}{2}}} E\sqrt{\int_0^T \phi_s^2 ds}$$

has a finite limit as $T \to 0$.

(H4) (H1) and (H2) hold and the terms

$$E\left(\int_0^T \frac{1}{u_s^2(T-s)}\phi_s \left(\int_s^T D_s\phi_r^2 dr\right) ds\right)$$

and

$$\frac{1}{T^2} E\left(\frac{1}{u_0}\int_0^T \left(\int_s^T D_s\phi_r^2 dr\right)^2 ds\right)$$

are well defined and tend to zero as $T \to 0$, where $u_t := \sqrt{\frac{1}{T-t}\int_t^T \phi_s^2 ds}$ and ϕ is defined as in (10.0.3).

Let us briefly discuss the above hypotheses. (H1) is a regularity condition in the Malliavin calculus sense that guarantees that M^T and $m^T \in \mathbb{L}^{1,2}$. (H2) is an integrability condition that, jointly with (H1), gives us that ϕ is also in $\mathbb{L}^{1,2}$. (H3) is a limit condition on the convergence of the 'volatility swap' corresponding to the underlying asset A. Finally, (H4) are technical conditions that we will need to prove that some terms in our analysis tend to zero.

Under (H1) and (H2), straightforward computations lead to

$$D_s M_t^T = E_t(D_s A),$$
$$D_s m_t^T = E_t(D_s D_t A),$$

and

$$D_s\phi_t = \frac{D_s m_t^T M_t^T - m_t^T D_s M_t^T}{(M_t^T)^2} = \frac{E_t(D_s D_t A)E_t(A) - E_t(D_t A)E_t(D_s A)}{(E_t(A))^2}, \quad (10.1.1)$$

for all $s < t$.

Remark 10.1.1 *In the case of VIX options, the above hypotheses are satisfied under some general regularity conditions on the volatility process that include the case of rough volatilities, as we will see in Section 10.2. In other cases, these hypotheses may not always hold. For example, in the case of realised variance options and under rough volatilities with Hurst parameter H, the term in (H3) tends to infinity, and what we have is that the term*

$$\frac{1}{T^H} E\sqrt{\int_0^T \phi_s^2 ds}$$

has a finite limit as $T \to 0$. Then, the corresponding ATM implied volatility blows up for short maturities (see Alòs, García-Lorite, and Muguruza (2018)).

The following result is a direct consequence of the same arguments in the proof of Theorem 6.5.5 in Chapter 6.

Theorem 10.1.2 *Assume that hypotheses (H1), (H2), and (H3) hold. Then,*

$$\lim_{T\to 0}\left(I_0(k_0^*) - E\sqrt{\frac{1}{T}\int_0^T \phi_s^2 ds} \right) = 0.$$

Proof. The same arguments as in Theorem 6.3.2 lead to

$$V_0 = E_t\left(BS\left(0, \ln(E(A)), k, u_0\right)\right) + \frac{1}{2}E_t\left(\int_0^T \frac{\partial G}{\partial x}(s, \ln(M_s^T), \ln(E(A)), u_s)\left(\phi_s \int_s^T D_s\phi_r^2 dr\right) .ds\right), \quad (10.1.2)$$

where $G := \left(\frac{\partial^2}{\partial x^2} - \frac{\partial}{\partial x}\right) BS$. Now, the proof follows the same ideas as in the analysis of the ATMI level in Chapter 6, so we only sketch it. Notice that

$$\begin{aligned} I_0(k_0^*) &= BS^{-1}(0, \ln(E(A)), k_0^*, V_0) \\ &= BS^{-1}(0, \ln(E(A)), k_0^*, \Gamma_T), \end{aligned}$$

where

$$\begin{aligned} \Gamma_r &= E\left(BS\left(0, \ln(E(A)), k, u_0\right)\right) \\ &+ \frac{1}{2} E\left(\int_0^r \frac{\partial G}{\partial x}(s, \ln(M_s^T), k_0^*, u_s)\left(\phi_s \int_s^T D_s \phi_r^2 dr\right) ds\right). \end{aligned}$$

This allows us to write

$$\begin{aligned} I_0 = E&\left(BS^{-1}(0, \Gamma_0)\right) \\ &+ E \int_0^T (BS^{-1})'(\ln(E(A)), \Gamma_s) \frac{\partial G}{\partial x}(s, \ln(E_s A), k_0^*, u_s) \\ &\left(\phi_s \int_s^T D_s \phi_r^2 dr\right) ds, \end{aligned} \tag{10.1.3}$$

where $(BS^{-1})'$ denotes the first derivative of BS^{-1} with respect to Γ. Now, it is easy to see that (H4) implies that the second term in (10.1.3) tends to zero. For the first term, we can write

$$\begin{aligned} &BS^{-1}(0, \Gamma_0) \\ &= BS^{-1}(0, \Lambda_0) \\ &= E(BS^{-1}(0, \Lambda_0)) \\ &= E(BS^{-1}(0, \Lambda_T) + BS^{-1}(0, \Lambda_0) - BS^{-1}(0, \Lambda_T)) \\ &= E(u_0) + E(BS^{-1}(0, \Lambda_0) - BS^{-1}(0, \Lambda_T)), \end{aligned}$$

where $\Lambda_r := E_r(BS\,(0, \ln(E(A)), k_0^*, u_0))$. As in the proof of Proposition 6.5.1, we can write

$$\Lambda_T = \Lambda_0 + \int_0^T U_s dW_s, \tag{10.1.4}$$

where

$$\begin{aligned} U_s &= E_s\left[D_s\left(BS\left(0, \ln(E(A)), \ln(E(A)), u_0\right)\right)\right] \\ &= E_s\left[AN'(d_+\left(\ln(E(A)), u_0\right)) \frac{\int_s^T D_s \phi_s^2 ds}{2\sqrt{T} u_0}\right]. \end{aligned} \tag{10.1.5}$$

Then, the classical Itô's formula and (10.1.4) lead to

$$
\begin{aligned}
&E\left(BS^{-1}(0,\Lambda_0)\right) - BS^{-1}(0,\Lambda_T) \\
&= -\frac{1}{2}E\left[\int_0^T \left(BS^{-1}\right)''(\ln(E(A)),\Lambda_r)\,U_r^2 dr\right]. \qquad (10.1.6)
\end{aligned}
$$

Now, (H4), 10.1.5, and the fact that

$$
\left(BS^{-1}\right)''(X_0,\Lambda_r) = \frac{BS^{-1}(X_0,\Lambda_r)}{4\left(\exp(X_t)N'(d_+(X_0,BS^{-1}(X_0,\Lambda_r)))\right)^2}
$$

allow us to prove that the last term in (10.1.6) tends to zero, which completes the proof. ■

10.1.2 The short-time skew of the ATMI volatility

Our main goal in this section is to study the short-time limit of the ATM implied volatility skew. We will need the following technical hypotheses:

(H1') (H1) holds, $A \in \mathbb{D}^{3,2}$, and $D^3 A \in L^p(\Omega; L^2([0,T]^3))$ for all $p > 1$.

(H5) There exists $H \in (0,1)$ such that

$$
E\left(\frac{\int_0^T \left(\int_s^T \phi_s D_s \phi_u^2 du\right) ds}{u_0^3 (T-t)^{\frac{3}{2}+H}}\right)
$$

has a finite limit as $T \to 0$.

(H6) The terms

$$
\frac{1}{\sqrt{T}} E\left(\int_0^T (u_s(T-s))^{-3}\left(\int_s^T \Theta_r dr\right)\Theta_s ds\right)
$$

and

$$
\frac{1}{\sqrt{T}} E\left(\int_0^T (u_s(T-s))^{-2}\left(\int_s^T D_s\Theta_r dr\right)\phi_s ds\right)
$$

tend to zero as $T \to 0$, where $\Theta_s := \phi_s \int_s^T D_s \phi_r^2 dr$.

Again, hypothesis (H1') is a Malliavin regularity condition that, jointly with (H2), guarantees that all the Malliavin derivatives in our computations are well defined. The term (H2) is linked to the short-end behaviour of the ATM skew leading term. (H6) imply that the higher-order terms in our analysis tend to zero.

The same arguments as in Chapter 7 allow us to prove the following result for the short-end of the ATM skew.

Theorem 10.1.3 *Consider a random variable A such that hypotheses (H1'), (H2), (H3), (H5), and (H6) hold. Then,*

$$\lim_{T\to 0} T^{\frac{1}{2}-H}\frac{\partial I_t}{\partial k}(k_0^*) = \frac{1}{2}\lim_{T\to 0} E\left(\frac{\int_0^T\left(\int_s^T \phi_s D_s\phi_u^2 du\right)ds}{u_0^3 T^{\frac{3}{2}+H}}\right). \tag{10.1.7}$$

Proof. This proof follows the same ideas as in the analysis of the ATM skew in Chapter 7, so we only sketch it. Taking partial derivatives with respect to k on the expression $V_0 = BS\left(0, \ln(E(A)), k, I_0(k)\right)$ we obtain

$$\begin{aligned}\frac{\partial V_0}{\partial k} &= \frac{\partial BS}{\partial k}(0,\ln(E(A)),k,I_0(k))) \\ &\quad + \frac{\partial BS}{\partial \sigma}(0,\ln(E(A)),k,I_0(k)))\frac{\partial I_0}{\partial k}(k).\end{aligned} \tag{10.1.8}$$

On the other hand, from the decomposition (10.1.2) in the proof of Theorem 10.1.2, we deduce that

$$\begin{aligned}\frac{\partial V_0}{\partial k} &= E_t\left(\frac{\partial BS}{\partial k}(0,\ln(E(A)),k,u_0)\right) \\ &\quad + E\left(\int_0^T \frac{\partial F}{\partial k}(s,\ln(E_s(A))),k,u_s)\Theta_s ds\right),\end{aligned} \tag{10.1.9}$$

where

$$F(s,x,k,\sigma) := \frac{1}{2}\frac{\partial G}{\partial x}(s,x,k,\sigma).$$

This implies that

$$\begin{aligned}&E_t\left(\frac{\partial BS}{\partial k}(0,\ln(E(A)),k,u_0)\right) + E\left(\int_0^T \frac{\partial F}{\partial k}(s,\ln(E_s(A))),k,u_s)\Theta_s ds\right) \\ &= \frac{\partial BS}{\partial k}(0,\ln(E(A)),k,I_0(k))) + \frac{\partial BS}{\partial \sigma}(0,\ln(E(A)),k,I_0(k)))\frac{\partial I_0}{\partial k}(k).\end{aligned} \tag{10.1.10}$$

After some algebra we get

$$E\left(\frac{\partial BS}{\partial k}(0,\ln(E(A)),\ln(E(A)),u_0)-\frac{\partial BS}{\partial k}(t,\ln(E(A)),\ln(E(A)),I_0)\right)$$
$$=\frac{1}{2}E\left(\int_0^T F(s,\ln(E_s(A))),\ln(E(A)),u_s)\Theta_s ds\right).$$

This, jointly with (10.1.10), implies that

$$\frac{\partial I_0}{\partial k}(k_0^*)=\frac{E\left(\int_0^T L(s,\ln(E_s(A))),k_0^*,u_s)\Theta_s ds\right)}{\frac{\partial BS}{\partial\sigma}(0,\ln(E(A)),k_0^*,I_0)},\tag{10.1.11}$$

where $L:=(\frac{1}{2}+\frac{\partial}{\partial k})F$. Now let us apply the Malliavin expansion technique explained in Remark 6.5.6 to study the numerator in (10.1.11). The anticipating Itô's formula allows us to write

$$E\left(\int_0^T L(s,\ln(E_s(A))),k_0^*,u_s)\Theta_s ds\right)$$
$$=E\left(L(0,\ln(E(A)),k_0^*,u_0)\int_0^T\Theta_s ds\right)$$
$$+\frac{1}{2}E\left(\int_0^T\left(\frac{\partial^3}{\partial x^3}-\frac{\partial^2}{\partial x^2}\right)L(s,\ln(E_s(A))),k_0^*,u_s)\left(\int_s^T\Theta_r dr\right)\Theta_s ds\right)$$
$$+E\left(\int_0^T\frac{\partial L}{\partial k}(s,\ln(E_s(A))),k_0^*,u_s)\left(\int_s^T D_s\Theta_r dr\right)\phi_s ds\right)$$
$$=:T_1+T_2+T_3.\tag{10.1.12}$$

Similarly as in the proof of Proposition 7.5.1, straightforward computations, Hypothesis (H6) and the fact that

$$\frac{\partial BS}{\partial\sigma}(0,x,x,\sigma)=\frac{\exp(x)\exp\left(-\frac{\sigma^2T}{8}\right)\sqrt{T}}{\sqrt{2\pi}}\tag{10.1.13}$$

imply that T_2+T_3 tend to zero as $T\to 0$. On the other hand,

$$L(0,\ln(E(A)),k_0^*,u_0)=\frac{1}{2}\frac{A\exp\left(-\frac{(u_0)^2T}{8}\right)}{\sqrt{2\pi}u_0\sqrt{T}}\left[\frac{1}{(u_0)^2T}-\frac{1}{2}\right].\tag{10.1.14}$$

This, jointly with (H5) and (10.1.13), implies that

$$\lim_{T\to 0} T^{\frac{1}{2}-H} T_1 = \frac{1}{2}\lim_{T\to 0} T^{\frac{1}{2}-H} E\left(\frac{\int_0^T \Theta_s ds}{u_0^3 T^2}\right)$$
$$= \frac{1}{2}\lim_{T\to 0} E\left(\frac{\int_0^T \left(\int_s^T \phi_s D_s \phi_u^2 du\right) ds}{u_0^3 T^{\frac{3}{2}+H}}\right). \quad (10.1.15)$$

Now the proof is complete. ■

10.2 VIX OPTIONS

This section is devoted to the application of the results in the previous section to the study of the ATMI level and skew of VIX options. In particular, we compute the ATM short-end skew and we characterise the class of stochastic volatility processes for which this skew is positive, as in real market data (see Figure 10.1).

10.2.1 The short-end level of the ATMI of VIX options

Let us write the result in Theorem 10.1.2 in the case of VIX options. From (10.0.6) we can write

$$\lim_{T\to 0} I_T^{VIX}(0) = \lim_{T\to 0} \frac{1}{T^{\frac{1}{2}}} E\sqrt{\int_0^T (\phi_s^{VIX})^2 ds}. \quad (10.2.1)$$

Now, (10.0.6) allows us to write

$$\lim_{T\to 0} \frac{1}{T^{\frac{1}{2}}} E\sqrt{\int_0^T (\phi_s^{VIX})^2 ds}$$
$$= \lim_{T\to 0} \frac{1}{T^{\frac{1}{2}}} E\sqrt{\int_0^T \left(\frac{1}{2\Delta M_s^{T,VIX}} E_s\left(\frac{1}{VIX_T}\int_T^{T+\Delta} (D_s(\sigma_u^2))du\right)\right)^2 ds}$$
$$= \frac{1}{2\Delta M_0^{0,VIX} VIX_0}\sqrt{\lim_{T\to 0}\frac{1}{T}\int_0^T \left(E_s\int_T^{T+\Delta} D_s\sigma_u^2 du\right)^2 ds}$$
$$= \frac{1}{2\Delta VIX_0^2}\sqrt{\lim_{T\to 0}\frac{1}{T}\int_0^T \left(E_s\int_T^{T+\Delta} D_s\sigma_u^2 du\right)^2 ds}. \quad (10.2.2)$$

As $D_s\sigma_u^2$ does not depend on T, the above limit is a finite quantity for most of the popular stochastic volatility models, as we see in the following examples.

Example 10.2.1 (The SABR model) *Consider the SABR model given by*

$$d\sigma_t = \alpha\sigma_t dW_t,$$

for some positive constant α. Then $\sigma^2 \in \mathbb{L}^{2,2}$ and $D_s\sigma_t^2 = 2\alpha\sigma_t^2$, for all $s < t$. This implies that (H1)–(H4) hold and

$$\begin{aligned}
&\lim_{T\to 0} I_T^{VIX}(0)\\
&= \frac{\alpha}{\Delta VIX_0^2}\sqrt{\lim_{T\to 0}\frac{1}{T}\int_0^T\left(E_s\int_T^{T+\Delta}\sigma_u^2 du\right)^2 ds}\\
&= \frac{\alpha}{VIX_0^2}\sqrt{VIX_0^4}\\
&= \alpha.
\end{aligned}$$

Example 10.2.2 (The Heston model) *Consider a Heston model as in Section 3.3.2. That is, assume that the volatility σ is given by $\sigma_t = \sqrt{\sigma_t^2}$, where*

$$d\sigma_t^2 = k(\theta - \sigma_t^2)dt + \nu\sqrt{\sigma_t^2}dW_t,$$

for some positive constants k, θ, ν. We have seen in Section 3.3.2 that a direct computation leads to $E_r(D_r\sigma_t^2) = \nu\sqrt{\sigma_r^2}\mathbf{1}_{[0,t]}(r)\exp(-k(t-r))$. Then,

$$\begin{aligned}
&\lim_{T\to 0} I_T^{VIX}(0)\\
&= \frac{\nu\sigma_0}{2\Delta VIX_0^2}\sqrt{\left(\int_0^\Delta \exp(-ku)du\right)^2}\\
&= \frac{\nu\sigma_0}{2VIX_0^2}\left(\frac{1-\exp(-k\Delta)}{k\Delta}\right).
\end{aligned}$$

Now, as

$$\begin{aligned}
VIX_0 &= \sqrt{\frac{1}{\Delta}E\int_0^\Delta \sigma_s^2 ds}\\
&= \sqrt{\theta + (\sigma_0^2 - \theta)\frac{1}{\Delta}\int_0^\Delta e^{-ks}ds}\\
&= \sqrt{\theta + (\sigma_0^2 - \theta)\frac{(1-\exp(-k\Delta))}{\Delta k}} \qquad (10.2.3)
\end{aligned}$$

we get the following explicit expression for the ATM short-end level of the VIX implied volatility

$$\lim_{T\to 0} I_T^{VIX}(0) = \frac{\nu\sigma_0\left(\frac{1-\exp(-k\Delta)}{k\Delta}\right)}{2\sqrt{\theta+(\sigma_0^2-\theta)\frac{(1-\exp(-k\Delta))}{\Delta k}}}.$$

Now, if $k\Delta$ is small enough, $\frac{1-\exp(-k\Delta)}{k\Delta}\approx 1$ and the above limit reads as

$$\lim_{T\to 0} I_T^{VIX}(0) \approx \frac{\nu}{2}.$$

Example 10.2.3 (Fractional volatilities) *Consider the rough Bergomi model as in Section 5.6.2, where the volatility σ is defined as*

$$\sigma_r^2 = \sigma_0^2 \exp\left(\nu\sqrt{2H}Z_r - \frac{1}{2}\nu^2 r^{2H}\right), r\in[0,T], \tag{10.2.4}$$

where

$$Z_r := \int_0^r (r-s)^{H-\frac{1}{2}} dW_s,$$

and for some positive real values σ_0^2, ν and for some $H<\frac{1}{2}$. We know (see Section 5.6.2) that

$$D_s\sigma_r^2 \quad = \quad \nu\sqrt{2H}\sigma_r^2(r-s)^{H-\frac{1}{2}}. \tag{10.2.5}$$

Then,

$$\begin{aligned}\lim_{T\to 0} I_T^{VIX}(0) &= \frac{\nu\sqrt{2H}\sigma_0^2}{2\Delta VIX_0^2}\sqrt{\left(\int_0^\Delta u^{H-\frac{1}{2}}du\right)^2} \\ &= \frac{\nu\sqrt{2H}\sigma_0^2\Delta^{H-\frac{1}{2}}}{2(H+\frac{1}{2})VIX_0^2}.\end{aligned}$$

Now, as $VIX_0^2 = \frac{1}{\Delta}E\int_0^\Delta \sigma_s^2 ds = \sigma_0^2$ the above limit reads as

$$\lim_{T\to 0} I_T^{VIX}(0) \quad = \quad \frac{\nu\sqrt{2H}\Delta^{H-\frac{1}{2}}}{2(H+\frac{1}{2})}.$$

10.2.2 The ATM skew of VIX options

The empirical VIX skew is positive, while some popular models are not able to replicate this property. For example, this VIX skew is negative for the Heston model, while it is flat for the SABR (see for example Fouque and Saporito (2015), Goutte, Ismail, and Pham (2017), de Marco (2018), or Guyon (2018)).

Our purpose in this section is to establish the class of stochastic volatility models that can reproduce a positive VIX skew. We have seen in Theorem 10.1.3 that

$$\lim_{T\to 0} T^{H-\frac{1}{2}} \frac{\partial I_t}{\partial k}(k_0^*) = \frac{1}{2} \lim_{T\to 0} E\left(\frac{\int_0^T \left(\int_s^T \phi_s D_s \phi_u^2 du \right) ds}{u_0^3 T^{\frac{3}{2}+H}} \right). \tag{10.2.6}$$

Notice that $D_s\phi_t^2 = 2\phi_t D_s\phi_t$. Then, in the short-end, $\phi_s D_s \phi_t^2 = 2\phi_s\phi_t D_s\phi_t \approx 2\phi_t^2 D_s\phi_t$ and the limit (10.2.6) is positive if $D_s\phi_u$ is positive for small values of s and u. Moreover, because of (10.1.1),

$$D_s\phi_t = \frac{D_s m_t^T M_t^T - m_t^T D_s M_t^T}{(M_t^T)^2},$$

which is positive if and only if $D_s m_t^T M_t^T - m_t^T D_s M_t^T$ is positive. This gives us a tool to identify those stochastic volatility models that can reproduce a short-time positive skew, as we see in the following examples.

Example 10.2.4 (The VIX skew in the SABR model) *Consider the SABR model given by $v = \sigma^2$, where*

$$d\sigma_t = \alpha\sigma_t dW_t,$$

for some positive constant α. Then, $v \in \mathbb{L}^{2,2}$ and $D_s\sigma_t^2 = 2\alpha\sigma_t^2$, $D_s D_r \sigma_t^2 = 4\alpha^2\sigma_t^2$, for all $r, s < t$. Notice that, as σ_t is log-normal, $E(\sigma_t^2) = \sigma_0^2 \exp(\alpha^2 t)$. Then,

$$\lim_{t,T-t\to 0} M_t^{T,VIX} = \frac{\sqrt{e^{\alpha^2\Delta} - 1}\sqrt{\sigma_0^2}}{\alpha\sqrt{\Delta}},$$

$$\lim_{s,t,T-t\to 0} D_s M_t^{T,VIX} = \frac{(e^{\alpha^2\Delta} - 1)\sigma_0^2}{VIX_0 \alpha\Delta} = \frac{\sqrt{e^{\alpha^2\Delta} - 1}\sqrt{\sigma_0^2}}{\sqrt{\Delta}},$$

and

$$
\begin{aligned}
& D_s m_t^{T,VIX} \\
&= \frac{1}{2\Delta} E_t \left(\frac{1}{VIX_T} E_T \int_T^{T+\Delta} (D_s D_t \sigma_r^2) dr \right) \\
&- \frac{1}{4\Delta} E_t \left(\frac{1}{\Delta (VIX_T)^3} E_T \left(\int_T^{T+\Delta} (D_t \sigma_r^2) dr \right) \left(E_T \int_T^{T+\Delta} (D_s \sigma_r^2) dr \right) \right) \\
&\to \frac{4\alpha^2 \sigma_0^2}{2VIX_0 \Delta} \frac{e^{\alpha^2 \Delta} - 1}{\alpha^2} - \frac{1}{4VIX_0^3 \Delta^2} \frac{4\alpha^2 \sigma_0^4 (e^{\alpha^2 \Delta} - 1)^2}{\alpha^4} \\
&= \frac{2\alpha \sqrt{\sigma_0^2} \sqrt{e^{\alpha^2 \Delta} - 1}}{\sqrt{\Delta}} - \frac{\alpha \sqrt{\sigma_0^2} \sqrt{e^{\alpha^2 \Delta} - 1}}{\sqrt{\Delta}} \\
&= \frac{\alpha \sqrt{\sigma_0^2} \sqrt{e^{\alpha^2 \Delta} - 1}}{\sqrt{\Delta}}
\end{aligned} \tag{10.2.7}
$$

in $L^p(\Omega)$, for all $p > 1$, from where we deduce that

$$
\begin{aligned}
& D_s m_t^{T,VIX} M_t^{T,VIX} - m_t^{T,VIX} D_s M_t^{T,VIX} \\
&\to \alpha \frac{\sqrt{\sigma_0^2} \sqrt{e^{\alpha^2 \Delta} - 1}}{\sqrt{\Delta}} \frac{\sqrt{e^{\alpha^2 \Delta} - 1} \sqrt{\sigma_0^2}}{\alpha \sqrt{\Delta}} - \left(\frac{\sqrt{e^{\alpha^2 \Delta} - 1} \sqrt{\sigma_0^2}}{\sqrt{\Delta}} \right)^2 = 0.
\end{aligned} \tag{10.2.8}
$$

That is, the SABR model generates a short-time flat VIX skew. We can observe this behaviour in Figure 10.2, corresponding to the VIX implied volatility with a maturity equal to 0.1 year for a SABR model with $\sigma_0 = 0.5$ and $\nu = 0.7$.

Example 10.2.5 (The sign of the VIX skew in the Heston model) For the Heston model, we have that

$$
d\sigma_t^2 = \kappa(\theta - \sigma_t^2) dt + \nu \sqrt{\sigma_t^2} dW_t,
$$

for some positive constants k, θ, and ν. We know (see Section 3.3.2 and Proposition 4.2 in Alòs and Ewald (2008)) that if $2k\theta > 3\nu^2$, the process $\sqrt{\sigma_t^2} \in \mathbb{L}^{2,2}$. Assume, for the sake of simplicity, that $\sigma_0^2 = \theta$. Notice that, for all $T < r$,

$$
E_T(\sigma_r^2) = \theta + (v_T - \theta) \exp(-k(r - T)).
$$

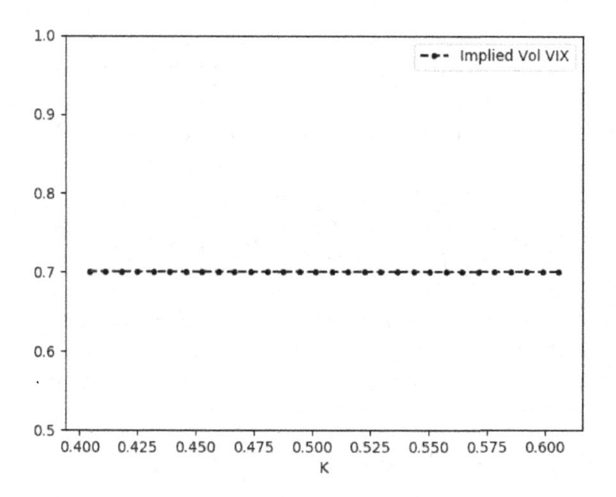

Figure 10.2 VIX implied volatility with maturity 0.1Y, corresponding to a SABR model with $T = 0.1, \sigma_0 = 0.5$, and $\nu = 0.7$.

This implies that, for all $t < T$

$$\begin{aligned} M_t^{T,VIX} &= E_t \sqrt{E_T \frac{1}{\Delta} \int_T^{T+\Delta} \sigma_r^2 dr} \\ &= E_t \sqrt{\theta + (v_T - \theta) \frac{1}{\Delta} \int_T^{T+\Delta} \exp(-k(r-T)) dr} \\ &= E_t \sqrt{\theta + (v_T - \theta) \frac{1 - \exp(-k\Delta)}{k\Delta}} \\ &\to E \sqrt{\theta + (\sigma_0^2 - \theta) \frac{1 - \exp(-k\Delta)}{k\Delta}} \end{aligned} \quad (10.2.9)$$

in $L^2(\Omega)$, as $T \to 0$. We know (see again Section 3.3.2) that $\sigma^2 \in \mathbb{D}^{1,2}$. Then, for all $s < t < T$

$$D_s M_t^{T,VIX} = E_t \left[\frac{1}{2VIX_T} E_T \left(\frac{1}{\Delta} \int_T^{T+\Delta} D_s \sigma_r^2 dr \right) \right].$$

The computations in Section 3.3.2 give us that

$$D_s \sigma_r^2 = \nu \exp(-k(r-s)) \sqrt{\sigma_s^2} + \nu \int_s^r \exp(-k(r-u)) D_s \sqrt{v_u} dW u. \quad (10.2.10)$$

Then, we get

$$D_s M_t^{T,VIX} \to \frac{\nu(1-\exp(-k\Delta))}{2k\Delta} \tag{10.2.11}$$

in $L^p(\Omega)$, for all $p > 1$, as $T \to 0$. On the other hand, (10.0.5) allows us to write

$$\begin{aligned} D_s m_t^{T,VIX} &= \frac{1}{2\Delta} E_t \left(\frac{1}{VIX_T} E_T \int_T^{T+\Delta} (D_s D_t \sigma_r^2) dr \right) \\ &- \frac{1}{4\Delta} E_t \left(\frac{1}{\Delta (VIX_T)^3} E_T \left(\int_T^{T+\Delta} (D_t \sigma_r^2) dr \right) \left(E_T \int_T^{T+\Delta} (D_s \sigma_r^2) dr \right) \right) \\ &=: T_1 + T_2. \end{aligned} \tag{10.2.12}$$

Let us first study the term T_1. It is easy to deduce from (10.2.10) that, for $s < t < r$

$$\begin{aligned} D_s D_t \sigma_r^2 &= \nu \exp(-k(r-t)) D_s \sqrt{\sigma_t^2} \\ &+ \nu \int_t^r \exp(-k(r-u)) D_t D_s \sqrt{v_u} dW u. \end{aligned} \tag{10.2.13}$$

The results in Section 3.3.2 give us that, as $s, T - t \to 0$, $D_s \sigma_t \to \frac{\nu}{2}$ in $L^p(\Omega)$, for all $p > 1$. This implies that

$$T_1 \to \frac{\nu^2(1-\exp(-k\Delta))}{4k\Delta\sqrt{\sigma_0^2}}. \tag{10.2.14}$$

Finally, we can write

$$T_2 \to -\frac{\nu^2(1-\exp(\Delta))^2}{4k^2\Delta^2\sqrt{\sigma_0^2}}. \tag{10.2.15}$$

Then (10.2.11), (10.2.12), (10.2.14), and (10.2.15) lead to

$$\begin{aligned} & D_s m_t^{T,VIX} M_t^{T,VIX} - m_t^{T,VIX} D_s M_t^{T,VIX} \\ & \to \frac{\nu^2(1-\exp(-k\Delta))}{4k\Delta} - \frac{\nu^2(1-\exp(-k\Delta))^2}{4k^2\Delta^2} \\ & - \left(\frac{\nu(1-\exp(-k\Delta)}{2k\Delta} \right)^2 \\ & = \frac{\nu^2(1-\exp(-k\Delta))}{4k\Delta} - 2\frac{\nu^2(1-\exp(-k\Delta))^2}{4k^2\Delta^2} \\ & = \frac{\nu^2(1-\exp(-k\Delta))}{4k\Delta} \left(1 - \frac{2(1-\exp(-k\Delta))}{k\Delta} \right), \end{aligned} \tag{10.2.16}$$

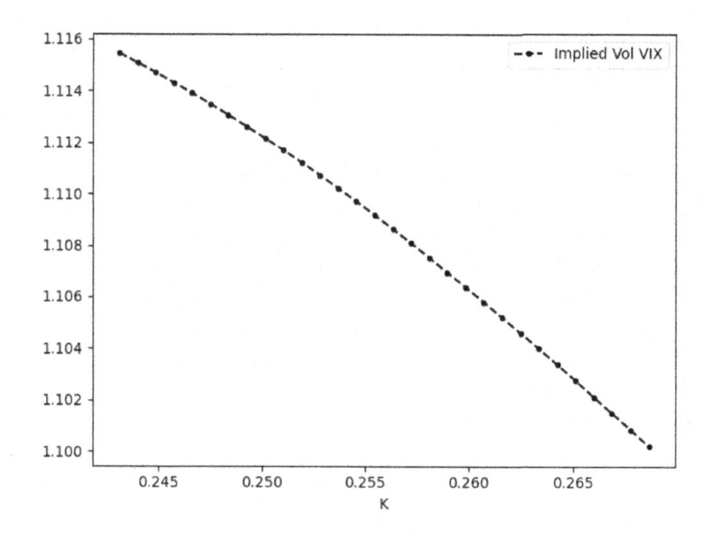

Figure 10.3 VIX implied volatility with maturity 0.1Y, corresponding to a Heston model with parameters $\sigma_0 = \sqrt{0.05}, \nu = 0.8, k = 3, \theta = 0.09$, and $\rho = -0.6$.

in $L^2(\Omega)$, as $T \to 0$. Under reasonable values of k, $f(\Delta)$ is negative and then the corresponding VIX skew is negative. We can observe the typical behaviour of the Heston VIX skew in Figure 10.3, corresponding to the VIX skew with maturity $T = 0.1$ for a Heston model with parameters $\sigma_0 = \sqrt{0.05}, \nu = 0.8, k = 3$, and $\theta = 0.09$. Notice that the Heston VIX skew can be positive if the model does not satisfy the Feller condition (see also Kokholm and Stisen (2015)), or for big values of the mean reversion k.

Example 10.2.6 (The VIX skew for a mixed Bergomi model) Consider a model given by

$$\sigma_t^2 = \sigma_0^2(\delta\mathcal{E}(\nu W_t) + (1-\delta)\mathcal{E}(\eta W_t))$$

for some positive constants $\delta \in [0,1]$ and $\nu, \eta > 0$ and where

$$\mathcal{E}(\Psi) = \exp(\Psi - \frac{1}{2}E(|\Psi|^2)).$$

In this model, the volatility is a weighted mean of two lognormal volatilities (see 2.5.1). A direct computation gives us that

$$D_s\sigma_u^2 = \sigma_0^2(\delta\nu\mathcal{E}(\nu W_u) + (1-\delta)\eta\mathcal{E}(\eta W_u))$$

and

$$D_r D_s \sigma_u^2 = \sigma_0^2(\delta \nu^2 \mathcal{E}(\nu W_u) + (1-\delta)\eta^2 \mathcal{E}(\eta W_u)),$$

for all $s < r < u$. Then, it is easy to see that

$$\lim_{t,T-t\to 0} M_t^{T,VIX} = \sigma_0,$$

$$\lim_{s,T-t\to 0} D_s M_t^{T,VIX} = \frac{\sigma_0}{2}(\delta\nu + (1-\delta)\eta),$$

and

$$\begin{aligned} \lim_{s,T-t\to 0} D_s m_t^{T,VIX} &= \frac{\sigma_0}{2}(\delta\nu^2 + (1-\delta)\eta^2) \\ &- \frac{\sigma_0}{4}(\delta\nu + (1-\delta)\eta)^2 \end{aligned} \tag{10.2.17}$$

in $L^p(\Omega)$, for all $p > 1$. Then,

$$\begin{aligned} & D_s m_t^{T,VIX} M_t^{T,VIX} - m_t^{T,VIX} D_s M_t^{T,VIX} \\ & \to \frac{\sigma_0^2}{2}\left((\delta\nu^2 + (1-\delta)\eta^2) - (\delta\nu + (1-\delta)\eta)^2\right) \\ & = \frac{\sigma_0^2}{2}\left(\delta\nu^2 + (1-\delta)\eta^2 - \delta^2\nu^2 - (1-\delta)^2\eta^2 - 2\delta(1-\delta)\nu\eta\right) \\ & = \frac{\sigma_0^2}{2}\left((\delta-\delta^2)\nu^2 + ((1-\delta) - (1-\delta)^2)\eta^2 - 2\delta(1-\delta)\nu\eta\right) \\ & = \frac{\sigma_0^2}{2}(1-\delta)\left(\delta\nu^2 + \delta\eta^2 - 2\delta\nu\eta\right) \\ & = \frac{\sigma_0^2}{2}(1-\delta)\delta\,(\nu-\eta)^2. \end{aligned} \tag{10.2.18}$$

If $\delta \neq 0, 1$ and $\nu \neq \eta$, the above limit is positive. This allows us to reproduce a positive VIX skew, according to the results in Bergomi (2008), where the idea of mixing several log-normal random variables with positive weights was proposed. We can observe this behaviour in Figure 10.4, corresponding to the VIX skew for a mixed Bergomi model with $\nu = 0.7, \eta = 0.8$, and $\delta = 0.5, 1$. Notice that the VIX skew is flat for the case $\delta = 1$ and positive for $\delta = 0.5$.

Example 10.2.7 (Fractional volatilities) *Consider again the rough Bergomi model as in Section 5.6.2, where the volatility σ is defined as*

$$\sigma_r^2 = \sigma_0^2 \exp\left(\nu\sqrt{2H} Z_r - \frac{1}{2}\nu^2 r^{2H}\right), r \in [0,T], \tag{10.2.19}$$

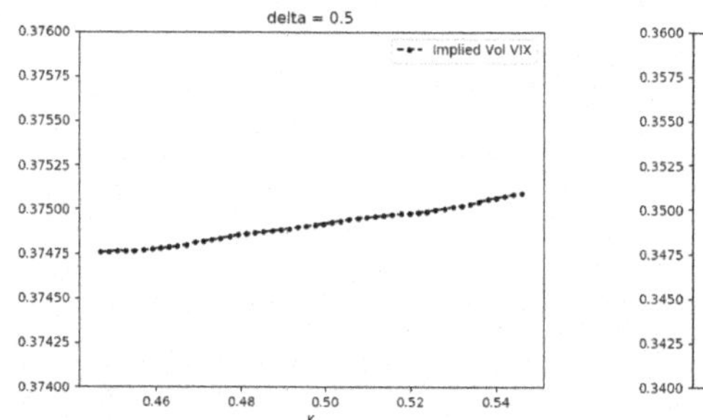

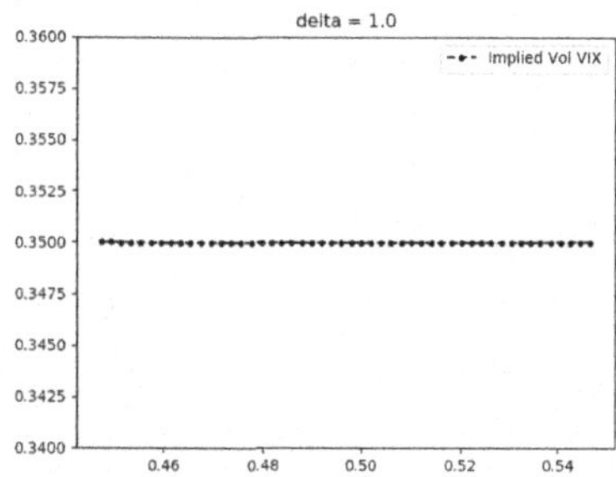

Figure 10.4 VIX implied volatility with maturity 0.1Y, corresponding to a mixed Bergomi model with parameters $\nu = 0.7, \eta = 0.8$, and $\delta = 0.5$ (left) and $\delta = 1$ (right).

with

$$Z_r := \int_0^r (r-s)^{H-\frac{1}{2}} dW_s,$$

and for some positive real values σ_0^2, ν *and for some* $H < \frac{1}{2}$. *Notice that* $E(\sigma_r^2) = \sigma_0^2$, *which implies that*

$$\lim_{t,T-t\to 0} M_t^{T,VIX} = \sigma_0.$$

Moreover, we know (see Section 5.6.2) that

$$D_s\sigma_r^2 \quad = \quad \nu\sqrt{2H}\sigma_r^2(r-s)^{H-\frac{1}{2}}. \tag{10.2.20}$$

Then,

$$\lim_{s,t,T-t\to 0} D_s M_t^{T,VIX} = \frac{\nu\sqrt{2H}\sigma_0^2\Delta^{H-\frac{1}{2}}}{2VIX_0(H+\frac{1}{2})} = \frac{\nu\sqrt{2H}\sigma_0\Delta^{H-\frac{1}{2}}}{2(H+\frac{1}{2})},$$

and

$$\begin{aligned}
D_s m_t^{T,VIX} &= \frac{1}{2\Delta} E_t\left(\frac{1}{VIX_T} E_T \int_T^{T+\Delta} (D_s D_t \sigma_r^2) dr\right) \\
&- \frac{1}{4\Delta} E_t\left(\frac{1}{\Delta(VIX_T)^3} E_T\left(\int_T^{T+\Delta} (D_t\sigma_r^2) dr\right)\left(E_T\int_T^{T+\Delta} (D_s\sigma_r^2) dr\right)\right) \\
&\to \frac{\nu^2\Delta^{2H-1}\sigma_0^2}{2VIX_0} - \frac{\nu^2 H\Delta^{2H-1}\sigma_0^4}{2(H+\frac{1}{2})^2 VIX_0^3}.
\end{aligned} \tag{10.2.21}$$

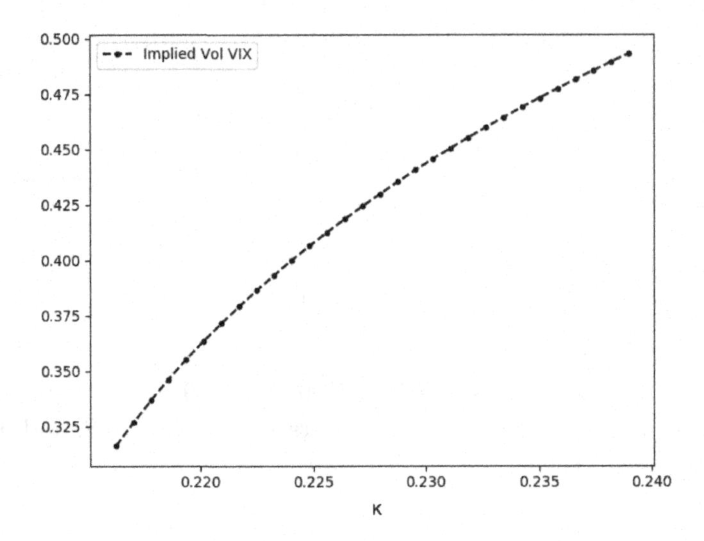

Figure 10.5 VIX implied volatility with maturity 0.1Y, corresponding to a rBergomi model with parameters $H = 0.3, \nu = 0.5$, and $\sigma_0 = \sqrt{0.05}$.

As $VIX_0 = \sigma_0$, it follows that

$$\begin{aligned} D_s m_t^{T,VIX} &\to \frac{\nu^2 \Delta^{2H-1}\sigma_0}{2} - \frac{\nu^2 H \Delta^{2H-1}\sigma_0}{2(H+\frac{1}{2})^2} \\ &= \frac{\nu^2 \Delta^{2H-1}\sigma_0}{2}\left(1 - \frac{H}{(H+\frac{1}{2})^2}\right). \end{aligned} \tag{10.2.22}$$

Then,

$$\begin{aligned} &D_s m_t^{T,VIX} M_t^{T,VIX} - m_t^{T,VIX} D_s M_t^{T,VIX} \\ &\to \frac{\nu^2 \Delta^{2H-1}\sigma_0^2}{2}\left(1 - \frac{H}{(H+\frac{1}{2})^2}\right) - \frac{\nu^2 H \sigma_0^2 \Delta^{2H-1}}{2(H+\frac{1}{2})^2} \\ &= \frac{\nu^2 \Delta^{2H-1}\sigma_0^2}{2}\left(1 - \frac{2H}{(H+\frac{1}{2})^2}\right). \end{aligned} \tag{10.2.23}$$

Notice that this limit is zero if $H = \frac{1}{2}$, while it is positive for any other value of $H \in (0,1)$. We can observe this behaviour in Figure 10.5, corresponding to the VIX skew for a rBergomi model with $H = 0.3, \nu = 0.5$, and $\sigma_0 = \sqrt{0.05}$.

10.3 CHAPTER'S DIGEST

The Clark-Ocone-Haussman formula allows us to study options on the VIX as options on a stock following a stochastic volatility model, where the volatility process may depend on maturity time T. Then, the same approach as in Chapters 6 and 7 can be applied to study the short-time behaviour of the ATMI level and skew for options on the VIX. In particular, we establish a criterion, based on Malliavin calculus, that identifies the class of models that are able to reproduce a positive VIX skew, as observed in real market data. As particular examples, we see that this skew is flat for the SABR model, negative for the Heston (under reasonable parameters), and positive for the mixed Bergomi and the rough Bergomi models.

Bibliography

[1] Ahn, D. and Gao, B. (1999). A parametric nonlinear model of term structure dynamics. Review of Financial Studies, 12, pp. 721–762.

[2] Aldrich, E. M., Heckenbach, I., and Laughlin, G. (2015). The random walk of high frequency trading, Working Paper, No. 722, University of California, Economics Department, Santa Cruz, CA, 113–123

[3] Alexander, C. and Nogueira, L. (2004). Hedging with stochastic local volatility. ICMA Centre Discussion Papers in Finance icma-dp2004-10, Henley Business School, Reading University, revised Dec 2004.

[4] Algieri, B. (2018). A journey through the history of commodity derivatives markets and the political economy of (de)regulation, ZEF Discussion Papers on Development Policy, No. 268, University of Bonn, Center for Development Research (ZEF), Bonn.

[5] Alòs, E. (2006). A generalization of the Hull and White formula with applications to option pricing approximation. Finance Stoch, 10, pp. 353–365.

[6] Alòs, E. and Ewald, C. (2008). Malliavin differentiability of the Heston volatility and applications to option pricing. Advances in Applied Probability, 40(1), pp. 144-162.

[7] Alòs, E. and Fukasawa, M. (2021). The asymptotic expansion of the regular discretization error of Itô's integrals. Mathematical Finance, 31(1), pp. 1–43.

[8] Alòs, E., García-Lorite, D., Muguruza, A. (2018). On smile properties of volatility derivatives and exotic products: understanding the VIX skew. Preprint available at arXiv:1808.03610, 2018.

[9] Alòs, E., Gatheral, J., and Radoičić, R. (2020). Exponentiation of conditional expectations under stochastic volatility. Quantitative Finance, 20(1), pp. 13–27.

[10] Alòs, E., Jacquier, A., and León, J.A. (2019). The implied volatility of Forward-Start options: ATM short-time level, skew and curvature, Stochastics, 91(1), pp. 37–51

[11] Alòs, E. and León, J. (2013). On the closed-form approximation of short-time random strike options. Working Paper 1347, UPF.

[12] Alòs, E. and León, J. (2016). On the short-maturity behaviour of the implied volatility skew for random strike options and applications to option pricing approximation. Quantitative Finance, 16(1), pp. 31–42.

[13] Alòs, E. and León, J. (2017). On the curvature of the smile in stochastic volatility models. SIAM Journal on Financial Mathematics, 8, pp. 373–399

[14] Alòs, E., León, J., and Vives, J. (2007). On the short-time behavior of the implied volatility for jump-diffusion models with stochastic volatility. Finance Stoch, 11, pp. 571–589.

[15] Alòs, E., León, J., Pontier, M., and Vives, J. (2008). A Hull and White formula for a general stochastic volatility jump-diffusion model with applications to the study of the short-time behavior of the implied volatility. Journal of Applied Mathematics and Stochastic Analysis. Article ID 359142.

[16] Alòs, E., Mazet, O., and Nualart, D. (2001). Stochastic calculus with respect to Gaussian processes. Annals of Probability, 29(2), pp. 766–801.

[17] Alòs, E. and Muguruza, A. (2020). A novel volatility swap representation formulae for stochastic volatility models. Work in progress.

[18] Alòs, E. and Yang, Y. (2017). A fractional Heston model with $H > \frac{1}{2}$. Stochastics, 89(1), pp. 384–399.

[19] Andersen, L. (2008). Simple and efficient simulation of the Heston stochastic volatility model. Journal of Computational Finance, 11(3), pp. 1+.

[20] Andersen, T. and Bollerslev, T. (1997). Heterogeneous information arrivals and returns volatility dynamics. Journal of Finance, 52, pp. 975–1005.

[21] Andersen, T.G., Bollerslev, T., Diebold, F.X., and Labys, P. (2003). Modeling and forecasting realized volatility. Econometrica, 71, pp. 579-625.

[22] Anis, A. A. and Lloyd, E. H. (1976). The Expected Value of the Adjusted Rescaled Hurst Range of Independent Normal Summands. Biometrika, 63(1), pp. 111-116

[23] Bachelier, L. (1900). Théorie de la spéculation. Annales scientifiques de l'Ecole Normale Supérieure , Sér. 3(17), pp. 21–86.

[24] Baillie, T. and Bollerslev, R.T. (1989). The message in daily exchange rates: a conditional variance tale. Journal of Business and Economic Statistics, 7(3), pp. 297–305

[25] Baldeaux, J. and Badran, A. (2014). Consistent modelling of VIX and equity derivatives using a 3/2 plus jumps model. Applied Mathematical Finance, 21(4), pp. 299–312.

[26] Banna, O., Mishura, Y., Ralchenko, K., and Shklyar, S. (2019). Fractional Brownian Motion: Approximations and Projections. ISTE Ltd. and Wiley.

[27] Bayer, C., Friz, P., and Gatheral, J. (2016). Pricing under rough volatility. Quantitative Finance, 16(6), pp. 887–904.

[28] Barndorff-Nielsen, O.E., Hansen, P.R., Lunde, A., Shephard, N. (2008). Designing realised kernels to measure the ex-post variation of equity prices in the presence of noise. Econometrica, 6, pp. 1481–1536.

[29] Barndorff-Nielsen, O.E., and Shephard, N. (2001). Modelling by Lévy processes for financial econometrics. In: Barndorff-Nielsen, O.E., Mikosch, T., and Resnick, S.I. (eds.) Lévy Processes: Theory and Applications, pp. 283–318. Birkhäuser, Basel.

[30] Barndorff-Nielsen, O.E., Shephard, N. (2002). Econometric analysis of realized volatility and its use in estimating stochastic volatility models. Journal of the Royal Statistical Society Series B (Statistical Methodology), 64, pp. 253–280.

[31] Bates, D.S. (1996). Jumps and stochastic volatility: exchange rate processes implicit in Deutsche Mark options. The Review of Financial Studies, 9, pp. 69–107.

[32] Benhamou, E. (January 2002). Malliavin calculus for Monte Carlo methods in finance. LSE Working Paper, Available at SSRN: https://ssrn.com/abstract=298084 or http://dx.doi.org/10.2139/ssrn.298084.

[33] Bennedsen, M., Lunde, A., and Pakkanen, M.S. (2017). Hybrid scheme for Brownian semistationary processes. Finance and Stochastics, 21(4), pp. 931–965.

[34] Bentata, A., and Cont (2009). Mimicking the marginal distributions of a semimartingale, Working Paper, arXiv:0910.3992v2.

[35] Benth, F.E., Di Nunno, G., Lokka, A., Oksendal, B., and Proske, F. (2003). Explicit representation of the minimal variance portfolio in markets driven by Levy processes. Available at SSRN: https://ssrn.com/abstract=371398

[36] Bergomi, L. (2005). Smile dynamics II. Risk October, pp. 67–73.

[37] Bergomi, L. (2009a). Smile dynamics III. Risk March, pp. 90–96.

[38] Bergomi, L. (2009b). Smile dynamics IV. Risk December, pp. 94–100.

[39] Bergomi, L. (2016). Stochastic Volatility Modeling. Chapman and Hall, CRC.

[40] Bergomi, L. Guyon, L. (2012). Stochastic volatility's orderly smiles. Risk May, pp. 60–66.

[41] Bernard, C., and Cui, Z. (2014). Prices and asymptotics for discrete variance swaps. Applied Mathematical Finance, 21(2), 140–173.

[42] Biagini, F., Hu, Y., Øksendal, B. and Zhang, T. (2008). Stochastic Calculus for Fractional Brownian Motion and Applications. Probability and its Applications. Springer-Verlag London, Ltd., London.

[43] Bollerslev, T. (1987). A conditionally heteroskedastic time series model for speculative prices and rates of return. The Review of Economics and Statistics, 69(3), pp. 542–547.

[44] Bossy, M. and Diop, A. (2004). An efficient discretization scheme for one dimensional SDEs with a diffusion coefficient function of the form $|x|^\alpha$, $\alpha \in [1/2, 1)$. INRIA working paper no. 5396.

[45] Breeden, D. and Litzenberger, R. (1978). Prices of state contingent claims implicit in option prices. Journal of Business, 51, pp. 621–651.

[46] Borodin, A.N. (2000). Versions of the Feynman-Kac formula. Journal of Mathematical Sciences, 99(2), pp. 1044–1052.

[47] Breidt, F.J., Crato, N., and de Lima, P. (1998). The detection and estimation of long memory in stochastic volatility. Journal of Econometrics, 83, pp. 325–348.

[48] Broadie, M. and Glasserman, P. (1996). Estimating security price derivatives using simulation. Management Science 42, pp. 269–285.

[49] Broadie, M., and A. Jain (2008). The effect of jumps and discrete sampling on volatility and variance swaps. International Journal of Theoretical and Applied Finance, 11(8), pp. 761–797.

[50] Bronzin, V. (1908). Theorie der Prämiengeschäfte; Franz Deuticke, Leipzig/Vienna.

[51] Brockhaus, O. and Long, D. (2000). Volatility swaps made simple. Risk Magazine, January.

[52] Brunick, G. and Shreve, S. (2013). Mimicking an Itô's process by a solution of a stochastic differential equation. Annals of Applied Probability, 23(4), pp. 1584–1628.

[53] Bühler, H. (2002). Applying stochastic volatility models for pricing and hedging derivatives. quantitative-research.de/dl/021118SV.pdf.

[54] Cai, N., Song, Y., and Chen, N. (2017). Exact simulation of the SABR model. Operations Research, 65(4), pp. 931–951.

[55] Cai, J., Fukasawa, M., Rosenbaum, M., and Tankov, P. (2016). Optimal discretization of hedging strategies with directional views. SIAM Journal on Financial Mathematics, 7(1), pp. 34–69.

[56] Carr, P. and Itkin, A. (2019). ADOL: Markovian approximation of a rough lognormal model. Risk.net, November.

[57] Carr, P. and Lee, R. (2008). Robust replication of volatility derivatives. In PRMIA award for Best Paper in Derivatives, MFA 2008 Annual Meeting.

[58] Carr, P., Geman, H., Madan, D.B., and Yor, M. (2005). Pricing options on realized variance. Finance Stochast, 9, pp. 453–475.

[59] Carr, P., Lee, R., and Loring, M. (2019). Pricing variance swaps on time-changed markov processes. arXiv:1705.01069v3 [q-fin.MF]

[60] Carr, P. and Madan, D. (1998). Towards a theory of volatility trading. In: Robert A. Jarrow (ed), Volatility: New Estimation Techniques for Pricing Derivatives, pp. 417–427. London: RISK Publications.

[61] Carr, P. and Madan, D. (2014). Joint modeling of VIX and SPX options at a single and common maturity with risk management applications. IIE Transactions, 46(11), pp. 1125–1131.

[62] Carr, P. and Wu, L. (2003), What type of process underlies options? A simple robust test. The Journal of Finance, 58, pp. 2581–2610.

[63] Chang, P. Fourier instantaneous estimators and the Epps effect. Plos One, 15(9): e0239415.

[64] Chen, R. (2019). The fourier transform method for volatility functional inference by asynchronous observations. Working Paper, arXiv:1911.02205.

[65] Cheridito, P. (2003). Arbitrage in fractional Brownian motion models. Finance Stochast, 7, pp. 533–553.

[66] Comte, F., Coutin, L., and Renault, E. (2012). Affine fractional stochastic volatility models. Annals of Finance, pp. 337–378.

[67] Comte, F. and Renault, E. (1998). Long memory in continuous-time stochastic volatility models Mathematical Finance, 8(4), pp. 291–323.

[68] Cont, R. (2005). Long range dependence in financial markets. In: Lévy-Véhel, J. and Lutton, E. (eds) Fractals in Engineering. Springer, London.

[69] Cont, R. (2007). Volatility clustering in financial markets: empirical facts and agent-based models. In: Teyssière, G. and Kirman, A.P. (eds) Long Memory in Economics. Springer, Berlin, Heidelberg.

[70] Cont, R. and Tankov, P. (2008). Financial modelling with Jump Processes. 2nd edition. Chapman and Hall/CRC Press.

[71] Cox, J.C. (1996). The constant elasticity of variance option pricing model. The Journal of Portfolio Management. A Tribute to Fischer Black, 23(5), pp. 15–17.

[72] Crownover, R.M. (1995). Introduction to Fractals and Chaos. Jones and Barlett Publishers.

[73] Czichowsky, C., Peyre, R., Schachermayer, W., and Yang, J. (2018). Shadow prices, fractional Brownian motion, and portfolio optimisation under transaction costs. Finance Stoch, 22, pp. 161–180.

[74] Dai, W. and Heyde, C.C (1996). It^o's formula with respect to fractional Brownian motion and its application. Journal of Applied Mathematics and Stochastic Analysis, 9, pp. 439–448.

[75] Da Prato, G. (2014). Introduction to Stochastic Analysis and Malliavin Calculus (Vol. 13). Springer.

[76] De Marco, E. (2018). VIX derivatives in rough forward variance models. Presentation, Bachelier Congress, Dublin.

[77] Decreusefond, L. and Ustünel, A. S. (1998). Fractional Brownian motion: theory and applications. ESAIM: Proceedings, 5, pp. 75–86.

[78] Decreusefond, L. and Ustünel, A.S. (1999). Stochastic analysis of the fractional Brownian motion. Potential Analysis, 10, pp. 177–214.

[79] Delbaen, F., and Schachermayer, W. (1994). A general version of the fundamental theorem of asset pricing. Mathematische Annalen, 300, 463–520.

[80] De Marco, S., Friz, P., and Gerhold, S., Risk, 26(2), 70.

[81] Derman, E., Kani, I., and Zou, J.Z. (1996). The local volatility surface: unlocking the information in index option prices. Financial Analysts Journal, 52(4), pp. 25–36.

[82] Detemple, J., Garcia, R., and Rindisbacher, M. (2005). Representation formulas for Malliavin derivatives of diffusion processes. Finance Stoch, 9, pp. 349–367.

[83] Dieker, T. (2002). Simulation of fractional Brownian motion. Master thesis. University of Twente, The Netherlands.

[84] Ding, Z. and Granger, C.W.J. (1996). Modeling volatility persistence of speculative returns: a new approach. Journal of Econometrics, Elsevier, 73(1), pp. 185–215.

[85] Di Nunno, G., Øksendal, B. K. and Proske, F. (2009). Malliavin Calculus for Lévy Processes with Applications to Finance (Vol. 2). Springer, Berlin.

[86] Dobrić, V. and Ojeda, F.M. (2006). Fractional Brownian fields, duality, and martingales. Institute of Mathematical Statistics Lecture Notes - Monograph Series. Institute of Mathematical Statistics, Beachwood, Ohio, USA.

[87] Duncan, T.E., Hu, Y., and Pasik-Duncan, B. (2000). Stochastic calculus for fractional Brownian motion I. Theory. SIAM Journal on Control and Optimization, 38(2), pp. 582–612.

[88] Dupire, B. (1994). Pricing with a smile. Risk, 7, pp. 18–20.

[89] Durrleman, V. (2008). Convergence of at-the-money implied volatilities to the spot volatility. Journal of Applied Probability, 45(2), pp. 542–550.

[90] Durrleman, V. (2010). From implied to spot volatilities. Finance Stoch, 14, pp. 157–177.

[91] El Euch, O. and Rosenbaum, M. (2018). Perfect hedging in rough Heston models. Annals of Applied Probability, 28, pp. 3813–3856

[92] Eraker, B., Johannes, M. and Polson, N. (2003). The impact of jumps in volatility and returns. The Journal of Finance, 58, pp. 1269–1300.

[93] Engelmann, B., Koster, F. and Oeltz, D. (2020). Calibration of the Heston stochastic local volatility model: a finite volume scheme. Available at SSRN: https://ssrn.com/abstract=1823769.

[94] El Euch, O., Fukasawa, M., and Rosenbaum, M. (2018). The microstructural foundations of leverage effect and rough volatility. Finance Stoch, 22, pp. 241–280.

[95] Falconer, K.J. (1990). Fractal Geometry. Wiley, New York.

[96] Fan. J. and Wang, Y. (2008). Spot volatility estimation for high frequency data. Statistics and its Interface, 1, pp. 279–288.

[97] Fama, E. and French, K. (1988). Permanent and temporary components of stock prices. Journal of Political Economy, 96, pp. 246–273.

[98] Fouque, J.P., Papanicolaou, G., and Sircar, K.R. (2000). Derivatives in Financial Markets with Stochastic Volatility. Cambridge University Press.

[99] Forde, M. and Jacquier, A. (2009). Small-time asymptotics for implied volatility under the Heston model. International Journal of Theoretical and Applied Finance, 12, pp. 861–876.

[100] Forde, M. and Jacquier, A. (2011). Small-time asymptotics for an uncorrelated local-stochastic volatility model. Applied Mathematical Finance, 18(6), pp. 517–535,

[101] Forde, M., Jacquier, A. and Lee, R. (2012). The small-time smile and term structure of implied volatility under the Heston model. SIAM Journal on Financial Mathematics, 3(1), pp. 690–708.

[102] Fournié, E., Lasry, J., Lebuchoux, J., Lions, J. P. and Touzi, N. (1999). Applications of Malliavin calculus to Monte Carlo methods in finance. Finance Stochast 3, pp. 391–412.

[103] Fournié, E., Lasry, J., Lebuchoux, J. and Lions, J.P. (2001). Applications of Malliavin calculus to Monte Carlo methods in finance II. Finance and Stochastics, 5(2), pp. 201–236.

[104] Friz, P., Gerhold, S., and Yor, M. (2014). How to make Du. pire's local volatility work with jumps Quantitative Finance, 14 (8), pp. 1327–1331

[105] Fouque, J.P, Papanicolaou, G., and Sircar, K. (2000). Derivatives in Financial Markets with Stochastic Volatility. Cambridge University Press.

[106] Fouque, J.-P., Papanicolaou, G., Sircar, R. and Solna, K. (2003). Singular perturbations in option pricing. SIAM Journal on Applied Mathematics, 63, pp. 1648–1665.

[107] Fouque, J.P., Papanicolaou, G., Sircar, R., and Sola, K. (2004). Maturity cycles in implied volatility. Finance and Stochastics, 8, pp. 451–477.

[108] Fouque, J.P and Saporito, Y.F. (2018). Heston stochastic vol-of-vol model for joint calibration of VIX and S&P 500 options. Quantitative Finance, 18(6), pp. 1003–1016,

[109] Fukasawa, M. (2011). Asymptotic analysis for stochastic volatility: martingale expansion. Finance Stoch, 15, 635–654.

[110] Fukasawa, M. (2014). Volatility derivatives and model-free implied leverage. IJTAF, 17(1), 1450002.

[111] Fukasawa, M. (2017). Short-time at-the-money skew and rough fractional volatility. Quantitative Finance, 17(2), pp. 189–198.

[112] Funahashi and Kijima (2017). A solution to the time-scale fractional puzzle in the implied volatility. Fractal and Fractional (1), pp. 14–31.

[113] Gangahar, A. (2006). Why volatility becomes an asset class. Financial Times, May 23, 2006.

[114] Gatheral, J., Jusselin, P., and Rosenbaum, M. (2020). The quadratic rough Heston model and the joint S&P 500/VIX smile calibration problem. arXiv:2001.01789v1 [q-fin.MF].

[115] Gatheral, J. (2006). The Volatility Surface: A Practicioner's Guide. John Wiley & Sons, Inc., Hoboken, New Jersey.

[116] Gatheral, J., Hsu, E.P., Laurence, P., Ouyang, C., and Wang, T.-H. (2012). Asymptotics of implied volatility in local volatility models. Mathematical Finance, 22, pp. 591–620.

[117] Gatheral, J., Jaisson, T., and Rosenbaum, M. (2018). Volatility is rough. Quantitative Finance, 18(6), pp. 933–949.

[118] Gerd, H., Lunde, A., Shephard, N., and Sheppard, K.K. (2009). Oxford-Man Institute's realized library, Oxford-Man Institute, University of Oxford. Library version: 0.3.

[119] Glasserman, P. and Yao, D.D. (1992). Some guidelines and guarantees for common random numbers. Management Science 38(6), pp. 884–908.

[120] Glynn, P.W. (1989). Optimization of stochastic systems via simulation, Proceedings of the 1989 Winter Simulation Conference, pp. 90–105.

[121] Göngy, I. (1986). Mimicking the one-dimensional marginal distributions of processes having an ito differential. Probability Theory and Related Fields, 71, pp. 501–516.

[122] Goutte, S., Ismail, A., and Pham, H. (2017). Regime-switching stochastic volatility model: estimation and calibration to VIX options. Applied Mathematical Finance, 24:1, pp. 38–75.

[123] Greene, M.T. and Fielitz, B.D. (1977). Long term dependence in common stock returns. Journal of Financial Economics, 4, pp. 339–349.

[124] Guasoni, P. (2006). Transaction costs, with fractional brownian motion and beyond mathematical finance, 16 no. 3, pp. 569–582.

[125] Guasoni, P., Nika, Z., and Rásonyi, M. (2019). Trading fractional Brownian motion. SIAM Journal on Financial Mathematics, 10(3), pp. 769–789.

[126] Guasoni, P., Rásonyi, M., and Schachermayer, W. (2010). The fundamental theorem of asset pricing for continuous processes under small transaction costs. Annals of Finance, 6(2), pp. 157–191.

[127] Guo, J. H. and Hung, M.W. (2008). A generalization of Rubinstein's pay now, choose later, Journal of Futures Markets, 28(5), pp. 488–515.

[128] Guyon, J. (2018). On the joint calibration of SPX and VIX options. Presentation, Bachelier Congress, Dublin.

[129] Guyon, J. and Henry-Labordère, P. (2012). Being particular about calibration. Risk Magazine, 25(1), pp. 92–96.

[130] Guyon, J. and Henry-Labordère, P. (2013). Nonlinear Option Pricing. Chapman and Hall/CRC Press Financial Mathematics Series.

[131] Hagan, P.S., Kumar, D., Lesniewski, A.S., and Woodward, D.E. (2002). Managing smile risk. Wilmott, September 2002, pp. 84–108.

[132] Hagan, P.S., Kumar, D., Lesniewski, A.S., and Woodward, D.E. (2014). Arbitrage-free SABR. Wilmott, 2014, pp. 60–75.

[133] Hairer, M. Advanced stochastic analysis. http://www.hairer.org/Course.pdf, 2016. date of last access: 11th November 2020.

[134] Guennoun, H., Jacquier, A., Roome, P., and Shi, F. (2018). Asymptotic behavior of the fractional Heston model. SIAM Journal on Financial Mathematics, 9(3), pp. 1017–1045.

[135] Harrison, M. and Pliska, S. (1981). Martingales and stochastic integrals in the theory of continuous trading. Stochastic Processes and their Applications, 11, pp. 215–260.

[136] Harvey, A.C. (1998). Long memory in stochastic volatility. In: Knight, J. and Satchell, S. (eds.) Forecasting Volatility in Financial markets. Butterworth-Heinemann, London.

[137] Hayashi, T. and Mykland, P. (2005). Evaluating hedging errors: an asymptotic approach, Math Finance, 15, pp. 309–343.

[138] Heston, S.L. (1993). A closed-form solution for options with stochastic volatility with applications to bond and currency options. Review of Financial Studies, 6, pp. 327–343.

[139] Hsieh, D.A. (1989). Modeling hetereskedasticity in daily foreign exchange rates. Journal of Business and Economic Statistics, 7, pp. 306–317.

[140] Hitsuda, M. (1972). Formula for Brownian partial derivatives. Proceedings of the 2nd Japan-USSR Symposium on Probability Theory. Communications in Mathematical Physics, 2, pp. 111–114.

[141] Hitsuda, M. (1979). Formula for Brownian partial derivatives. Publ. Fac. of Integrated Arts and Sciences Hiroshima Univ., 3, pp. 1–15.

[142] Horvath, B., Jacquier, A., and Tankov, P. (2019). Volatility options in rough volatility models. arXiv:1802.01641v2 [q-fin.PR]

[143] Hu, Y. and Øksendal, B. (2003). Fractional white noise calculus and applications to finance. Infinite Dimensional Analysis, Quantum Probability and Related Topics, 6(1), pp. 1–32.

[144] Hull, J. (2016). Options, Futures and Other Derivatives. 9th edition. Pearson Education.

[145] Hurst, E. (1951). Long-term storage capacity of reservoirs. Transactions of the American Society of Civil Engineers, 116, pp. 770–808.

[146] Itkin, A. (2020). Fitting Local Volatility. World Scientific, Singapore.

[147] Jacobsen, B. (1996). Long term dependence in stock returns. Journal of Empirical Finance, 3(4), pp. 393-417.

[148] Jacod, J., Li, Y., Mykland, P.A., Podolskij, M., and Vetter, M. (2009). Microstructure noise in the continuous case: the preaveraging approach. Stochastic Processes and Their Applications, 119(7), pp. 2249–2276.

[149] Kennedy, E. Mitra, S., and Pham, D. (2012). On the approximation of the SABR model: a probabilistic approach. Applied Mathematical Finance, 19(6), pp. 553–586.

[150] Jacquier, A. and Roome, P. (2013). The small-maturity Heston forward smile. SIAM Journal on Financial Mathematics, 4(1), pp. 831–856.

[151] Jacquier, A. and Roome, P. (2015). Asymptotics of forward implied volatility. SIAM Journal on Financial Mathematics, 6(1), pp. 307–351.

[152] Jacquier, A. and Roome, P. (2016). Large-maturity regimes of the Heston forward smile. Stochastic Processes and their Applications, 126(4), pp. 1087–1123.

[153] Kijima, M. (2002). Stochastic Processes with Applications to Finance. Chapman and Hall, London.

[154] Kokholm, T. and Stisen, M. (2015). Joint pricing of VIX and SPX options with stochastic volatility and jump models. The Journal of Risk Finance, 16(1), pp. 27–48.

[155] Kolmogorov, A. N. (1940). Wienersche Spiralen und einige andere interessante Kurven im Hilbertschen Raum. C. R. (Doklady). Acad. Sci. URSS (N.S.) 26, pp. 115–118

[156] Kreps, D.M. (1981). Arbitrage and equilibrium in economics with infinitely many commodities. Journal of Mathematical Economics, 8, pp. 15–35.

[157] Kristensen, D. (2010). Nonparametric filtering of the realized spot volatility: a kernel based approach. Econometric Theory, 26, pp. 60–93.

[158] Kruse, S. and Nögel, U. (2005). On the pricing of forward starting options in Heston's model on stochastic volatility. Finance and Stochastics, 9(2), pp. 233–250.

[159] Kurtz, T.G. and Stockbridge, R.H. (1998). Existence of Markov controls and characterization of optimal Markov controls. Finance and Stochastics, 36, pp. 609–653.

[160] Lamberton, D. and Lapeyre, B. (2008). Introduction to Stochastic Calculus Applied to Finance. Chapman and Hall/CRC Financial Mathematics Series.

[161] L'Ecuyer, P. and Perron, G. (1994). On the convergence rates of IPA and FDC derivative estimators for finite-horizon stochastic simulations. Operations Research, 42(4), pp. 643–656.

[162] Lee, R. (2001). Implied and local volatilities under stochastic volatility. International Journal of Theoretical Applied Finance, 4(1), pp. 45–89,

[163] Lee, R. (2005). Implied volatility: statics, dynamics, and probabilistic interpretation. In: Baeza-Yates, R., Glaz, J., Gzyl, H., Hüsler, J., and Palacios, J.L. (eds) Recent Advances in Applied Probability. Springer, Boston, MA.

[164] Leontsinis, S. and Alexander, C. (2017). Arithmetic variance swaps. Quantitative Finance, 17(4), pp. 551–569.

[165] Leung, T. and Lorig, M. (2016). Optimal static quadratic hedging. Quantitative Finance, 16(9), pp. 1341–1355.

[166] Lewis, A.L. (2000). Option Valuation Under Stochastic Volatility with Mathematica Code. Finance Press, Newport Beach.

[167] Lim, S.C. and Sithi, V.M. (1995). Asymptotic properties of the fractional Brownian motion of Riemann-Liouville type. Physics Letters A, 206(5–6), pp. 311–317.

[168] Lin, S.J. (1995). Stochastic analysis of fractional Brownian motions. Stochastics and Stoch. Reports, 55, pp. 121–140.

[169] Liptser, R. and Shiryaev, A. (1986). Theory of Martingales. Nauka, Moscow. (In Russian.)

[170] Lo, A.W. (1991). Long-term memory in stock market prices. Econometrica, 59(5), 1279–1313.

[171] Lobato, I.N. and Velasco, C. (2000). Long memory in stock market trading volume. Journal of Business and Economic Statistics, 18, pp. 410–427.

[172] Lucic, V. (2003). Forward-start options in stochastic volatility models. Wilmott Magazine, September.

[173] Malliavin, P. (1997). Stochastic Analysis. Springer Verlag.

[174] Malliavin, P. and Mancino, M.E. (2002). Fourier series method for measurement of multivariate volatilities. Finance and Stochastics, 4, pp. 49–61.

[175] Malliavin, P. and Thalmaier, A. (2006). Stochastic calculus of variations in mathematical finance. Springer Science and Business Media. Springer, Berlin, Heidelberg.

[176] Mancino, M.E., Recchioni, M.C., and Sanfelici, S. (2017). Fourier-Malliavin Volatility Estimation. Theory and Practice. Springer.

[177] Mancino, M.E. and Sanfelici, S. (2008). Robustness of Fourier estimator of integrated volatility in the presence of microstructure noise. Computational Statistics and Data Analysis, 52(6), pp. 2966–2989

[178] Mandelbrot, B. (1963). The variation of certain speculative prices. The Journal of Business, 36, pp. 394–419.

[179] Mandelbrot, B. (1971). When can price be arbitraged efficiently? A limit to the validity of the random walk and martingale models. The Review of Economics and Statistics, 53, pp. 225–236.

[180] Mandelbrot, B. (1972). Statistical methodology for nonperiodic cycles: from the covariance to R/S analysis. Annals of Economic and Social Measurement, 1, pp. 259–290.

[181] Mandelbrot, B. (1982). The Fractal Geometry of Nature. Freeman, San Francisco.

[182] Mandelbrot, B. (2001). Scaling in financial prices: I. Tails and dependence. Quantitative Finance, pp. 113–123.

[183] Mandelbrot, B. (2005). The inescapable need for fractal tools in finance. Annals of Finance, 1, pp. 193–195.

[184] Mandelbrot, B.; van Ness, J.W. (1968). Fractional Brownian motions, fractional noises and applications. SIAM Review, 10 (4), pp. 422–437.

[185] Marinucci, D. and Robinson, P.M. (1999). Alternative forms of fractional Brownian motion. Journal of Statistical Planning and Inference, 80(1–2), pp. 111–122.

[186] McCurdy, T. H. and Morgan, I. G. (1988). Testing the martingale hypothesis in deutsch mark futures with models specifying the form of heteroscedasticity, Journal of Applied Econometrics, 3, pp. 187–202.

[187] Medvedev, A. and Scaillet, O. (2007). Approximation and calibration of short-term implied volatilities under jump-diffusion stochastic volatility. Review of Financial Studies, 20(2), pp. 427–459.

[188] Merton, R.C. (1976). Option pricing when underlying stock returns are discontinuous. Journal of Financial Economics, 3(1–2), pp. 125-144,

[189] Mikosch, T. and Stărică, C. (2000). Long-range dependence effects and ARCH modeling, in Theory and applications of long-range dependence, Birkhäuser Boston, Boston, MA, 2003, pp. 439–459.

[190] Mishura, Y.S. (2008). Stochastic Calculus for Fractional Brownian Motion and Related Processes, Lecture Notes in Mathematics, 1929. Springer-Verlag, Berlin.

[191] Molchan, G., and Golosov, J. (1969). Gaussian stationary processes with asymptotic power spectrum. Soviet Mathematics Doklady, 10(1), pp. 134–137.

[192] Musiela, M. and Rutkowski, M. (2005). Martingale Methods in Financial Modelling. 2nd edition. Springer-Verlag, Berlin, Heidelberg.

[193] Norros, I., Valkeila, E., and Virtamo, J. (1999). An elementary approach to a Girsano formula and other analytical results on fractional Brownian motion. Bernoulli, 5, pp. 571–587.

[194] Nourdin, I. (2012). Selected Aspects of Fractional Brownian Motion, Bocconi and Springer Series, Vol. 4. Springer, Milan; Bocconi University Press, Milan.

[195] Nualart, D. (2006). The Malliavin Calculus and Related Topics. Probability and its Applications (New York). 2nd edition. Springer-Verlag, Berlin

[196] Nualart, D. and Nualart, E. (2018). Introduction to Malliavin Calculus (Vol. 9). Cambridge University Press.

[197] Nualart, D. and Pardoux, E. (1988). Stochastic calculus with anticipating integrands. Probability Theory and Related Fields, 78, pp. 535–581.

[198] Pascucci, A. and Mazzon, A. (2017). The forward smile in local-stochastic volatility models. Risk, 20(3), pp. 1–29.

[199] Picard, J. (2011). Representation formulae for the fractional Brownian motion. In: Donati-Martin, C., Lejay, A., and Rouault,

A. (eds) Séminaire de Probabilités XLIII. Lecture Notes in Mathematics, vol 2006. Springer, Berlin, Heidelberg.

[200] Pigato, P. (2019). Extreme at-the-money skew in a local volatility model. Finance and Stochastics, 23, pp. 827–859.

[201] Poterba, J.M. and Summers, L. H. (1988). Mean reversion in stock prices: evidence and implications. Journal of Financial Economics, 22(1), pp. 27–59

[202] Rebonato, R. (1999). Volatility and Correlation in the Pricing of Equity, FX and Interest Rate Options. Wiley, New York.

[203] Renault, E. (1997). Econometric models of option pricing errors. In: Kreps, D. and Wallis, K. (eds) Advances in Economics and Econometrics: Theory and Applications: Seventh World Congress (Econometric Society Monographs, pp. 223–278). Cambridge University Press, Cambridge.

[204] Renault, E. and Touzi, N. (1996). Option hedging and implied volatilities in a stochastic volatility model. Mathematical Finance, 6, pp. 279–302.

[205] Revusz, D. and Yor, M. (1999). Continuous Martingales and Brownian Motion. Springer-Verlag, Berlin, Heidelberg.

[206] Romano, M. and Touzi, N. (1997). Contingent claims and market completeness in a stochastic volatility model. Mathematical Finance, 7, pp. 399–410.

[207] Rouah, F.D. (1993). The Heston Model and Its Extensions in Matlab and C#. Wiley Online Library.

[208] Stoll, H. (1969). The relationship between call and put option prices. Journal of Finance, 23, pp. 801–824.

[209] Sanz-Solé, M. (2005). Malliavin Calculus: With Applications to Stochastic Partial Differential Equations. EPFL Press.

[210] Schachermayer, W. and Teichmann, J. (2008). How close are the option pricing formulas of Bachelier and Black-Merton-Scholes?. Mathematical Finance, 18, pp. 155-170.

[211] Schürger, K. (2002). Laplace transforms and suprema of stochastic processes. In: Sandmann, K. and Schönbucher, P. J. (ed) Advances in Finance and Stochastics, pp. 287-293. Springer, Berlin.

[212] Skorohod, A.V. (1975). On a generalization of a stochastic integral. Theory of Probability and Its Applications, 20, 219–233.

[213] Sevljakov, A. Ju. (1981). The Itô's formula for the extended stochastic integral. Theory of Probability and Mathematical Statistics, 22, pp. 163–174.

[214] Shevchenko, G. (2015). Fractional Brownian motion in a nutshell. 7th Jagna International Workshop (2014) International Journal of Modern Physics: Conference Series Vol. 36, 1560002.

[215] Todorov, V. and Tauchen, G. (2011), Volatility jumps. Journal of Business and Economic Statistics, 29(3), p. 356–371.

[216] Sekiguchi, T. and Shiota, Y. (1985). L^2-theory of noncausal stochastic integrals. Mathematics Reports Toyama University, 8, pp. 119–195.

[217] Serinaldi, F. (2010). Use and misuse of some Hurst parameter estimators applied to stationary and non-stationary financial time series. Physica A: Statistical Mechanics and its Applications, 389(14), pp. 2770–2781.

[218] Ustunel, A.S. (1986). La formule de changement de variable pour l'intégrale anticipante de Skorohod. Comptes Rendus de l'Académie des Sciences, Series I, 303, pp. 329–331.

[219] Teverovsky, V., Taqqu, M.S., and Willinger, W. (1999). A critical look at Lo's modified R/S statistic, Journal of Statistical Planning and Inference, 80, pp. 211–227.

[220] Van Der Stoep, A.A., Grzelak, l., and Oosterlee, C. (2014). The Heston stochastic-local volatility model: efficient Monte Carlo simulation, International Journal of Theoretical and Applied Finance, 17, pp. 1–30.

[221] Wang, F. (2014). Optimal design of Fourier estimator in the presence of microstructure noise. Computational Statistics and Data Analysis, 76, pp. 708–722.

[222] Weron, R. (2002). Estimating long-range dependence: finite sample properties and confidence intervals. Physica A: Statistical Mechanics and its Applications, 312(1–2), pp. 285–299.

[223] Willard, G.A. (1997). Calculating prices and sensitivities for path-independent securities in multifactor models. The Journal of Derivatives, 5, pp. 45–61.

[224] Willinger, W., Taqqu, M., and Teverovsky, V. (1999). Stock market prices and long-range dependence. Finance Stochast, 3, pp. 1–13.

[225] Zhang, L. (2006). Efficient estimation of stochastic volatility using noisy observations: a multi-scale approach. Bernoulli, 12(6), pp. 1019–1043.

[226] Zhang, L., Mykland, P.A., and Ait-Sahalia, Y. (2005). A tale of two time scales: determining integrated volatility with noisy high-frequency data. Journal of the American Statistical Association, 100(472), pp. 1394–1411.

[227] Zähle, M. (1998). Integration with respect to fractal functions and stochastic calculus I. Probability Theory and Related Fields, 111(3), pp. 333–374.

[228] Zähle, M. (1999). Integration with respect to fractal functions and stochastic calculus II. Mathematische Nachrichten, 225, pp. 145–183.

[229] Zhou, B. (1996). High frequency data and volatility in foreign-exchange rates. Journal of Business and Economic Statistics, 14(1), pp. 45–52.

Index

3/2 Heston volatility, 71

A

Anticipating Itô formula, 94–97
Arithmetic variance swaps, 106
ATMI approximation, 165
  for diffusions, 168–172
  for fractional and rough volatilities, 176–178
  for local volatilities, 172–176
  numerical experiments, 178–181
ATMI curvature, 215
  blow-up, 215
  in the correlated case, 227–238
  for diffusions, 241–243
  empirical facts, 215–216
  for fractional and rough volatilities, 243–245
  limit results, 220–223, 230–238
  for local volatilities, 239–241
  in the uncorrelated case, 216–220
ATMI short-time level, 133
  in the correlated case, 152–165
  rate of convergence, 152, 163
  in the uncorrelated case, 146–152
ATMI skew, 185
  blow-up, 185–186, 198
  in the correlated case, 190–191
  for diffusion volatilities, 197, 198–200
  for fractional and rough volatilities, 203–206
  for local volatilities, 201–202
  short-end limit, 192–197
  in the uncorrelated case, 189–190

B

Bachelier model, 16
Basket options, 6
Bates model, 208–212
  skew slope, 211–212
Black-Scholes model, 14–16
  hedging in the, 76
  Malliavin derivative of, 58

C

Call option, 5
Call-put parity, 8
CEV model, 36
  ATMI approximation, 173–174, 178–179
  ATM skew slope, 201
  curvature, 241
  forward smile, 264, 278
  Malliavin derivative of, 67–68
  martingale representation, 75
  option price decomposition, 145
Clark-Ocone-Haussman formula, 73–75

Complete markets, 10
Computation of the Greeks, 84
  in the Black-Scholes model, 84–91
  in stochastic volatility models, 91–94
Convexity adjustment, 81

D
Decomposition formula for option prices, 138–143
  for the CEV model, 144
  for local volatilities, 144
  for the rough Bergomi model, 143
  for the SABR model, 143
Decompostion formulas for implied volatilities, 145–146
Delta, 12
  delta hedging, 16
  delta-gamma-vega relationship, 98
Derivatives, 4
Divergence operator, 61
Dupire formula, 34–35
Dynamic replication, 8

F
Floating strike Asian options, 249
Forward ATMI curvature, 272–280
Forward ATMI skew, 264–272
  for local volatility models, 269
  for stochastic volatilities, 269
Forward contract, 4
Forward implied volatility, 254
  ATM limit, 258–264
Forward price, 5
Forward start options, 6, 249
  option price decomposition, 256–258
  as random strike options, 253–255
Fourier estimation of volatility 26–29
Fractional Brownian motion, 47, 111–117
  in finance, 124–126
  integration with respect to, 117–122
  Malliavin derivative, 126–128
  simulation, 122–124
Fractional Heston model, 129
Fractional Ornstein-Uhlenbeck volatilities, 127
  ATMI approximation, 176–178
  curvature, 225–226, 243–244
  Malliavin derivative of, 127
  option price decomposition, 176
  skew 203–204,
Fractional volatilities, 126
  ATMI approximation, 176
  curvature, 216, 243–246
  forward skew, 271
  Markov approximation, 205
  skew, 203
  VIX ATMI, 292, 298–300
Fundamental Theorems of Asset Pricing, 10
Futures contract, 5

G
Gamma, 17
Gamma swaps, 53
  leverage swaps, 104
Gyöngy's lemma, 32–34

H
Heston model, 39

ATMI approximation, 172, 178–179
curvature, 223–224, 242–243
delta and gamma, 91
forward skew, 270
leverage swap, 105
option price decomposition, 143
skew, 200
VIX ATMI, 293, 296–299
Heston volatility, 69
Malliavin derivative of, 69
martingale representation, 80
Hull and White formula, 135
from conditional expectations, 135
extension to the correlated case, 138–145
from the Itô formula, 135–136

I
Implied volatility, 18–19
convexity of, 238
term structure of the, 185
see ATMI approximation; ATMI curvature; ATMI skew
Integrated variance, 25
Integrated volatility, 25
martingale representation, 79
Integration by parts, 59, 84
In-the-money, 14
Intrinsic value, 6

L
Leverage swaps, 104
Local volatility model, 32
ATMI approximation, 172
curvature, 239–241
option price decomposition for, 144
skew, 201–202
one-half rule, 202
Log-moneyness, 18
Long and short positions, 5
Long memory, 31

M
Malliavin derivative operator, 57
basic properties, 59–60
computation, 62–65
representation formulas, 65–66
Martingale representation formula, 73
Maturity time, 4
Mimicking processes, 33
Mixed log-normal distribution, 97
law of an asset price as a perturbation of, 97–100
Moments of log-prices, 101–104
Multifactor volatilities,
curvature, 222
skew, 196

O
Ornstein-Uhlenbeck volatilities
skew, 199
curvature, 224
Options, 5
American and European, 5
vanilla, 5
Out-of-the-money, 6

P
Path dependent options, 5
Payoff, 4
Put option, 5

R
Random strike options, 250–253
Realized variance, 25

Replicating portfolio, 4
Riemann-Liouville fractional Brownian motion, 117
  simulation, 122–124
Risk-neutral probability, 11
Rough Bergomi model, 48
  ATMI approximation, 176–177, 180–181
  forward smile, 263–264, 278–279
  Malliavin derivative of, 127–128
  option price decomposition, 143
  skew, 204–205
    curvature, 227, 244–245
Rough volatilities, 47

S

SABR model, 41
  arithmetic variance swap, 108
  ATMI approximation, 171, 178–179
  curvature 224, 242
  delta and gamma, 92
  forward smile, 263, 270, 278
  Malliavin derivative of, 61
  martingale representation, 79–80
  option price decomposition, 143
  skew slope, 199
  variance and skewness, 101
  VIX ATMI, 291
  VIX ATMI skew, 293–294
Self-financing portfolio, 10
Short memory, 31
  vs long memory, 31, 206–207
Short-time behaviour of the implied volatility, 146
Skew effect, 19
Skorohod integral, 61
Smile effect, 19
Spot volatility, 19, 25
  empirical properties, 28–31
  martingale representation, 79
Spread options, 6, 249
Static replication, 8
Stochastic-local volatilities, 45
  ATM skew slope, 203
Stochastic volatilities, 38–39
Strike price, 5

T

Time-varying coefficients model, 186
  forward skew, 271
  skew slope, 198
Two-factor Bergomi skew, 200
  curvature, 225

V

Variance swap, 5, 49
Vega, 17
VIX, 49
  martingale representation, 83, 283
VIX implied volatility, 283
  ATMI short-time limit, 290
  ATMI skew, 293
Volatility clustering, 30
Volatility derivatives, 49
Volatility swap, 5, 51
  and curvature, 223
  difference with variance swap, 81–82, 149
  Laplace transform of, 56
  model-free approximation, 196–197

W

Weighted variance swaps, 53

Z

Zero-vanna implied volatility, 163

Printed in the United States
by Baker & Taylor Publisher Services